Sequences and Series

Ana Alves de Sá • Bento Louro

Sequences and Series

Theory and Practice

Ana Alves de Sá
Department of Mathematics
Nova School of Science and
Technology-NOVA University Lisbon
(Faculdade de Ciências e
Tecnologia-Universidade Nova de
Lisboa)
Lisboa, Portugal

Bento Louro
Department of Mathematics
Nova School of Science and
Technology-NOVA University Lisbon
(Faculdade de Ciências e
Tecnologia-Universidade Nova de
Lisboa)
Lisboa, Portugal

ISBN 978-3-031-67201-9   ISBN 978-3-031-67202-6 (eBook)
https://doi.org/10.1007/978-3-031-67202-6

Mathematics Subject Classification: 97I30, 40A05, 40, 40A30, 40D15, 40C15

The translation was done with the help of an artificial intelligence machine translation tool. A subsequent human revision was done primarily in terms of content.

Translation from the Portuguese language edition: "Sucessões e Séries: Teoria e Prática" by Ana Alves de Sá and Bento Louro, © Ana Maria de Souza Alves de Sá and Bento José Carrilho Miguens Louro 2014. Published by Escolar Editora. All Rights Reserved.

© The Editor(s) (if applicable) and The Author(s), under exclusive license to Springer Nature Switzerland AG 2024

This work is subject to copyright. All rights are solely and exclusively licensed by the Publisher, whether the whole or part of the material is concerned, specifically the rights of reprinting, reuse of illustrations, recitation, broadcasting, reproduction on microfilms or in any other physical way, and transmission or information storage and retrieval, electronic adaptation, computer software, or by similar or dissimilar methodology now known or hereafter developed.
The use of general descriptive names, registered names, trademarks, service marks, etc. in this publication does not imply, even in the absence of a specific statement, that such names are exempt from the relevant protective laws and regulations and therefore free for general use.
The publisher, the authors and the editors are safe to assume that the advice and information in this book are believed to be true and accurate at the date of publication. Neither the publisher nor the authors or the editors give a warranty, expressed or implied, with respect to the material contained herein or for any errors or omissions that may have been made. The publisher remains neutral with regard to jurisdictional claims in published maps and institutional affiliations.

This Springer imprint is published by the registered company Springer Nature Switzerland AG
The registered company address is: Gewerbestrasse 11, 6330 Cham, Switzerland

If disposing of this product, please recycle the paper.

Preface

This book is primarily intended for students of Science and Engineering. Although the topics were not always addressed on the same subject, we decided to group them here due to their scientific coherence. We have strived to create a self-contained text, presenting theoretical exposition, illustrated with various examples, solved exercises of different natures and degrees of difficulty, and proposed exercises. Thus, the readers can test their knowledge and level of understanding of the material.

We rigorously present the statements of definitions and results without delving too much into the theoretical exposition and without demonstrating a large part of the Theorems. Many examples highlight aspects that could go unnoticed by readers, mainly those inexperienced in reading mathematical texts, for whom this book is primarily intended.

Any reader who has completed High School should be able to comprehend Chap. 1, and the same holds true for Chap. 2, with the possible exception of the Integral Test. For those unfamiliar with integral calculus, skipping this section will not detract from their understanding of the rest of the chapter. To understand Chap. 3, the reader must have basic knowledge of differential and integral calculus. We go beyond the typical introduction to function series by delving into the study of Fourier series. This topic will prove to be especially useful when studying Differential Equations, but no additional knowledge is required beyond what is already necessary for a general understanding of function series.

At the end of each chapter, there are two lists of exercises: one titled "Solved Exercises" followed by the respective solutions, and another titled "Proposed Exercises," which are intended for the reader's practice and whose answers can be found at the end of the book.

We strongly encourage students to attempt the "Solved Exercises" on their own before referring to the provided solution. It is important to note that the written solution does not include the reasoning, drafts, or false starts made in arriving at this point. By comparing their solution to the one presented in the book, readers can more accurately assess their understanding. Reviewing the solution without first attempting the problem independently may not effectively facilitate learning.

The exercises featured in this book were used during practical classes and exams for the subjects we taught. We extend our sincere appreciation to the students whose inquiries aided in enhancing this publication, as well as to our fellow educators who assisted us in teaching Calculus courses at Nova School of Science & Technology. Their thorough review of the text and copious suggestions were invaluable contributions.

In the margins of the text, we have provided concise biographies of the mentioned mathematicians. For those interested in learning more about these individuals, among many others, we suggest consulting the following web page: https://mathshistory.st-andrews.ac.uk/Biographies/.

Despite our efforts toward meticulous writing and thorough proofreading, we know some typos persisted. We extend our apologies to our readership for any such errors and express our gratitude in advance for their assistance in identifying and correcting them for future editions.

Lisboa, Portugal   Ana Alves de Sá
Lisboa, Portugal   Bento Louro
2024

Notes to the Reader

This section consists of a list of notations. They are quite common. Our aim is to avoid ambiguity and to help the reader who, here or there, may be used to a different notation.

$\mathbb{N} = \{1, 2, 3, 4, \dots\}$: the set of natural numbers

$\mathbb{N}_0 = \mathbb{N} \cup \{0\}$

$\mathbb{Z} = \{\dots, -4, -3, -2, -1, 0, 1, 2, 3, 4, \dots\}$: the set of integers

$\mathbb{Q} = \{x \in \mathbb{R} : x = p/q, \ p, q \in \mathbb{Z}, \ q \neq 0\}$: the set of rationals

$\mathbb{R}$: the set of real numbers (remark that $\mathbb{N} \subset \mathbb{Z} \subset \mathbb{Q} \subset \mathbb{R}$)

$\mathbb{R}^+ = \{x \in \mathbb{R} : x > 0\}$: the set of positive real numbers

$\mathbb{R}_0^+ = \mathbb{R}^+ \cup \{0\}$

triangle inequality: $|x + y| \leq |x| + |y|, \ \forall x, y \in \mathbb{R}$

intervals on $\mathbb{R}$: (suppose $a, b \in \mathbb{R}$, with $a < b$) are the following sets:

$[a, b] = \{x \in \mathbb{R} : a \leq x \leq b\}$; $]-\infty, b] = \{x \in \mathbb{R} : x \leq b\}$;

$]a, b[= \{x \in \mathbb{R} : a < x < b\}$; $]-\infty, b[= \{x \in \mathbb{R} : x < b\}$;

$[a, b[= \{x \in \mathbb{R} : a \leq x < b\}$; $[a, +\infty[= \{x \in \mathbb{R} : a \leq x\}$;

$]a, b] = \{x \in \mathbb{R} : a < x \leq b\}$; $]a, +\infty[= \{x \in \mathbb{R} : a < x\}$;

$[a, a] = \{a\}$; $]-\infty, +\infty[= \mathbb{R}$

ε **neighborhood of** a: is the set $V_\varepsilon(x) =]a - \varepsilon, a + \varepsilon[$ (where $a \in \mathbb{R}$ and $\varepsilon > 0$)

V **is a neighborhood of** a: there is an $\varepsilon > 0$ such that $V_\varepsilon(a) \subset V$

$$\sum_{i=n}^{n} a_i = a_n, \quad a_n \in \mathbb{R} \text{ and } n \in \mathbb{N}$$

$$\sum_{i=m}^{n} a_i = a_m + a_{m+1} + \cdots + a_n, \quad a_i \in \mathbb{R}, i = m, m+1, \ldots, n \text{ and } m, n \in \mathbb{N} \text{ with } n > m$$

factorial of $n \in \mathbb{N}_0$: $0! = 1, 1! = 1$ and $n! = n \times (n-1)!$

(in a suggestive way, $n! = n \times (n-1) \times \cdots \times 3 \times 2 \times 1$)

number of p-combinations of n distinct objects $(n, p \in \mathbb{N}_0, n \geq p)$: $\binom{n}{p} = \dfrac{n!}{p! \times (n-p)!}$

Newton Binomial: if $a, b \in \mathbb{R}$, $n \in \mathbb{N}$ then $(a+b)^n = \sum_{k=0}^{n} \binom{n}{k} a^k b^{n-k}$

Method of Mathematical Induction: to prove that a proposition $P(n)$ is true for all $n \in \mathbb{N}$ proceed as follows:

1) prove that $P(1)$ is true
2) prove hereditary, that is, assume that the Induction Hypothesis $P(n)$ is true, and then prove the Induction Thesis $P(n+1)$ is true

integer part of $x \in \mathbb{R}^+$: $\text{Int}(x)$, the largest integer less than or equal to x

upper bound of $X \subset \mathbb{R}$: a real number z such that $x \leq z, \forall x \in X$

$X \subset \mathbb{R}$ is **bounded from above**: there exists an upper bound of X

supremum of $X \subset \mathbb{R}$: a real number w such that w is an upper bound of X and $w \leq z$ for all z upper bound of X; we write $w = \sup(X)$; if X is bounded from above, $\sup(X)$ exists

maximum of $X \subset \mathbb{R}$: if $\sup(X) \in X$ it is called the maximum of X and is denoted by $\max(X)$

lower bound of $X \subset \mathbb{R}$: a real number y such that $y \leq x, \forall x \in X$

$X \subset \mathbb{R}$ is **bounded from below**: there exists a lower bound of X

infimum of $X \subset \mathbb{R}$: a real number u such that u is a lower bound of X and $u \geq y$ for all y lower bound of X; we write $u = \inf(X)$; if X is bounded from below, $\inf(X)$ exists

minimum of $X \subset \mathbb{R}$: if $\inf(X) \in X$ it is called the minimum of X and is denoted by $\min(X)$

$X \subset \mathbb{R}$ is **bounded**: X is bounded from below and from above

a **function of the set** X **into the set** Y is a relation between X and Y such that, to each element $x \in X$ corresponds an unique element $y \in Y$, and we will write $f : X \to Y$. X is the **domain** of f and the set $\{y \in Y : \exists x \in X, y = f(x)\}$ is the **range** of f

if $a \in \mathbb{R}^+$, $f : [-a, a] \to \mathbb{R}$ is an **odd function** if $f(x) = -f(-x), \forall x \in [-a, a]$ and is an **even function** if $f(x) = f(-x), \forall x \in [-a, a]$

function of class C^1 on $X \subset \mathbb{R}$: function with continuous derivative on X

function of class C^p on $X \subset \mathbb{R}$: function with all continuous derivatives until order p on X

indeterminate forms: $\dfrac{0}{0}, \dfrac{\infty}{\infty}, \infty - \infty, 0 \times \infty, 1^\infty, 0^0; \infty^0$

L'Hôpital's Rule: Let I be an interval, $a \in I$, f and g two differentiable functions on I such that:

a) $g'(x) \neq 0, \forall x \in I \setminus \{a\}$.

b) $\lim\limits_{x \to a} f(x) = \lim\limits_{x \to a} g(x) = 0$ or $\lim\limits_{x \to a} f(x) = \lim\limits_{x \to a} g(x) = +\infty$.

If there exists (real or ∞) $\lim\limits_{x \to a} \dfrac{f'(x)}{g'(x)}$, then $\lim\limits_{x \to a} \dfrac{f'(x)}{g'(x)} = \lim\limits_{x \to a} \dfrac{f(x)}{g(x)}$.

L'Hôpital's Rule (another version): Let $b \in \mathbb{R}$, f and g two differentiable functions on $]b, +\infty[$ such that:

a) $g'(x) \neq 0, \forall x \in]b, +\infty[$.

b) $\lim\limits_{x \to +\infty} f(x) = \lim\limits_{x \to +\infty} g(x) = 0$ or $\lim\limits_{x \to +\infty} f(x) = \lim\limits_{x \to +\infty} g(x) = +\infty$.

If there exists (real or ∞) $\lim\limits_{x \to +\infty} \dfrac{f'(x)}{g'(x)}$, then $\lim\limits_{x \to +\infty} \dfrac{f'(x)}{g'(x)} = \lim\limits_{x \to +\infty} \dfrac{f(x)}{g(x)}$.

Contents

Preface   v

Notes to the Reader   vii

1 Sequences of Real Numbers   1
   1.1 Sequences of Real Numbers 1
   1.2 Solved Exercises 38
   1.3 Proposed Exercises 64

2 Numerical Series   67
   2.1 Generalization of the Addition Operation 67
   2.2 Definition of Series: Convergence – General Properties 69
   2.3 Alternating Series 82
   2.4 Absolute Convergence 87
   2.5 Series of Nonnegative Terms 90
   2.6 Products of Series 113
   2.7 Solved Exercises 116
   2.8 Proposed Exercises 161

3 Series of Functions   167
   3.1 Introduction: Sequences of Functions 167
   3.2 Pointwise and Uniform Convergence of Series of Functions 172
   3.3 Power Series 186
   3.4 Taylor Series and Maclaurin Series 193
   3.5 Introduction to Fourier Series 201
   3.6 Solved Exercises 229
   3.7 Proposed Exercises 320

Contents xi

Answers to Proposed Exercises 327

Bibliography 337

Index 339

CHAPTER 1: Sequences of Real Numbers

In everyday language, the word "sequence" refers to a collection of objects or events arranged in a specific order. For instance, the daily stock quotes of a particular company on the stock exchange market can be considered a sequence. In this chapter, we shall explore sequences of real numbers and establish precise definitions for the terms "sequence" and "convergence."

1.1 Sequences of Real Numbers

Definition 1.1.1 *A sequence of real numbers is a function of $\mathbb{N}$ into $\mathbb{R}$.*

Note: The elements of the range of a sequence u are called **terms of the sequence**, and they are usually denoted by u_n, instead of the notation $u(n)$ which is generally used for functions. A sequence u is represented by $(u_n)_{n \in \mathbb{N}}$ or, more simply, (u_n). As a function, its graph is the set formed by ordered pairs of the form (n, u_n), $n \in \mathbb{N}$. For example, in Fig. 1.1 is represented the graph of the sequence $u_n = n/(n+1)$. Of course, it is only possible to represent part of a sequence graph.

Definition 1.1.2 *The designatory expression that defines the sequence is called **general term** of the sequence.*

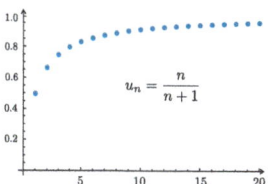

Figure 1.1: The graph of the sequence $u_n = n/(n+1)$

Example 1.1.1 The sequences of general terms $a_n = n^2$ and $b_n = \cos(n)$ are illustrated in Fig. 1.2.

Notes:

1. We can define a sequence without specifying a general term. This is the case for sequences defined recursively. A **recursive relation** is an expression that relates the nth element of a sequence to the previous element or elements. For example,

   $u_1 = 1,\ u_2 = 2,\ u_{n+2} = u_{n+1} + u_n,\ \forall n \in \mathbb{N}$ (Fibonacci sequence).

   In this sequence, we obtain the term u_n as the sum of the two immediately preceding terms: $u_3 = u_2 + u_1 = 3$, $u_4 = u_3 + u_2 = 5$, $u_5 = u_4 + u_3 = 8,\ \ldots$

2. Sometimes only a few terms of the sequence are given, leading the reader to "infer" the rest, for example,

   $$1, 1, 2, 1, 2, 3, 1, 2, 3, 4, \ldots$$

3. There are sequences that are not defined for a finite number of values of $n \in \mathbb{N}$. For instance, the sequence with general term $u_n = \dfrac{1}{n-3}$ is only defined for $n \neq 3$.

Leonardo Fibonacci, also known as Leonardo of Pisa (1170–1250), was born in Pisa, Italy and received his education in North Africa. He is primarily known for his work on number theory and algebra. There are copies of some of his works, one of the most important being the *Liber Abaci*, which introduced Arabic numerals to Europe. It is also in this book that the famous Fibonacci sequence appears: $1, 1, 2, 3, 5, 8, 13, 21, 34, \ldots$, where each number is the sum of the two previous numbers. (Source of image: I benefattori dell'umanità; vol. VI, Firenze, Ducci, 1850)

> **Definition 1.1.3** *A sequence is* **bounded from above** *(or bounded above) if the set of its terms has an upper bound; it is* **bounded from below** *(or bounded below) if the set of its terms has a lower bound; and it is* **bounded** *if the set of its terms is bounded, that is, if the set has an upper bound and a lower bound.*

Example 1.1.2 The sequence $a_n = n^2$ is bounded below but not above (see Fig. 1.2). Specifically, $x = 1$ is a lower bound; however, there are no upper bounds since there is no number L such that $n^2 \leq L,\ \forall n \in \mathbb{N}$.

Example 1.1.3 The sequence $b_n = \cos(n)$ is bounded (see Fig. 1.2). It is enough to note that $-1 \leq \cos(x) \leq 1,\ \forall x \in \mathbb{R}$.

Example 1.1.4 The sequence $c_n = -n$ is bounded above but not below (see Fig. 1.3). In fact, $x = -1$ is an upper bound, but there are no lower bounds because there is no number L such that $-n \geq L,\ \forall n \in \mathbb{N}$.

1.1. Sequences of Real Numbers

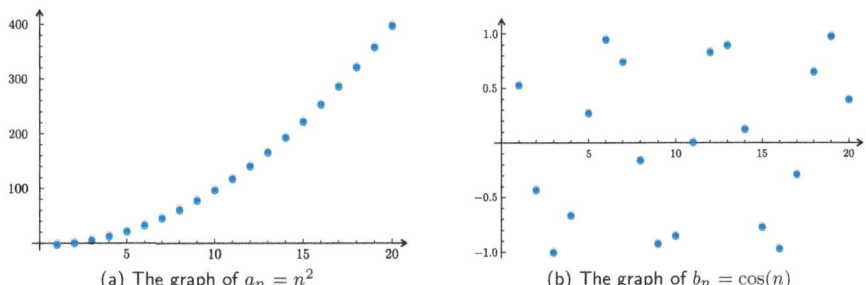

Figure 1.2: A bounded below sequence (**a**) and a bounded sequence (**b**)

Example 1.1.5 The sequence $d_n = (-1)^n n$ is not bounded above or below (see Fig. 1.3).

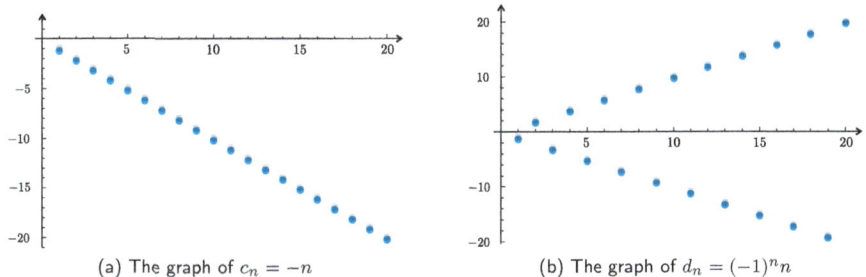

Figure 1.3: A bounded above and unbounded below sequence (**a**) and a sequence not bounded above or below (**b**)

Theorem 1.1.1 *A sequence u is bounded if and only if there exists $M \in \mathbb{R}^+$ such that $|u_n| \leq M$, $\forall n \in \mathbb{N}$.*

Proof: Let u be a bounded sequence. According to Definition 1.1.3, u has upper and lower bounds. Suppose α is a lower bound and β is an upper bound. This means that $\alpha \leq u_n \leq \beta$, $\forall n \in \mathbb{N}$. Let us consider $M = \max(|\alpha|, |\beta|)$. Then

$$|u_n| \leq M, \quad \forall n \in \mathbb{N}.$$

Conversely, suppose there exists $M \in \mathbb{R}^+$ such that
$$|u_n| \leq M, \quad \forall n \in \mathbb{N}.$$
We have $-M \leq u_n \leq M$, $\forall n \in \mathbb{N}$. Thus, $-M$ is a lower bound and M is an upper bound of the sequence. By Definition 1.1.3, u is a bounded sequence. ∎

Example 1.1.6 The sequence $u_n = (-1)^n \dfrac{n+2}{n}$ is bounded (see Fig. 1.4):
$$|u_n| = \left|(-1)^n \frac{n+2}{n}\right| = 1 + \frac{2}{n} \leq 3, \quad \forall n \in \mathbb{N}.$$

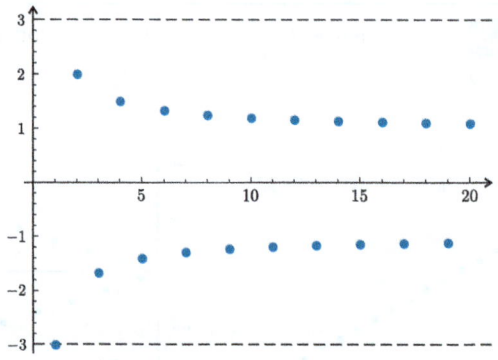

Figure 1.4: A bounded sequence

Definition 1.1.4 *Given two sequences of real numbers, u and v, we call* **sum**, **difference**, *and* **product** *of u and v to the sequences $u+v$, $u-v$, and uv with general terms, respectively, $u_n + v_n$, $u_n - v_n$, and $u_n v_n$. If $v_n \neq 0$, $\forall n \in \mathbb{N}$, we define the* **quotient** *sequence of u and v as the sequence u/v with general term u_n/v_n.*

Definition 1.1.5 *A sequence u is* **increasing** *if $u_n \leq u_{n+1}$, $\forall n \in \mathbb{N}$; it is* **strictly increasing** *if $u_n < u_{n+1}$, $\forall n \in \mathbb{N}$; it is* **decreasing** *if $u_n \geq u_{n+1}$, $\forall n \in \mathbb{N}$; it is* **strictly decreasing** *if $u_n > u_{n+1}$, $\forall n \in \mathbb{N}$; it is* **monotonic** *if it is increasing or decreasing; and it is* **strictly monotonic** *if it is strictly increasing or decreasing.*

1.1. Sequences of Real Numbers

Example 1.1.7 The sequence $u_n = \dfrac{2n}{3n+7}$ is strictly increasing. We can prove this by computing and simplifying the expression $u_{n+1} - u_n$:

$$u_{n+1} - u_n = \frac{2(n+1)}{3(n+1)+7} - \frac{2n}{3n+7} = \frac{(2n+2)(3n+7) - 2n(3n+10)}{(3n+10)(3n+7)}$$

$$= \frac{14}{(3n+10)(3n+7)} > 0, \quad \forall n \in \mathbb{N}.$$

Example 1.1.8 The sequence $u_n = \dfrac{5n}{n^2+1}$ is strictly decreasing. In fact,

$$u_{n+1} - u_n = \frac{5(n+1)}{(n+1)^2+1} - \frac{5n}{n^2+1}$$

$$= \frac{5(n+1)(n^2+1) - 5n(n^2+2n+2)}{(n^2+1)(n^2+2n+2)}$$

$$= -\frac{5(n^2+n-1)}{(n^2+1)(n^2+2n+2)} < 0, \quad \forall n \in \mathbb{N}.$$

Example 1.1.9 The sequence $u_n = n^2$ is strictly increasing because

$$u_{n+1} - u_n = (n+1)^2 - n^2 = n^2 + 2n + 1 - n^2 = 2n + 1 > 0, \quad \forall n \in \mathbb{N}.$$

Example 1.1.10 The sequence $u_n = -n$ strictly decreases because

$$u_{n+1} - u_n = -(n+1) + n = -n - 1 + n = -1 < 0, \quad \forall n \in \mathbb{N}.$$

Example 1.1.11 The sequence $u_n = (-n)^n$ is not monotonic. Indeed,

$$u_{n+1} - u_n = \bigl(-(n+1)\bigr)^{n+1} - (-n)^n = (-1)^{n+1}\bigl((n+1)^{n+1} + n^n\bigr),$$

and this difference is positive if n is odd and negative if n is even.

Example 1.1.12 The sequence $u_n = \dfrac{n + (-1)^n}{n^2}$ is not monotonic (see Fig. 1.5).

$$u_{n+1} - u_n = \begin{cases} \dfrac{n+1-1}{(n+1)^2} - \dfrac{n+1}{n^2} = \dfrac{n^3 - (n+1)^3}{n^2(n+1)^2} < 0, & \text{if } n \text{ is even} \\[2ex] \dfrac{n+1+1}{(n+1)^2} - \dfrac{n-1}{n^2} = \dfrac{n^2+n+1}{n^2(n+1)^2} > 0, & \text{if } n \text{ is odd} \end{cases}$$

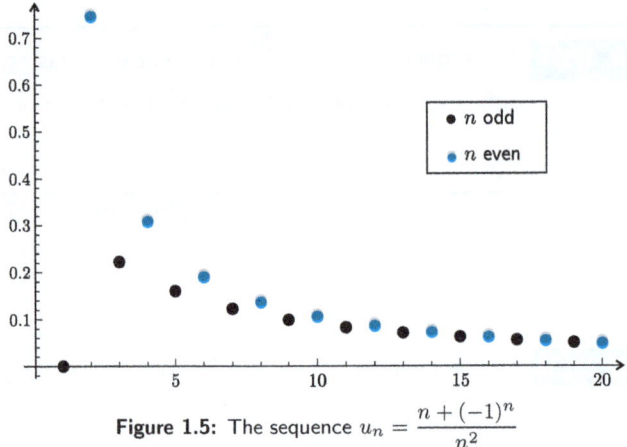

Figure 1.5: The sequence $u_n = \dfrac{n + (-1)^n}{n^2}$

• Arithmetic and Geometric Sequences

We will now discuss two distinct types of sequences that hold significant importance due to their properties: arithmetic sequences (also known as arithmetic progressions) and geometric sequences (also known as geometric progressions).

An **arithmetic sequence** (a_n) is a sequence defined by the recursive relation:

$$\begin{cases} a_1 = k, \\ a_{n+1} = a_n + r, \quad \forall n \in \mathbb{N}, \end{cases}$$

where $k, r \in \mathbb{R}$, and r is called the **common difference** of the arithmetic sequence.

According to this expression, it is easy to conclude that the arithmetic sequence is:

- Strictly increasing if $r > 0$
- Strictly decreasing if $r < 0$
- Constant if $r = 0$

As an example, $1, 4, 7, 10, 13, \ldots$ is an arithmetic sequence with $r = 3$.

Let (a_n) be an arithmetic sequence with common difference r. Then

1.1. Sequences of Real Numbers

$$a_2 = a_1 + r$$
$$a_3 = a_2 + r = a_1 + 2r$$
$$a_4 = a_3 + r = a_1 + 3r$$
$$\vdots$$
$$a_n = a_{n-1} + r = a_1 + (n-1)r$$
$$\vdots$$

The **general term of the arithmetic sequence** with common difference r is given by the expression

$$a_n = a_1 + (n-1)r, \quad \forall n \in \mathbb{N}.$$

At first glance, the sum of the first terms of an arithmetic sequence may seem like a time-consuming problem to solve.

However, according to legend, Gauss, one of the greatest mathematicians of all time, solved the question of calculating the sum of the first 100 natural numbers in a few minutes when he was still a child. What reasoning would have allowed him to arrive so quickly at the result?

If we write the integers from 1 to 100, we find that:

- The sum of the first and last numbers is 101.
- The sum of the second and penultimate numbers is 101.
- The sum of the third and third-to-last numbers is 101, and so on.

There are precisely 50 pairs of numbers whose sum is a constant of value 101, so it is concluded that the sum of the first 100 natural numbers is $101 \times 50 = 5050$.

Carl Friedrich Gauss (1777–1855) was a German mathematician. He worked in a wide variety of areas in both Mathematics and Physics. He was a gifted child. His first mathematical discoveries were made in his teens. His magnum opus, *Disquisitiones Arithmeticae*, was completed in 1798, when Gauss was only 21 years old. (Source of image: Oil portrait by Gottlieb Wilhelm Emil Biermann)

In general, S_n, **the sum of the first n consecutive terms of an arithmetic sequence** (a_n) with common difference r, is given by the formula

$$S_n = \frac{a_1 + a_n}{2} \times n.$$

Let us prove this using the method of induction.
For $n = 1$, the equality is trivial. The property is proven if it is hereditary.
If we admit that equality is valid for n, then

$$\begin{aligned}
S_{n+1} = S_n + a_{n+1} &= \frac{a_1 + a_n}{2} \times n + a_{n+1} \\
&= \frac{a_1 + a_1 + (n-1)r}{2} \times n + a_1 + nr \\
&= n\,a_1 + \frac{(n-1)r}{2} \times n + a_1 + nr \\
&= (n+1)a_1 + \left(\frac{n-1}{2} + 1\right) \times nr \\
&= (n+1)a_1 + \frac{n+1}{2} \times nr \\
&= (n+1)\left(a_1 + \frac{nr}{2}\right) \\
&= (n+1)\frac{a_1 + a_1 + nr}{2} \\
&= (n+1)\frac{a_1 + a_{n+1}}{2},
\end{aligned}$$

which confirms that the equality holds for $n+1$.

A **geometric sequence** (a_n) is the sequence defined by the recursive relation

$$\begin{cases} a_1 = k \\ a_{n+1} = a_n \times r, \quad \forall n \in \mathbb{N}, \end{cases}$$

where $k, r \in \mathbb{R} \setminus \{0\}$, and r is known as the **common ratio** of the geometric sequence.

For example, $\frac{1}{10}, \frac{1}{100}, \frac{1}{1000}, \ldots$ is a geometric sequence with $r = \frac{1}{10}$.

Similar to arithmetic sequences, it is possible to calculate any term of the sequence knowing the first term and the common ratio. Considering the definition of a geometric sequence with common ratio r, we obtain the following:

$$\begin{aligned}
a_2 &= a_1 \times r \\
a_3 &= a_2 \times r = a_1 \times r^2 \\
a_4 &= a_3 \times r = a_1 \times r^3 \\
&\vdots \\
a_n &= a_1 \times r^{n-1} \\
&\vdots
\end{aligned}$$

1.1. Sequences of Real Numbers

The **general term of a geometric sequence** with ratio r is expressed as follows:
$$a_n = a_1 \times r^{n-1}, \quad \forall n \in \mathbb{N}.$$

According to this expression, it is easy to conclude that the monotonicity of a geometric sequence is:

- Increasing if $a_1 > 0$ and $r > 1$ or if $a_1 < 0$ and $0 < r < 1$
- Decreasing if $a_1 > 0$ and $0 < r < 1$ or if $a_1 < 0$ and $r > 1$
- Constant if $r = 1$
- Not monotonic if $r < 0$

There is a well-known legend that a Hindu prince, fascinated with the chess game, decided to reward its inventor, who asked for this reward to be given in wheat, receiving one grain for the first square of the chessboard, two grains for the second square, four grains for the third, and so on, doubling the number of grains from the last square considered. We conclude that the number of wheat grains is represented by the sum of the first 64 terms of a geometric sequence with common ratio 2.

$$S_{64} = 1 + 2 + 2^2 + 2^3 + \ldots + 2^{63}.$$

How can we determine the value of this expression?
We can calculate the sum in the more general case of a geometric sequence, which involves finding the sum of the first n terms. Let

$$S_n = a_1 + a_2 + a_3 + \ldots + a_n \tag{1.1}$$

be the sum of the first n terms. Multiplying S_n by the ratio r, we obtain

$$r\,S_n = r\,a_1 + r\,a_2 + r\,a_3 + \ldots + r\,a_n = a_2 + a_3 + a_4 + \ldots + a_{n+1}. \tag{1.2}$$

Subtracting (1.2) from (1.1), we obtain

$$S_n - r\,S_n = (a_1 + a_2 + a_3 + \ldots + a_n) - (a_2 + a_3 + a_4 + \ldots + a_{n+1}),$$

that is,

$$S_n = \frac{a_1 - a_{n+1}}{1 - r} = \frac{a_1 - a_1 r^n}{1 - r} = a_1 \frac{1 - r^n}{1 - r}.$$

Returning to the inventor of the game of chess, the requested reward is

$$S_{64} = 2^{64} - 1 = 18\ 446\ 744\ 073\ 709\ 551\ 615$$

wheat grains (which not even the richest prince would be able to pay).

Example 1.1.13 Let us consider Fig. 1.6, and let (A_n) be the sequence of the areas of the semicircles delimited by the lines $C_1, C_2, C_3, \ldots$. We show that (A_n) is a geometric sequence and express its general term as follows:

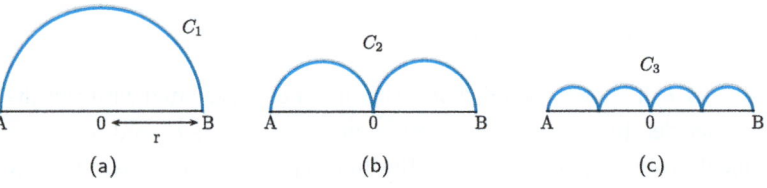

Figure 1.6: Semicircles

Each semicircle has an area $\dfrac{1}{2}\pi r^2$, where r is the radius of the respective circumference. In Fig. 1.6a, the radius of the circumference is r; in Fig. 1.6b, each semicircle has a radius of $\dfrac{r}{2}$; in Fig. 1.6c, each semicircle has a radius of $\dfrac{r}{4}$; and so on. Therefore, we can write the first terms of the sequence A_n.

$$A_1 = \frac{1}{2}\pi r^2$$
$$A_2 = \frac{1}{2}\pi \left(\frac{r}{2}\right)^2 \times 2 = \pi \frac{r^2}{4}$$
$$A_3 = \frac{1}{2}\pi \left(\frac{r}{4}\right)^2 \times 4 = \pi \frac{r^2}{8}.$$

By induction, it can be proven that $A_{n+1} = \dfrac{1}{2} \times A_n$, that is, it is a geometric sequence with common ratio $\dfrac{1}{2}$, so the general term of the sequence is $A_n = \pi \dfrac{r^2}{2^n}$.

1.1. Sequences of Real Numbers

- **Limits of Sequences**

> **Definition 1.1.6** *A sequence u tends to $+\infty$ (and we write $u_n \to +\infty$ or $\lim_{n\to+\infty} u_n = +\infty$ or $\lim u_n = +\infty$), if*
>
> $$\forall L \in \mathbb{R}^+ \ \exists p \in \mathbb{N}: \ n > p \Rightarrow u_n > L.$$
>
> *A sequence u tends to $-\infty$ (and we write $u_n \to -\infty$ or $\lim_{n\to+\infty} u_n = -\infty$ or $\lim u_n = -\infty$), if*
>
> $$\forall L \in \mathbb{R}^+ \ \exists p \in \mathbb{N}: \ n > p \Rightarrow u_n < -L.$$

Example 1.1.14 We prove that $u_n = n^2 \to +\infty$. Given $L > 0$, there exists $p \in \mathbb{N}$ such that $p > \sqrt{L}$; if $n > p$, then $n^2 > p^2 > L$. It is evident that p depends on L; for instance, if $L = 100$, it is sufficient to consider $p = 10$, but if $L = 200$, we must consider $p = 14$. This example is illustrated in Fig. 1.7.

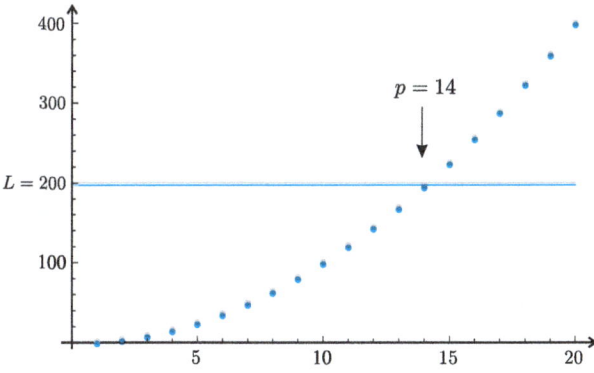

Figure 1.7: The sequence $u_n = n^2$ tends to $+\infty$

Similarly, it can be shown that $v_n = n^\alpha$, $\alpha \in \mathbb{R}^+$, tends to $+\infty$, that $y_n = -n$ tends to $-\infty$, and that if $w_n = (-n)^n$, then $|w_n| = n^n \to +\infty$.

Notes:

1. Sometimes the following concept is also used: A sequence (u_n) has **general infinite limit** if $|u_n| \to +\infty$. For example, the sequence

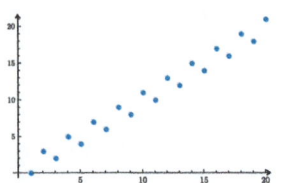

Figure 1.8: The sequence $u_n = n + (-1)^n$

$u_n = (-1)^n n$ has general infinite limit despite $u_n \not\to +\infty$ and $u_n \not\to -\infty$.

2. The fact that $u_n \to +\infty$ does not imply that u is increasing (nor that there is an order from which it is increasing). For instance, the sequence $u_n = n + (-1)^n$, illustrated in Fig. 1.8, tends to $+\infty$, but it is not monotonic.

3. If u is such that $u_n \to +\infty$, $u_n \to -\infty$, or $|u_n| \to +\infty$, then u is unbounded. The converse is not true. For example, the sequence

$$u_n = \begin{cases} n, & \text{if } n \text{ is even} \\ \dfrac{1}{n}, & \text{if } n \text{ is odd} \end{cases}$$

is unbounded and $u_n \not\to +\infty$, $u_n \not\to -\infty$, $|u_n| \not\to +\infty$.

From Definition 1.1.6, the next result follows immediately.

> **Theorem 1.1.2** *Let u and v be sequences such that from a certain order, $u_n \leq v_n$. Then,*
>
> a) $u_n \to +\infty \Rightarrow v_n \to +\infty$.
>
> b) $v_n \to -\infty \Rightarrow u_n \to -\infty$.

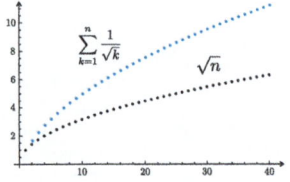

Figure 1.9: An illustration of Theorem 1.1.2

Example 1.1.15 Let us consider the sequence

$$u_n = \sum_{k=1}^{n} \frac{1}{\sqrt{k}} = 1 + \frac{1}{\sqrt{2}} + \frac{1}{\sqrt{3}} + \cdots + \frac{1}{\sqrt{n}}.$$

Because we have inequality

$$1 + \frac{1}{\sqrt{2}} + \frac{1}{\sqrt{3}} + \cdots + \frac{1}{\sqrt{n}} \geq n \times \frac{1}{\sqrt{n}} = \sqrt{n}$$

and $\sqrt{n} \to +\infty$, we can state that $u_n \to +\infty$ (see Fig. 1.9).

1.1. Sequences of Real Numbers

Theorem 1.1.3 *Let $u_n \to +\infty$, $v_n \to +\infty$, and $w_n \to -\infty$. Then*

a) $\lim (u_n + v_n) = +\infty$.

b) $\lim (u_n v_n) = +\infty$.

c) $\lim (u_n w_n) = -\infty$.

d) $\lim (u_n)^p = +\infty$, $\forall p \in \mathbb{N}$.

e) $\lim |u_n| = \lim |v_n| = \lim |w_n| = +\infty$.

Definition 1.1.7 *Let u be a sequence and $a \in \mathbb{R}$. We say that u converges to a (or that tends to a or, still, that the limit of the sequence is a), and we write $u_n \to a$ or $\lim_{n \to +\infty} u_n = a$ or $\lim u_n = a$, if*

$$\forall \varepsilon > 0 \; \exists p \in \mathbb{N}: \; n > p \Rightarrow |u_n - a| < \varepsilon.$$

We say that a sequence u is convergent if there is $a \in \mathbb{R}$ such that u converges to a; otherwise, u is divergent.

In other words, u is convergent if for every $\varepsilon > 0$, we can choose p such that from the order p all terms u_n are in the interval $]a - \varepsilon, a + \varepsilon[$.

Taking the sequence $u_n = \dfrac{5n}{9n+1}$, we illustrate this definition in Fig. 1.10 for two values of ε.

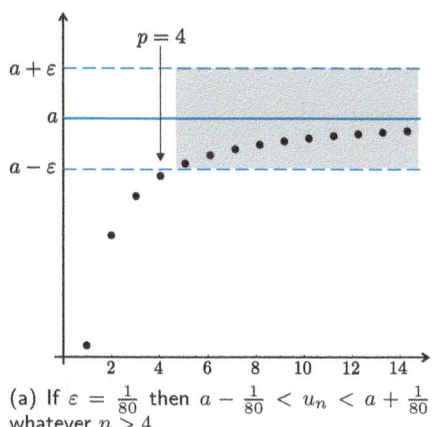

(a) If $\varepsilon = \frac{1}{80}$ then $a - \frac{1}{80} < u_n < a + \frac{1}{80}$ whatever $n > 4$

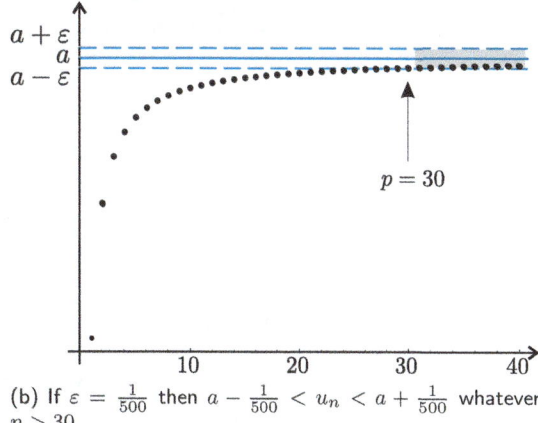

(b) If $\varepsilon = \frac{1}{500}$ then $a - \frac{1}{500} < u_n < a + \frac{1}{500}$ whatever $n > 30$

Figure 1.10: The value of p varies with the value of ε

The limit of this sequence is $a = \dfrac{5}{9}$. To prove this fact using the definition, we proceed as follows: Let $\varepsilon > 0$; we must show that there exists a natural number, p, such that

$$\text{if } n > p \text{ then } \left| \dfrac{5n}{9n+1} - \dfrac{5}{9} \right| < \varepsilon.$$

Note that
$$\left| \dfrac{5n}{9n+1} - \dfrac{5}{9} \right| = \left| \dfrac{-5}{9(9n+1)} \right| = \dfrac{5}{9(9n+1)}.$$

Solving inequality
$$\dfrac{5}{9(9n+1)} < \varepsilon$$

for n, we obtain $n > \dfrac{5 - 9\varepsilon}{81\varepsilon}$.

If we choose a natural number p greater than or equal to $\dfrac{5 - 9\varepsilon}{81\varepsilon}$, then

$$n > p \Rightarrow n > \dfrac{5 - 9\varepsilon}{81\varepsilon} \Rightarrow \left| \dfrac{5n}{9n+1} - \dfrac{5}{9} \right| < \varepsilon.$$

Note that if $\varepsilon \geq \dfrac{5}{9}$, we can choose $p = 1$. If $\varepsilon < \dfrac{5}{9}$, we can choose $p = \text{Int}\left(\dfrac{5-9\varepsilon}{81\varepsilon} \right)$. This number $p = \text{Int}\left(\dfrac{5-9\varepsilon}{81\varepsilon} \right)$ is the smallest number that makes the statement true.

Example 1.1.16 Let us prove that $u_n = \dfrac{1}{n} \to 0$. Given an arbitrary $\varepsilon > 0$, we must show that there exists a natural number, p, such that

$$\text{if } n > p \text{ then } \left| \dfrac{1}{n} \right| = \dfrac{1}{n} < \varepsilon.$$

Let p be a natural number greater than or equal to $\dfrac{1}{\varepsilon}$. Then, for $n > p$, we have $\dfrac{1}{n} < \dfrac{1}{p} < \varepsilon$. If we choose $p = \text{Int}\left(\dfrac{1}{\varepsilon} \right)$, this will be the smallest order from which the proposition is true for a given value of ε (see Fig. 1.11).

Similarly, we can prove that the sequence $v_n = \dfrac{1}{n^\alpha}$, $\alpha \in \mathbb{R}^+$, converges to zero.

1.1. Sequences of Real Numbers

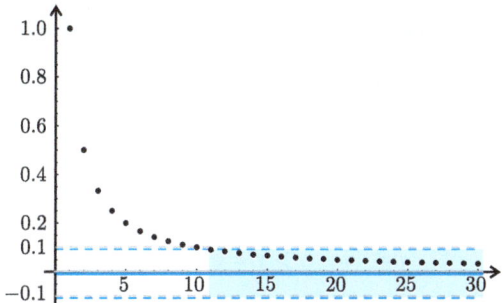

Figure 1.11: If $\varepsilon = 0.1$, then $-\varepsilon < \frac{1}{n} < \varepsilon$ if $n > 10$

Notes:

1. In the language of neighborhoods, Definition 1.1.7 is equivalent to
$$\forall \varepsilon > 0\ \exists p \in \mathbb{N}:\ n > p \Rightarrow u_n \in V_\varepsilon(a).$$

2. We could also write equivalently
$$\forall \varepsilon > 0\ \exists p \in \mathbb{N}: |u_n - a| < \varepsilon,\ \forall n > p.$$

3. Consider the set $\overline{\mathbb{R}} = \mathbb{R} \cup \{-\infty, +\infty\}$, where $-\infty$ and $+\infty$ are two distinct mathematical objects that are not real numbers. We can introduce an order relation in this set as follows:

    i) If $x, y \in \mathbb{R}$, $x < y$ on $\overline{\mathbb{R}}$ if and only if $x < y$ on $\mathbb{R}$.

    ii) $-\infty < x < +\infty$, $\forall x \in \mathbb{R}$.

    The set $\overline{\mathbb{R}}$, with this order relation, is called **extended real line**.

    We can extend the notion of neighborhood to $\overline{\mathbb{R}}$. Let $\varepsilon > 0$. If $a \in \mathbb{R}$, we call the ε neighborhood of a the set $V_\varepsilon(a) =]a - \varepsilon, a + \varepsilon[$ (which coincides with the neighborhood in $\mathbb{R}$). The set $V_\varepsilon(+\infty) =]\frac{1}{\varepsilon}, +\infty]$ is called the ε neighborhood of $+\infty$. The set $V_\varepsilon(-\infty) = [-\infty, -\frac{1}{\varepsilon}[$ is called the ε neighborhood of $-\infty$. With the definitions given previously, we can unify from a formal viewpoint Definitions 1.1.6 and 1.1.7: If $a \in \overline{\mathbb{R}}$,
$$u_n \to a\ \text{if and only if}\ \forall \varepsilon > 0\ \exists p \in \mathbb{N}:\ n > p \Rightarrow u_n \in V_\varepsilon(a).$$

Theorem 1.1.4 (Uniqueness of the limit) *Let $a, b \in \mathbb{R}$. If $u_n \to a$ and $u_n \to b$, then $a = b$.*

Proof: (By contradiction)

If $a \neq b$, we can consider $\varepsilon = \dfrac{|b-a|}{3}$. By the definition of a convergent sequence,
$$\exists p \in \mathbb{N}: n > p \Rightarrow |u_n - a| < \varepsilon$$
and
$$\exists q \in \mathbb{N}: n > q \Rightarrow |u_n - b| < \varepsilon.$$

Let $m > \max\{p, q\}$. Then,
$$\varepsilon = \frac{|b-a|}{3} = \frac{|b - u_m + u_m - a|}{3} \leq \frac{|b - u_m| + |u_m - a|}{3}$$
$$< \frac{\varepsilon + \varepsilon}{3} = \frac{2\varepsilon}{3} < \varepsilon$$

and we conclude that $\varepsilon < \varepsilon$, which is impossible. ∎

Theorem 1.1.5 *If u and v are convergent sequences, then:*

a) $\lim(u_n + v_n) = \lim u_n + \lim v_n$.

b) $\lim(u_n \cdot v_n) = \lim u_n \cdot \lim v_n$.

c) $\lim(u_n)^p = (\lim u_n)^p$, $p \in \mathbb{N}$.

d) $\lim \dfrac{u_n}{v_n} = \dfrac{\lim u_n}{\lim v_n}$, *if $v_n \neq 0$, $\forall n \in \mathbb{N}$, and $\lim v_n \neq 0$.*

e) $\lim(u_n)^{1/p} = (\lim u_n)^{1/p}$ *(if p is even, it must be $u_n \geq 0$, $\forall n \in \mathbb{N}$).*

f) $\lim\left((u_n)^{v_n}\right) = (\lim u_n)^{\lim v_n}$, *if $u_n > 0$, $\forall n \in \mathbb{N}$, and the limits of the sequences are not both zero.*

g) $\lim |u_n| = |\lim u_n|$.

h) $(\exists p \in \mathbb{N} \; \forall n \geq p, \; u_n \geq 0) \Rightarrow \lim u_n \geq 0$.

i) $(\exists p \in \mathbb{N} \; \forall n \geq p, \; u_n \geq v_n) \Rightarrow \lim u_n \geq \lim v_n$.

Proof: We will prove only items a) and g).

1.1. Sequences of Real Numbers

a) Let a and b be the limits of sequences (u_n) and (v_n), respectively. Let $\varepsilon > 0$ be arbitrary. By definition,
$$\exists p_1 \in \mathbb{N}: n > p_1 \Rightarrow |u_n - a| < \frac{\varepsilon}{2}$$
and
$$\exists p_2 \in \mathbb{N}: n > p_2 \Rightarrow |v_n - b| < \frac{\varepsilon}{2}.$$
Let $p = \max\{p_1, p_2\}$; then, if $n > p$,
$$|(u_n + v_n) - (a+b)| = |u_n + v_n - a - b| = |u_n - a + v_n - b|$$
$$\leq |u_n - a| + |v_n - b| < \frac{\varepsilon}{2} + \frac{\varepsilon}{2} = \varepsilon.$$
We thus conclude that $u_n + v_n \to a + b$.

g) Let us denote $a = \lim u_n$. Let ε be an arbitrary positive real number. By definition,
$$\exists p \in \mathbb{N}: n > p \Rightarrow |u_n - a| < \varepsilon.$$
But $||u_n| - |a|| \leq |u_n - a|$, so $||u_n| - |a|| < \varepsilon$, $\forall n > p$. We conclude that $|u_n| \to |a|$. ∎

Definition 1.1.8 *A sequence u is a* **null sequence** *if $u_n \to 0$.*

Notes:

1. It is evident from the definitions that (u_n) converges to a if and only if $(u_n - a)$ is a null sequence.

2. Considering the definitions, it is easy to verify that if u is a null sequence, $|1/u_n| \to +\infty$, and, reciprocally, if $|v_n| \to +\infty$, then $1/v_n \to 0$.

Theorem 1.1.6 *If $u_n \to 0$ and v is a bounded sequence, then $u_n \cdot v_n \to 0$.*

Proof: Let $M > 0$ such that $|v_n| \leq M$, $\forall n \in \mathbb{N}$. Given an arbitrary $\varepsilon > 0$, let $p \in \mathbb{N}$, such that $|u_n| < \frac{\varepsilon}{M}$, $\forall n > p$. Then
$$|u_n \cdot v_n| < \varepsilon, \; \forall n > p. \; \blacksquare$$

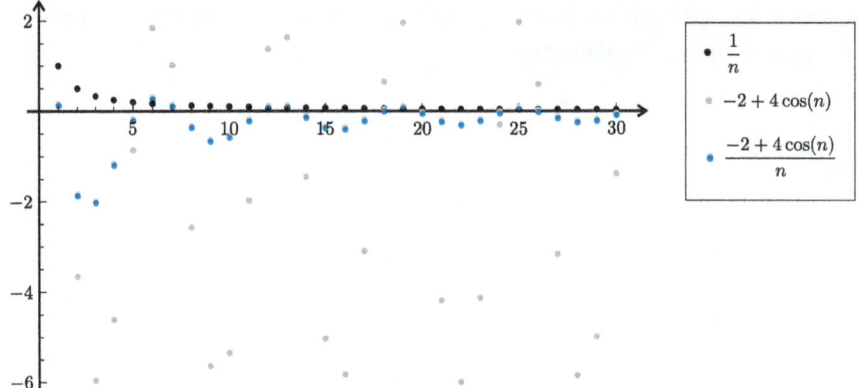

Figure 1.12: An illustration of Theorem 1.1.6

Example 1.1.17 Let us calculate the limit of the sequence
$$a_n = \frac{-2 + 4\cos(n)}{n}.$$
We know that $-1 \leq \cos(n) \leq 1$, $\forall n \in \mathbb{N}$; therefore,
$$-6 \leq -2 + 4\cos(n) \leq 2, \quad \forall n \in \mathbb{N}.$$
In other words, the sequence $b_n = -2 + 4\cos(n)$ is bounded (see Fig. 1.12). As Example 1.1.15 shows, the sequence with general term $\dfrac{1}{n}$ is a null sequence. By Theorem 1.1.6 we can assert that
$$\lim \frac{-2 + 4\cos(n)}{n} = 0.$$

From Definition 1.1.7 it is easy to show the following theorem:

Theorem 1.1.7 *Every convergent sequence is bounded.*

Note: The converse is not true. For example, the sequence
$$u_n = \cos(n\pi) = (-1)^n$$
is bounded but does not converge.

1.1. Sequences of Real Numbers

Theorem 1.1.8 (Squeeze Theorem) *If $u_n \to a$, $v_n \to a$ and, from a certain order, $u_n \leq w_n \leq v_n$, then $w_n \to a$.*

Proof: Let ε be any positive real number. Then
$$\exists p_1 \in \mathbb{N}: n > p_1 \Rightarrow a - \varepsilon < u_n < a + \varepsilon,$$
$$\exists p_2 \in \mathbb{N}: n > p_2 \Rightarrow a - \varepsilon < v_n < a + \varepsilon,$$
$$\exists p_3 \in \mathbb{N}: n > p_3 \Rightarrow u_n \leq w_n \leq v_n.$$
Let $p = \max\{p_1, p_2, p_3\}$. If $n > p$, then
$$a - \varepsilon < u_n \leq w_n \leq v_n < a + \varepsilon. \blacksquare$$

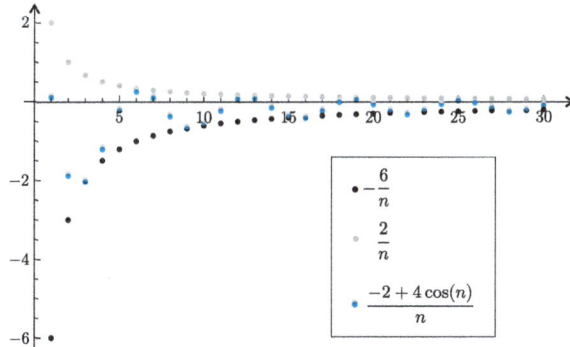

Figure 1.13: An illustration of the Squeeze Theorem

Example 1.1.18 Let us calculate the limit of the sequence
$$a_n = \frac{-2 + 4\cos(n)}{n}$$
from the previous example, now applying Theorem 1.1.8.
We know that $-1 \leq \cos(n) \leq 1$, $\forall n \in \mathbb{N}$, which implies that
$$-\frac{6}{n} \leq \frac{-2 + 4\cos(n)}{n} \leq \frac{2}{n}, \quad \forall n \in \mathbb{N}.$$
Considering that $\frac{1}{n} \to 0$, we conclude that $\lim \frac{-2 + 4\cos(n)}{n} = 0$ (see Fig. 1.13).

Theorem 1.1.9 *Let $P(x) = a_0 x^p + \cdots + a_p$ and $Q(x) = b_0 x^q + \cdots + b_q$ be two polynomial functions with real coefficients, $p, q \in \mathbb{N}$, $a_0 \neq 0, b_0 \neq 0$, and v be a sequence, $v_n \to +\infty$. Then*

$$\lim \frac{P(v_n)}{Q(v_n)} = \begin{cases} \dfrac{a_0}{b_0}, & \text{if } p = q \\ +\infty, & \text{if } p > q \land \dfrac{a_0}{b_0} > 0 \\ -\infty, & \text{if } p > q \land \dfrac{a_0}{b_0} < 0 \\ 0, & \text{if } p < q. \end{cases}$$

Proof: Let $p = q$. Dividing by v_n^p both members of the fraction

$$\frac{P(v_n)}{Q(v_n)} = \frac{a_0 v_n^p + a_1 v_n^{p-1} + \cdots + a_{p-1} v_n + a_p}{b_0 v_n^p + b_1 v_n^{p-1} + \cdots + b_{p-1} v_n + b_p},$$

we obtain

$$\frac{P(v_n)}{Q(v_n)} = \frac{a_0 + \dfrac{a_1}{v_n} + \cdots + \dfrac{a_{p-1}}{v_n^{p-1}} + \dfrac{a_p}{v_n^p}}{b_0 + \dfrac{b_1}{v_n} + \cdots + \dfrac{b_{p-1}}{v_n^{p-1}} + \dfrac{b_p}{v_n^p}}.$$

By hypothesis, $v_n \to +\infty$; then $\dfrac{1}{v_n^s} \to 0$ for all $s \in \mathbb{N}$. We conclude by Theorem 1.1.5 that

$$\frac{P(v_n)}{Q(v_n)} \to \frac{a_0}{b_0}.$$

Let $p \neq q$. Putting v_n^p in evidence in the numerator and v_n^q in the denominator, we obtain the equalities

$$\frac{P(v_n)}{Q(v_n)} = \frac{a_0 v_n^p + a_1 v_n^{p-1} + \cdots + a_{p-1} v_n + a_p}{b_0 v_n^q + b_1 v_n^{q-1} + \cdots + b_{q-1} v_n + b_q}$$

$$= \frac{v_n^p \left(a_0 + \dfrac{a_1}{v_n} + \cdots + \dfrac{a_{p-1}}{v_n^{p-1}} + \dfrac{a_p}{v_n^p} \right)}{v_n^q \left(b_0 + \dfrac{b_1}{v_n} + \cdots + \dfrac{b_{q-1}}{v_n^{q-1}} + \dfrac{b_q}{v_n^q} \right)}$$

$$= v_n^{p-q} \frac{a_0 + \dfrac{a_1}{v_n} + \cdots + \dfrac{a_{p-1}}{v_n^{p-1}} + \dfrac{a_p}{v_n^p}}{b_0 + \dfrac{b_1}{v_n} + \cdots + \dfrac{b_{q-1}}{v_n^{q-1}} + \dfrac{b_q}{v_n^q}}.$$

1.1. Sequences of Real Numbers

We saw earlier that

$$\frac{a_0 + \dfrac{a_1}{v_n} + \cdots + \dfrac{a_{p-1}}{v_n^{p-1}} + \dfrac{a_p}{v_n^p}}{b_0 + \dfrac{b_1}{v_n} + \cdots + \dfrac{b_{q-1}}{v_n^{q-1}} + \dfrac{b_q}{v_n^q}} \to \frac{a_0}{b_0}.$$

If $p > q$, $v_n^{p-q} \to +\infty$, so the limit of $\dfrac{P(v_n)}{Q(v_n)}$ will be $+\infty$ if $\dfrac{a_0}{b_0} > 0$ and $-\infty$ if $\dfrac{a_0}{b_0} < 0$.

If $p < q$, $v_n^{p-q} \to 0$, so $\dfrac{P(v_n)}{Q(v_n)} \to 0 \cdot \dfrac{a_0}{b_0} = 0$. ∎

Theorem 1.1.10 *Let $a \in \mathbb{R}$. Then*

$$\lim a^n = \begin{cases} +\infty, & \text{if } a > 1 \\ 0, & \text{if } |a| < 1 \\ 1, & \text{if } a = 1. \end{cases}$$

If $a \leq -1$, the limit does not exist. If $a < -1$, $|a^n| \to +\infty$.

Proof: Let $a \in \mathbb{R}$ such that $|a| > 1$. We intend to prove that $|a|^n \to +\infty$. Consider an arbitrary $L > 0$. Since $\log(|a|)$ is positive when $|a| > 1$, we have

$$|a|^n > L \Leftrightarrow e^{n \log(|a|)} > e^{\log(L)} \Leftrightarrow n \log(|a|) > \log(L) \Leftrightarrow n > \frac{\log(L)}{\log(|a|)}.$$

Therefore, it is sufficient to consider in Definition 1.1.6, for example, $p = \text{Int}\left(\frac{\log(L)}{\log(|a|)}\right) + 1$. Thus, we have shown that if $|a| > 1$, then $|a^n| \to +\infty$.

If $a > 1$, $|a^n| = a^n$, so $a^n \to +\infty$. If $a < -1$, $|a^n| \to +\infty$ (note that, in this case, a^n is negative when n is odd and positive when n is even).

If $a = 0$, it is obvious that $a^n = 0 \to 0$.

Let a be such that $|a| < 1$, $a \neq 0$, and take an arbitrary $\varepsilon > 0$. As $\log(|a|) < 0$ because $|a| < 1$, we get

$$|a^n| = |a|^n < \varepsilon \Leftrightarrow n \log(|a|) < \log(\varepsilon) \Leftrightarrow n > \frac{\log(\varepsilon)}{\log(|a|)}.$$

If $\varepsilon < 1$, it is enough to take in Definition 1.1.7, for instance, $p = \text{Int}\left(\frac{\log(\varepsilon)}{\log(|a|)}\right) + 1$; if $\varepsilon \geq 1$, we can consider $p = 1$.

If $a = 1$, then $a^n = 1$, $\forall n \in \mathbb{N}$, so $a^n \to 1$.

If $a = -1$, then $a^n = (-1)^n$, and this sequence does not converge. ∎

Example 1.1.19 Let us calculate the limit of the sequence

$$\left(\left(\frac{4n}{2n+1}\right)^n\right)_{n \in \mathbb{N}}.$$

Given $n \in \mathbb{N}$,

$$3n \geq 2n+1 \Rightarrow \frac{4}{3} = \frac{4n}{3n} \leq \frac{4n}{2n+1} \Rightarrow \left(\frac{4}{3}\right)^n \leq \left(\frac{4n}{2n+1}\right)^n.$$

But by Theorem 1.1.10, $\lim\left(\frac{4}{3}\right)^n = +\infty$. Considering the last inequality, we conclude by Theorem 1.1.2 that $\lim\left(\frac{4n}{2n+1}\right)^n = +\infty$.

Example 1.1.20 Let us evaluate the limit of the sequence of general term

$$(a_n)^n = \left(\frac{n}{5n+1}\right)^n.$$

The sequence (a_n) is monotonically increasing and has a limit of $\frac{1}{5}$, so

$$\frac{1}{6} \leq \frac{n}{5n+1} \leq \frac{1}{5} \Rightarrow \left(\frac{1}{6}\right)^n \leq \left(\frac{n}{5n+1}\right)^n \leq \left(\frac{1}{5}\right)^n.$$

But by Theorem 1.1.10, $\lim\left(\frac{1}{5}\right)^n = \lim\left(\frac{1}{6}\right)^n = 0$. From the previous inequalities, we conclude by Theorem 1.1.8 that $\lim\left(\frac{n}{5n+1}\right)^n = 0$.

1.1. Sequences of Real Numbers 23

> **Theorem 1.1.11**
>
> a) Let $a \in \overline{\mathbb{R}}$. If $u_n \to a$, then $\dfrac{u_1 + \cdots + u_n}{n} \to a$.
>
> b) If $b \in \mathbb{R}^+$, then $\sqrt[n]{b} \to 1$.
>
> c) Let $c \in \overline{\mathbb{R}}$. If $u_n > 0$, $\forall n \in \mathbb{N}$, and $\dfrac{u_{n+1}}{u_n} \to c$, then $\sqrt[n]{u_n} \to c$.

<u>Proof</u>: a) Let $a \in \mathbb{R}$ and $\varepsilon > 0$ be arbitrary. We want to prove that

$$\exists s \in \mathbb{N}: \; n > s \Rightarrow \left| \frac{u_1 + u_2 + \cdots + u_n}{n} - a \right| < \varepsilon,$$

that is,

$$\exists s \in \mathbb{N}: \; n > s \Rightarrow \left| \frac{u_1 - a + u_2 - a + \cdots + u_n - a}{n} \right| < \varepsilon.$$

We know that $\exists p \in \mathbb{N}: \; n > p \Rightarrow |u_n - a| < \frac{\varepsilon}{2}$. Additionally,

$$\left| \frac{u_1 - a + u_2 - a + \cdots + u_p - a}{n} \right| \leq$$
$$\leq \frac{|u_1 - a| + |u_2 - a| + \cdots + |u_p - a|}{n} \leq \frac{Mp}{n},$$

where $M = \max\{|u_1 - a|, |u_2 - a|, \cdots, |u_p - a|\}$. Let $q \in \mathbb{N}$ such that $n > q \Rightarrow \frac{Mp}{n} < \frac{\varepsilon}{2}$. Let $s = \max\{p, q\}$. If $n > s$, we have

$$\left| \frac{u_1 - a + u_2 - a + \cdots + u_n - a}{n} \right| \leq$$
$$\leq \frac{|u_1 - a| + |u_2 - a| + \cdots + |u_p - a|}{n} + \frac{|u_{p+1} - a| + \cdots + |u_n - a|}{n}$$
$$\leq \frac{\varepsilon}{2} + \frac{\frac{\varepsilon}{2}(n-p)}{n} \leq \varepsilon.$$

Thus, if $a \in \mathbb{R}$, the statement is proved.

Now, we show the theorem for $a = +\infty$. Let L be an arbitrary positive real number. We know that there exists $p \in \mathbb{N}: \; n > p \Rightarrow u_n > 2L$. Given that

$$\frac{u_1 + u_2 + \cdots + u_p}{n} \to 0,$$

then

$$\exists q \in \mathbb{N}: \; n > q \Rightarrow \frac{u_1 + u_2 + \cdots + u_p}{n} > \frac{-L}{2}.$$

Additionally,
$$\frac{u_{p+1}+u_{p+1}+\cdots+u_n}{n} > \frac{2L(n-p)}{n}.$$

As $\frac{n-p}{n} \to 1$, $\exists r \in \mathbb{N}: n > r \Rightarrow \frac{n-p}{n} > \frac{3}{4}$. Let $s = \max\{p,q,r\}$. If $n > s$, we have
$$\frac{u_1+u_2+\cdots+u_n}{n} > \frac{-L}{2} + \frac{2L(n-p)}{n} > \frac{-L}{2} + 2L\frac{3}{4} = L.$$

Thus, the statement is proved if $a = +\infty$.

For $a = -\infty$, the procedure is similar but with obvious adaptations.

b) Let $b > 1$. In this case $\sqrt[n]{b} > 1$, so we can write $\sqrt[n]{b} = 1 + h_n$, where, for each n, $h_n > 0$. Then, by Newton's Binomial,
$$b = (1+h_n)^n = \sum_{k=0}^{n}\binom{n}{k}h_n^k > 1 + n\,h_n,$$

so $n\,h_n < b-1$, from which $0 < h_n < \frac{b-1}{n}$, and we deduce that $h_n \to 0$, that is, $\sqrt[n]{b} \to 1$.

If $b < 1$, we take $a = \frac{1}{b} > 1$. Then
$$\sqrt[n]{b} = \sqrt[n]{\frac{1}{a}} = \frac{1}{\sqrt[n]{a}} \to \frac{1}{1} = 1.$$

If $b = 1$, $\sqrt[n]{b} = \sqrt[n]{1} = 1$ for all n, which is the limit of a constant sequence.

c) Let $c \in \mathbb{R}^+$ and $0 < \varepsilon < c$ be arbitrary. We know that
$$\exists p \in \mathbb{N}: n > p \Rightarrow c - \frac{\varepsilon}{2} < \frac{u_{n+1}}{u_n} < c + \frac{\varepsilon}{2}.$$

Taking $n > p$, we obtain
$$c - \frac{\varepsilon}{2} < \frac{u_{p+2}}{u_{p+1}} < c + \frac{\varepsilon}{2},\ c - \frac{\varepsilon}{2} < \frac{u_{p+3}}{u_{p+2}} < c + \frac{\varepsilon}{2}, \ldots, c - \frac{\varepsilon}{2} < \frac{u_n}{u_{n-1}} < c + \frac{\varepsilon}{2},$$

and multiplying term by term these inequalities, we obtain
$$\left(c-\frac{\varepsilon}{2}\right)^{n-p-1} < \frac{u_n}{u_{p+1}} < \left(c+\frac{\varepsilon}{2}\right)^{n-p-1},$$

which implies
$$\frac{u_{p+1}}{\left(c-\frac{\varepsilon}{2}\right)^{p+1}}\left(c-\frac{\varepsilon}{2}\right)^n < u_n < \frac{u_{p+1}}{\left(c+\frac{\varepsilon}{2}\right)^{p+1}}\left(c+\frac{\varepsilon}{2}\right)^n.$$

1.1. Sequences of Real Numbers

Applying roots,

$$\sqrt[n]{\frac{u_{p+1}}{(c-\frac{\varepsilon}{2})^{p+1}}}\left(c-\frac{\varepsilon}{2}\right) < \sqrt[n]{u_n} < \sqrt[n]{\frac{u_{p+1}}{(c+\frac{\varepsilon}{2})^{p+1}}}\left(c+\frac{\varepsilon}{2}\right).$$

By item b),

$$\sqrt[n]{\frac{u_{p+1}}{(c-\frac{\varepsilon}{2})^{p+1}}} \to 1 \quad \text{and} \quad \sqrt[n]{\frac{u_{p+1}}{(c+\frac{\varepsilon}{2})^{p+1}}} \to 1;$$

therefore,

$$\sqrt[n]{\frac{u_{p+1}}{(c-\frac{\varepsilon}{2})^{p+1}}}\left(c-\frac{\varepsilon}{2}\right) \to c-\frac{\varepsilon}{2} \quad \text{and} \quad \sqrt[n]{\frac{u_{p+1}}{(c+\frac{\varepsilon}{2})^{p+1}}}\left(c+\frac{\varepsilon}{2}\right) \to c+\frac{\varepsilon}{2}.$$

As $c - \frac{\varepsilon}{2} \neq c + \frac{\varepsilon}{2}$, we cannot apply the Squeeze Theorem. However, by applying the definition of limit, we know that there exists $q \in \mathbb{N}$ such that if $n > q$, then

$$c - \varepsilon = \left(c-\frac{\varepsilon}{2}\right) - \frac{\varepsilon}{2} < \sqrt[n]{\frac{u_{p+1}}{(c-\frac{\varepsilon}{2})^{p+1}}}\left(c-\frac{\varepsilon}{2}\right) < \left(c-\frac{\varepsilon}{2}\right) + \frac{\varepsilon}{2} = c,$$

and there exists $r \in \mathbb{N}$ such that if $n > r$, then

$$c = \left(c+\frac{\varepsilon}{2}\right) - \frac{\varepsilon}{2} < \sqrt[n]{\frac{u_{p+1}}{(c+\frac{\varepsilon}{2})^{p+1}}}\left(c+\frac{\varepsilon}{2}\right) < \left(c+\frac{\varepsilon}{2}\right) + \frac{\varepsilon}{2} = c+\varepsilon;$$

taking $s = \max\{p, q, r\}$, if $n > s$, then

$$c-\varepsilon < \sqrt[n]{\frac{u_{p+1}}{(c-\frac{\varepsilon}{2})^{p+1}}}\left(c-\frac{\varepsilon}{2}\right) < \sqrt[n]{u_n} < \sqrt[n]{\frac{u_{p+1}}{(c+\frac{\varepsilon}{2})^{p+1}}}\left(c+\frac{\varepsilon}{2}\right) < c+\varepsilon,$$

and it follows from the definition of limit that $\lim \sqrt[n]{u_n} = c$.

If $c = 0$, we follow the same reasoning, considering that $u_n > 0$ and, therefore, $\sqrt[n]{u_n} > 0$. We show that, from a certain order,

$$0 < \sqrt[n]{u_n} < \frac{\varepsilon}{2}\sqrt[n]{\frac{u_{p+1}}{(\frac{\varepsilon}{2})^{p+1}}}.$$

From this, we deduce that there is an order from which $0 < \sqrt[n]{u_n} < \varepsilon$ and, as before, we conclude that $\lim \sqrt[n]{u_n} = 0$.

Let $c = +\infty$. Let L be an arbitrary positive number. We know that

$$\exists p \in \mathbb{N} : n > p \Rightarrow \frac{u_{n+1}}{u_n} > 2L.$$

Taking $n > p$, we have
$$\frac{u_{p+2}}{u_{p+1}} > 2L, \ \frac{u_{p+3}}{u_{p+2}} > 2L, \ldots, \ \frac{u_n}{u_{n-1}} > 2L,$$
and multiplying term by term these inequalities yields
$$\frac{u_n}{u_{p+1}} > (2L)^{n-p-1},$$
which implies $u_n > u_{p+1}(2L)^{n-p-1} = (2L)^n \frac{u_{p+1}}{(2L)^{p+1}}$. Applying roots,
$$\sqrt[n]{u_n} > 2L \sqrt[n]{\frac{u_{p+1}}{(2L)^{p+1}}}.$$

By item b), $\sqrt[n]{\frac{u_{p+1}}{(2L)^{p+1}}} \to 1$; therefore,
$$\exists q \in \mathbb{N}: \ n > q \Rightarrow \sqrt[n]{\frac{u_{p+1}}{(2L)^{p+1}}} > \frac{1}{2}.$$
Let $s = \max\{p, q\}$. If $n > s$, then $\sqrt[n]{u_n} > 2L \frac{1}{2} = L$. ∎

Notes:

1. It is easy to see, using item c) of the previous theorem, that $\sqrt[n]{n} \to 1$.

2. The converse of item c) of the previous theorem is not true, that is, $\sqrt[n]{u_n} \to c$ does not imply that $\frac{u_{n+1}}{u_n} \to c$ ($u_n > 0, \ \forall n \in \mathbb{N}$). For example, consider the sequence $u_n = e^{-n-(-1)^n}$:
$$\sqrt[n]{u_n} = e^{-1-\frac{(-1)^n}{n}} \to e^{-1},$$
and
$$\frac{u_{n+1}}{u_n} = e^{-1+2(-1)^n} = \begin{cases} e^{-3}, & \text{if } n \text{ is odd} \\ e, & \text{if } n \text{ is even} \end{cases}$$
has no limit.

Example 1.1.21 Let us calculate the limit of the sequence with general term $\sqrt[n]{2^n + 3^n}$. The sequence $u_n = 2^n + 3^n$ is positive and
$$\lim_{n \to +\infty} \frac{u_{n+1}}{u_n} = \lim_{n \to +\infty} \frac{2^{n+1} + 3^{n+1}}{2^n + 3^n} = \lim_{n \to +\infty} \frac{\left(\frac{2}{3}\right)^{n+1} + 1}{\left(\frac{2}{3}\right)^n \cdot \frac{1}{3} + \frac{1}{3}} = 3.$$

1.1. Sequences of Real Numbers

By item c) of Theorem 1.1.11, we can conclude that

$$\lim_{n \to +\infty} \sqrt[n]{2^n + 3^n} = 3.$$

Theorem 1.1.12 *Every bounded monotonic sequence (u_n) is convergent and*
$$\lim u_n = \begin{cases} \sup\{u_n : n \in \mathbb{N}\}, & \text{if } (u_n) \text{ is increasing} \\ \inf\{u_n : n \in \mathbb{N}\}, & \text{if } (u_n) \text{ is decreasing.} \end{cases}$$

Proof: For example, suppose that the sequence (u_n) is increasing. The set of terms of the sequence is bounded; therefore, it has a supremum, which we denote by S. Let $\varepsilon > 0$ be arbitrary. There is an element of the sequence, u_p, such that $S - \varepsilon < u_p \leq S$. However, the sequence is increasing; therefore, $u_p \leq u_n, \forall n > p$. We then have $S - \varepsilon < u_n \leq S, \forall n > p$, so $u_n \to S$. ∎

Note: The converse is not true; that is, there are convergent sequences that are not monotonic. For example, the sequence $u_n = (-1)^n \dfrac{1}{n}$ converges to 0 and is not monotonic (Fig. 1.14).

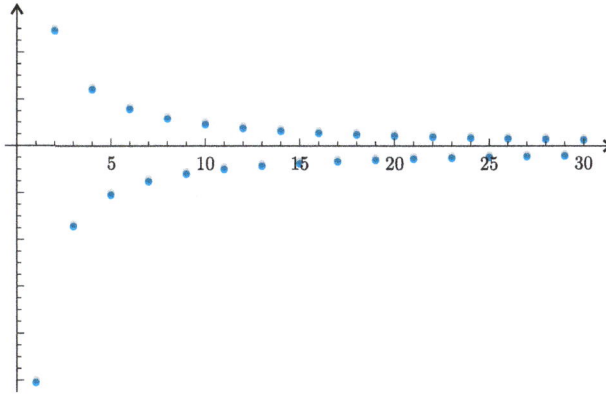

Figure 1.14: The sequence $u_n = \dfrac{(-1)^n}{n}$

Sequences of Real Numbers

- **The Sequence** $\left(\left(1+\dfrac{1}{n}\right)^n\right)_{n\in\mathbb{N}}$

Let us consider the sequence with general term $u_n = \left(1+\dfrac{1}{n}\right)^n$. According to Newton's Binomial, it can be written as follows:

$$\begin{aligned}
u_n &= \left(1+\frac{1}{n}\right)^n = \sum_{p=0}^{n} \binom{n}{p}\left(\frac{1}{n}\right)^p \\
&= \binom{n}{0}\left(\frac{1}{n}\right)^0 + \binom{n}{1}\left(\frac{1}{n}\right)^1 + \binom{n}{2}\left(\frac{1}{n}\right)^2 + \cdots + \binom{n}{n}\left(\frac{1}{n}\right)^n \\
&= 1 + \frac{n!}{1!(n-1)!}\cdot\frac{1}{n} + \frac{n!}{2!(n-2)!}\cdot\frac{1}{n^2} + \frac{n!}{3!(n-3)!}\cdot\frac{1}{n^3} + \\
&\quad + \cdots + \frac{n!}{(n-1)!(n-(n-1))!}\cdot\frac{1}{n^{n-1}} + \frac{n!}{n!(n-n)!}\cdot\frac{1}{n^n} \\
&= 1 + \frac{1}{1!}\cdot\frac{n!}{(n-1)!}\cdot\frac{1}{n} + \frac{1}{2!}\cdot\frac{n!}{(n-2)!}\cdot\frac{1}{n^2} + \frac{1}{3!}\cdot\frac{n!}{(n-3)!}\cdot\frac{1}{n^3} + \\
&\quad + \cdots + \frac{1}{(n-1)!}\cdot\frac{n!}{(n-(n-1))!}\cdot\frac{1}{n^{n-1}} + \frac{1}{n!}\cdot\frac{n!}{(n-n)!}\cdot\frac{1}{n^n} \\
&= 1 + \frac{1}{1!} + \frac{1}{2!}\cdot\frac{n(n-1)}{n^2} + \frac{1}{3!}\cdot\frac{n(n-1)(n-2)}{n^3} + \\
&\quad + \cdots + \frac{1}{(n-1)!}\cdot\frac{n!}{n^{n-1}} + \frac{1}{n!}\cdot\frac{n!}{n^n} \\
&= 1 + \frac{1}{1!} + \frac{1}{2!}\left(1-\frac{1}{n}\right) + \frac{1}{3!}\left(1-\frac{1}{n}\right)\left(1-\frac{2}{n}\right) + \\
&\quad + \cdots + \frac{1}{n!}\left(1-\frac{1}{n}\right)\left(1-\frac{2}{n}\right)\cdots\left(1-\frac{n-1}{n}\right).
\end{aligned}$$

Since $0 < 1 - \dfrac{p}{n} < 1$, for any $1 \leq p \leq n-1$, we can write

$$2 \leq u_n \leq 1 + 1 + \frac{1}{2} + \frac{1}{3!} + \cdots + \frac{1}{n!}.$$

It is easy to prove by induction that

$$\frac{1}{n!} < \frac{1}{2^{n-1}}, \quad \forall n \geq 3,$$

which allows us to write, taking into account the formula for the sum of the terms of a geometric sequence,

$$2 \leq u_n \leq 1 + 1 + \frac{1}{2} + \frac{1}{2^2} + \cdots + \frac{1}{2^{n-1}} = 1 + \frac{1-\left(\frac{1}{2}\right)^n}{1-\frac{1}{2}} < 1 + \frac{1}{1-\frac{1}{2}} = 3.$$

1.1. Sequences of Real Numbers

We have just proved that the sequence is bounded. Let us see that it is monotonic:

$u_{n+1} - u_n =$

$$= 1 + \frac{1}{1!} + \frac{1}{2!}\left(1 - \frac{1}{n+1}\right) + \frac{1}{3!}\left(1 - \frac{1}{n+1}\right)\left(1 - \frac{2}{n+1}\right) +$$
$$+ \cdots + \frac{1}{n!}\left(1 - \frac{1}{n+1}\right)\left(1 - \frac{2}{n+1}\right)\cdots\left(1 - \frac{n-1}{n+1}\right) +$$
$$+ \frac{1}{(n+1)!}\left(1 - \frac{1}{n+1}\right)\left(1 - \frac{2}{n+1}\right)\cdots\left(1 - \frac{n}{n+1}\right) -$$
$$- \left[1 + \frac{1}{1!} + \frac{1}{2!}\left(1 - \frac{1}{n}\right) + \frac{1}{3!}\left(1 - \frac{1}{n}\right)\left(1 - \frac{2}{n}\right) + \right.$$
$$\left. + \cdots + \frac{1}{n!}\left(1 - \frac{1}{n}\right)\left(1 - \frac{2}{n}\right)\cdots\left(1 - \frac{n-1}{n}\right)\right].$$

By associating terms, we obtain

$$u_{n+1} - u_n = \frac{1}{2!}\left[\left(1 - \frac{1}{n+1}\right) - \left(1 - \frac{1}{n}\right)\right] +$$
$$+ \frac{1}{3!}\left[\left(1 - \frac{1}{n+1}\right)\left(1 - \frac{2}{n+1}\right) - \left(1 - \frac{1}{n}\right)\left(1 - \frac{2}{n}\right)\right] +$$
$$+ \cdots + \frac{1}{n!}\left[\left(1 - \frac{1}{n+1}\right)\left(1 - \frac{2}{n+1}\right)\cdots\left(1 - \frac{n-1}{n+1}\right) - \right.$$
$$\left. - \left(1 - \frac{1}{n}\right)\left(1 - \frac{2}{n}\right)\cdots\left(1 - \frac{n-1}{n}\right)\right] +$$
$$+ \frac{1}{(n+1)!}\left(1 - \frac{1}{n+1}\right)\left(1 - \frac{2}{n+1}\right)\cdots\left(1 - \frac{n}{n+1}\right).$$

Since $1 - \frac{p}{n+1} > 1 - \frac{p}{n} > 0$ and the last term is positive, we have $u_{n+1} - u_n > 0$, for all $n \in \mathbb{N}$. The sequence

$$u_n = \left(1 + \frac{1}{n}\right)^n$$

is bounded and monotonic. By Theorem 1.1.12 we conclude that it is convergent. The limit of this sequence is an irrational number denoted by

Leonhard Euler (1707–1783) was a Swiss mathematician who studied philosophy and mathematics at the University of Basel. He was a professor in St. Petersburg (Russia) and Berlin. His contributions are important in almost all areas of mathematics, namely, number theory, differential equations, analysis, calculus of variations, and rational mechanics, as well as in other scientific areas, such as calculation of planetary orbits, artillery and ballistics, shipbuilding, and navigation. In 1771, he became blind, but this did not prevent him from continuing an extraordinary scientific production. After his death, St. Petersburg Academy continued to publish his unpublished works for 50 years. Euler introduced the notation $f(x)$ for a function, e for the base of natural logarithms, i for the root of -1, π, $\sum$ for summation, Δy for finite differences, and many others. He was the most prolific mathematician of all time. (Source of image: Oil portrait by Jakob Emanuel Handmann (1718–1781), Deutsches Museum)

the letter e:
$$\lim\left(1+\frac{1}{n}\right)^n = e.$$
This number is usually referred to as Neper's number or Euler's constant.

Theorem 1.1.13 *Let $x \in \mathbb{R}$. Then $\lim\left(1+\dfrac{x}{n}\right)^n = e^x$.*

Theorem 1.1.14 *Let $x \in \mathbb{R}$ and u be a sequence such that $\lim u_n = +\infty$ or $\lim u_n = -\infty$. Then $\lim\left(1+\dfrac{x}{u_n}\right)^{u_n} = e^x$.*

Example 1.1.22 We know that $x_n = n+5 \to +\infty$. For every $n \in \mathbb{N}$,
$$a_n = \left(\frac{n+3}{n+5}\right)^{2n+1} = \left(1-\frac{2}{n+5}\right)^{2n+1} = \left[\left(1-\frac{2}{n+5}\right)^{n+5}\right]^{\frac{2n+1}{n+5}}.$$

Using the previous theorem, we obtain the following:
$$\lim\left(1-\frac{2}{n+5}\right)^{n+5} = e^{-2}.$$

Since $\dfrac{2n+1}{n+5} \to 2$, then
$$\lim a_n = (e^{-2})^2 = e^{-4}.$$

John Napier (or Jhone Neper) (1550–1617) was a Scottish mathematician. He has dedicated himself to studies in theology and mathematics. He is essentially known as the inventor of logarithms, although he also has works on spherical triangles. He invented a calculating rule, known as "Napier's bones" because it was made of ivory, which allowed multiplication, division, and the mechanical calculation of square and cubic roots. (Source of image: Engraving by Samuel Freeman (1773–1857))

- **Subsequences**

Let f and g be two functions. The composition $f \circ g$ is only possible if the range of g is contained in the domain of f. Sequences are functions of a natural variable; thus, the composition of two sequences, $u \circ v$, is only possible if the range of v is a subset of $\mathbb{N}$.

Consider two sequences u and v, where v is a sequence of natural numbers. The composition $u \circ v$ is still a sequence, of general term u_{v_n}. For example,

1.1. Sequences of Real Numbers 31

if u is the sequence $1, 2, 1, 3, 1, 4, \ldots$ and $v_n = 2n - 1$, then $u_{v_n} = 1$; if $z_n = 2n$, then $u_{z_n} = n + 1$; and if $s_n = 4$, then $u_{s_n} = 3$.

A subsequence of a sequence can be obtained by omitting some of its terms and keeping the remaining terms in the original order. Let us now consider a more formal definition.

> **Definition 1.1.9** *Given two sequences u and w, we say that w is a **subsequence** of u if there exists a strictly increasing sequence of natural numbers v, such that $w = u \circ v$.*

Example 1.1.23 From the sequences considered earlier, $u \circ v$ and $u \circ z$ are subsequences of u, but $u \circ s$ is not a subsequence of u.

Notes:

1. Every subsequence of a bounded sequence is bounded.

2. An unbounded sequence may have bounded subsequences. For example,
$$u_n = \begin{cases} n, & \text{if } n \text{ is even} \\ \dfrac{1}{n}, & \text{if } n \text{ is odd.} \end{cases}$$

3. Every subsequence of a monotonic sequence is monotonic.

4. A non-monotonic sequence can have monotonic subsequences, as shown by the sequence of Example 1.1.12. The same can be verified in the sequence whose graph is illustrated in Fig. 1.14.

> **Theorem 1.1.15** *Every subsequence of a convergent sequence converges to the same limit.*

Proof: Let u be a sequence that converges to a. Let $\varepsilon > 0$. By definition, we know that
$$\exists p \in \mathbb{N} : n > p \Rightarrow |u_n - a| < \varepsilon.$$

Let w be a subsequence of u. By Definition 1.1.9, there exists a sequence of natural numbers, strictly increasing, v, such that $w = u \circ v$. Since $v_n \geq n, \forall n \in \mathbb{N}$, then if $n > p$, also $v_n > p$ and

$$|w_n - a| = |u_{v_n} - a| < \varepsilon,$$

which proves that $w_n \to a$. ∎

Note: This theorem is mainly used as a negative test: If two subsequences with different limits are found, the sequence diverges.

Theorem 1.1.16 *Let u be a sequence. Let (u_{v_n}) and (u_{z_n}) be two subsequences such that the union of their indices is $\mathbb{N}$. If (u_{v_n}) and (u_{z_n}) converge to the same limit a, then u converges to a.*

Proof: Let $\varepsilon > 0$ be arbitrary. By definition, we know that

$$\exists p_1 \in \mathbb{N} : v_n > p_1 \Rightarrow |u_{v_n} - a| < \varepsilon$$

and

$$\exists p_2 \in \mathbb{N} : z_n > p_2 \Rightarrow |u_{z_n} - a| < \varepsilon.$$

Let $p = \max\{p_1, p_2\}$. Observing that every natural number is a term of one of the sequences (v_n) or (z_n), then $|u_n - a| < \varepsilon$ if $n > p$. ∎

Note: Theorem 1.1.16 remains valid for a finite number of subsequences with the same limit, provided that the union of the indices is $\mathbb{N}$.

Lemma 1.1.1 *Every sequence of real numbers has monotonic subsequences.*

Proof: Let
$$\mathcal{M} = \{p \in \mathbb{N} : \ u_p < u_n, \ \forall n > p\}.[1]$$

If $\mathcal{M}$ is infinite, that is, if there exists $p_1 < p_2 < \cdots < p_n < \cdots$ belonging to $\mathcal{M}$,

$$u_{p_1} < u_{p_2} < u_{p_3} < \cdots < u_{p_n} < \cdots,$$

---

[1] If, for example, (u_n) is strictly increasing, $\mathcal{M} = \mathbb{N}$, if (u_n) is decreasing, $\mathcal{M} = \varnothing$, and if $u_n = (-1)^n \dfrac{1}{n}$, $\mathcal{M}$ is the set of odd numbers.

1.1. Sequences of Real Numbers

and the subsequence (u_{p_n}) is monotonic (increasing).

If $\mathcal{M}$ is finite or empty, we will show that there exists a decreasing subsequence (u_{q_n}). If $\mathcal{M}$ is empty, we consider $q_1 = 1$. If $\mathcal{M}$ is finite, let $q_1 = \max(\mathcal{M}) + 1$. It is clear that $q_1 \notin \mathcal{M}$. By the definition of $\mathcal{M}$, there exists $q_2 > q_1$ such that $u_{q_2} \leq u_{q_1}$. Again, by the definition of $\mathcal{M}$, there exists $q_3 > q_2$ such that $u_{q_3} \leq u_{q_2}$. Similarly, for each $n \in \mathbb{N}$, there exists $q_n > q_{n-1}$ such that $u_{q_n} \leq u_{q_{n-1}}$. Therefore, there exists a subsequence (u_{q_n}) satisfying $u_{q_1} \geq u_{q_2} \cdots \geq u_{q_n} \geq u_{q_{n+1}} \geq \cdots$, that is, the subsequence (u_{q_n}) is decreasing. ∎

Theorem 1.1.17 *Every bounded sequence has convergent subsequences.*

Proof: By Lemma 1.1.1, the sequence (u_n) admits at least one monotonic subsequence (u_{p_n}). Because (u_{p_n}) is monotonic and bounded, then by Theorem 1.1.12 it is convergent. ∎

Definition 1.1.10 *A number $a \in \mathbb{R}$ is a **sublimit** of the sequence u if there exists a subsequence of u that converges to a.*

Example 1.1.24 Consider the sequence $u_n = (-1)^n + \dfrac{1}{n}$. The sublimits of (u_n) are -1 and 1 because 1 is the limit of subsequence $u_{2n} = 1 + \dfrac{1}{2n}$ and -1 is the limit of subsequence $u_{2n-1} = -1 + \dfrac{1}{2n-1}$ (see Fig. 1.15).

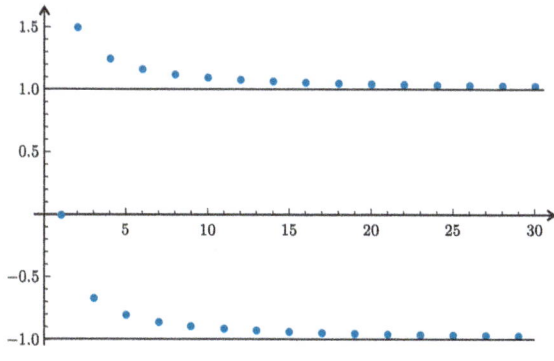

Figure 1.15: Sublimits of sequence $u_n = (-1)^n + \dfrac{1}{n}$

Notes: Let S be the set of sublimits of the sequence u.

1. By Theorem 1.1.17, if u is bounded, $S \neq \varnothing$.

2. S can be empty. As an example, $u_n = n$, which is not bounded, does not have convergent subsequences.

3. If u is convergent, then S is a unit set, that is, it has only one element.

4. S can be a unit set, and u can be divergent; for example, we have
$$u_n = \begin{cases} \dfrac{1}{n}, & \text{if } n \text{ is even} \\ n, & \text{if } n \text{ is odd}. \end{cases}$$

5. S can be an infinite set; for example, given the sequence
$$1, 1, 2, 1, 2, 3, 1, 2, 3, 4, 1, 2, 3, 4, 5, \ldots,$$
then $S = \mathbb{N}$.

Theorem 1.1.18 *The set of sublimits of a bounded sequence has a maximum and a minimum.*

Proof: Let $M > 0$, such that $-M \leq u_n \leq M, \forall n \in \mathbb{N}$. Let S be the set of sublimits of sequence u, which is not empty because the sequence is bounded. By Theorem 1.1.5, if $a \in S$, then $-M \leq a \leq M$, so S is bounded. Therefore, S has an infimum, α, and a supremum, β. We will show that $\beta \in S$, that is, $\beta = \max S$. By definition of supremum, there exists $a_1 \in S$ such that $\beta - 1/2 < a_1 \leq \beta$; by definition of sublimit, there exists u_{p_1} such that $a_1 - 1/2 < u_{p_1} < a_1 + 1/2$; and then,

$$\beta - 1 < a_1 - \frac{1}{2} < u_{p_1} < a_1 + \frac{1}{2} < \beta + 1.$$

Again, by definition of supremum, there exists $a_2 \in S$ such that
$$\beta - 1/4 < a_2 \leq \beta;$$
by definition of sublimit, there exists u_{p_2} such that
$$a_2 - 1/4 < u_{p_2} < a_2 + 1/4;$$
and then,
$$\beta - \frac{1}{2} < a_2 - \frac{1}{4} < u_{p_2} < a_2 + \frac{1}{4} < \beta + \frac{1}{2}.$$

1.1. Sequences of Real Numbers

Repeating this reasoning, for each $n \in \mathbb{N}$, there exists u_{p_n} such that
$$\beta - \frac{1}{n} < u_{p_n} < \beta + \frac{1}{n}.$$
We have constructed a subsequence (u_{p_n}) that converges to β, so $\beta \in S$. Similarly, it can be shown that $\alpha \in S$. ∎

> **Definition 1.1.11** Let u be a bounded sequence and S be the set of sublimits of u. The maximum of S is called the **superior limit** or **upper limit** of u and is represented by $\overline{\lim} \, u_n = \limsup u_n = \max(S)$. The minimum of S is called the **inferior limit** or **lower limit** of u and is represented by $\underline{\lim} \, u_n = \liminf u_n = \min(S)$.
> If u is not bounded above, we define $\overline{\lim} \, u_n = +\infty$. If u is not bounded below, we define $\underline{\lim} \, u_n = -\infty$.
> If $u_n \to +\infty$, then we define $\underline{\lim} \, u_n = \overline{\lim} \, u_n = +\infty$. If $u_n \to -\infty$, we define $\underline{\lim} \, u_n = \overline{\lim} \, u_n = -\infty$.

> **Theorem 1.1.19** A bounded sequence is convergent if and only if $\underline{\lim} \, u_n = \overline{\lim} \, u_n$.

Proof: If u is convergent, with limit a, every subsequence of u has limit a (Theorem 1.1.15), which implies that $S = \{a\}$ and
$$\underline{\lim} \, u_n = \min(S) = a = \max(S) = \overline{\lim} \, u_n.$$
Suppose u is bounded and $\underline{\lim} \, u_n = \overline{\lim} \, u_n = a$. If $u_n \not\to a$, then
$$\exists \varepsilon > 0 \, \forall p \in \mathbb{N} \, \exists n > p : |u_n - a| \geq \varepsilon.$$
Thus, there exists n_1 such that $|u_{n_1} - a| \geq \varepsilon$; there exists $n_2 > n_1$ such that $|u_{n_2} - a| \geq \varepsilon$; and there exists $n_m > n_{m-1}$ such that $|u_{n_m} - a| \geq \varepsilon$. Because it is bounded, the subsequence (u_{n_m}) has a convergent subsequence $(u_{n_{m_s}})$: $u_{n_{m_s}} \to b \neq a$, since $|u_{n_{m_s}} - a| \geq \varepsilon$ for all terms. Therefore, we have shown that S has an element b different from a, which is incompatible with the hypothesis of $\underline{\lim} \, u_n = \overline{\lim} \, u_n = a$. ∎

> **Definition 1.1.12** A sequence u is a **Cauchy sequence** if
> $$\forall \varepsilon > 0 \, \exists p \in \mathbb{N} : \, m, n > p \Rightarrow |u_n - u_m| < \varepsilon.$$

Augustin Louis Cauchy (1789–1857) was a French engineer and mathematician. He was a pioneer of Mathematical Analysis and initiated a project to express and rigorously prove the theorems of calculus, making significant contributions to Complex Analysis. He wrote about 800 works covering all areas of Mathematics and Mathematical Physics. (Source of image: Charles H. Reutlinger, Smithsonian Institution Libraries)

Example 1.1.25 We prove that the sequence $u_n = \dfrac{1}{n}$ is a Cauchy sequence. Let $m, n > p$; then

$$\left|\frac{1}{n} - \frac{1}{m}\right| \leq \frac{1}{n} + \frac{1}{m} < \frac{1}{p} + \frac{1}{p} = \frac{2}{p}.$$

Let ε denote an arbitrary positive number; to conclude, it is enough to take $p > \dfrac{2}{\varepsilon}$.

Note: In the definition of a convergent sequence, we introduced an element external to the sequence, the limit. The sequence converges if, from a certain order, all elements of the sequence "are close" to the limit. In the definition of Cauchy sequence, we compare only the elements of the sequence with each other. We say that the sequence is Cauchy if, from a certain order, all elements of the sequence "are close" to each other.

Theorem 1.1.20 *A real sequence is convergent if and only if it is a Cauchy sequence.*

Proof: Let u be a convergent sequence and a be its limit. Let $\varepsilon > 0$ be arbitrary. By definition,

$$\exists p \in \mathbb{N} : |u_n - a| < \frac{\varepsilon}{2}, \forall n > p.$$

Let $m, n > p$ be arbitrary. Then,

$$|u_m - u_n| = |u_m - a + a - u_n| \leq |u_m - a| + |a - u_n| \leq \frac{\varepsilon}{2} + \frac{\varepsilon}{2} = \varepsilon,$$

so u is a Cauchy sequence.

We must now prove the converse, that is, every Cauchy sequence is convergent.

Let u be a Cauchy sequence. We begin by demonstrating that u is bounded. In the definition, we can take, for example, $\varepsilon = 1$: There exists $p \in \mathbb{N}$ such that $|u_n - u_m| < 1, \forall m, n > p$; in particular,

$$|u_n| - |u_{p+1}| \leq |u_n - u_{p+1}| < 1, \forall n > p,$$

so

$$|u_n| < |u_{p+1}| + 1, \forall n > p.$$

1.1. Sequences of Real Numbers

Considering $M = \max\{|u_1|, |u_2|, \ldots, |u_p|, |u_{p+1}| + 1\}$, we obtain

$$|u_n| \leq M, \ \forall n \in \mathbb{N}.$$

Because sequence u is bounded, it has by Theorem 1.1.17 a convergent subsequence (u_{v_n}), with limit a. Let $\varepsilon > 0$ be arbitrary. There exists $q \in \mathbb{N}$ such that
$$|u_{v_n} - a| < \frac{\varepsilon}{2}, \ \forall v_n > q.$$

By definition of Cauchy sequence, there exists $r \in \mathbb{N}$ such that
$$|u_n - u_m| < \frac{\varepsilon}{2}, \ \forall m, n > r.$$

Let $p = \max\{q, r\}$. Let $n > p$ and $v_n > p$; then,

$$|u_n - a| = |u_n - u_{v_n} + u_{v_n} - a| \leq |u_n - u_{v_n}| + |u_{v_n} - a| < \frac{\varepsilon}{2} + \frac{\varepsilon}{2} = \varepsilon.$$

We conclude, therefore, that $u_n \to a$. ∎

Note: This theorem allows us to show that a sequence converges without having to calculate its limit. Consider the following sequence:

$$u_n = 1 + \frac{1}{2^2} + \frac{1}{3^2} + \cdots + \frac{1}{n^2}.$$

We can take, without loss of generality, $n > m$; thus, we obtain

$$|u_n - u_m| = \left| \frac{1}{(m+1)^2} + \frac{1}{(m+2)^2} + \cdots + \frac{1}{n^2} \right|$$
$$= \frac{1}{(m+1)^2} + \frac{1}{(m+2)^2} + \cdots + \frac{1}{n^2}$$
$$\leq \frac{1}{m(m+1)} + \frac{1}{(m+1)(m+2)} + \cdots + \frac{1}{(n-1)n}$$
$$= \left(\frac{1}{m} - \frac{1}{m+1}\right) + \left(\frac{1}{m+1} - \frac{1}{m+2}\right) + \cdots + \left(\frac{1}{n-1} - \frac{1}{n}\right)$$
$$= \frac{1}{m} - \frac{1}{n} \leq \frac{1}{m}.$$

If $p > \frac{1}{\varepsilon}$ and $n \geq m > p$, we obtain $|u_n - u_m| < \varepsilon$, so the sequence is Cauchy and therefore converges.

1.2 Solved Exercises

1. Evaluate the limits of the following sequences and provide an argument for the calculations:

   a) $\dfrac{n}{n^2+3} + \sqrt[n]{n}$

   b) $\dfrac{n+1}{2n+5}$

   c) $\dfrac{n^3+3n^2+5}{n^2+2}$

   d) $\dfrac{35n^7+1}{71n^7+5n^6+1}$

   e) $\dfrac{\sqrt[3]{n^2+n}+n}{\sqrt[4]{2n^4+1}+\sqrt{n}}$

   f) $\dfrac{\sqrt{n^3+2n^4+1}-n}{-2n^2+\sqrt[3]{n^2+3}}$

   g) $\dfrac{(1+2\sqrt{n})\sqrt{n+1}}{n+\sqrt[3]{n}}$

   h) $\dfrac{\sqrt[3]{1-27n^3}}{1+4n}$

   i) $\dfrac{n((-1)^n+\sqrt{n})}{2+\sqrt{n^3+1}}$

   j) $\dfrac{n\sqrt[3]{n^2+2}}{n^2+(-1)^n\,n}$

   k) $\dfrac{2n \cdot e^{1/n}}{(-1)^n+\sqrt{n^2+5}}$

2. Calculate the limits of the following sequences:

   a) $\left(\dfrac{n^2-1}{n^2}\right)^n$

   b) $\left(\dfrac{4^n-5}{4^n+3}\right)^{2n}$

   c) $\left(\dfrac{n+2}{n+4}\right)^{n+1}$

   d) $\left(\dfrac{2+n}{5+5n}\right)^n$

   e) $\left(\dfrac{3n+1}{3n+2}\right)^n$

   f) $\left(3-\dfrac{2n+1}{n}\right)^{4n-2}$

   g) $\left(\dfrac{2n+5}{2n+1}\right)^{n+4}$

   h) $\left(\dfrac{n^2+3}{2n^2+1}\right)^n e^{\arctan(n)}$

3. Find the limits of the following sequences and justify the calculations:

   a) $\dfrac{3^n \sin(2^{3n}+1)}{2^{3n}+1}$

   b) $\dfrac{1}{n} \cdot \cos(n+1) \cdot \log(n)$

   c) $\dfrac{n^2+3}{n\sqrt{n^3+2}} \cdot \cos\left(\sqrt{n^3+2}\right)$

   d) $\dfrac{1}{n}\sqrt[n]{n!}$

   e) $\sqrt[n]{n^2 e^{-n}} - \left(\dfrac{n^4}{n^4+1}\right)^{n^4}$

   f) $\dfrac{n\sin(n)}{2^n\sqrt{5n^3+1}}$

   g) $\sqrt{n^2+2n}-n$

   h) $\left(\dfrac{(n+1)^{n+2}}{(n+2)^{n+1}} - \dfrac{n}{3}\right) \cdot \sin\left(\dfrac{1}{n}\right)$

   i) $\dfrac{3^n-5}{5^n+3}$

4. Find the limits of the following sequences and provide a clear explanation of the reasoning behind each calculation:

   a) $\displaystyle\sum_{k=1}^{n-1} \dfrac{\sin^2(n)}{n^2+3k^2}$

   b) $\displaystyle\sum_{k=1}^{n} \dfrac{5n}{\sqrt{n^4+k}}$

   c) $\displaystyle\sum_{k=1}^{n} \dfrac{\sqrt[3]{2n}}{\sqrt[3]{n^4+k}}$

5. Evaluate the supremum and infimum of the set of terms, and the limit of the sequence
   $$a_n = n + \dfrac{1}{2n} - \sqrt{n^2+1}.$$

6. a) Find the limit of the sequence
   $$a_n = \left(\sqrt{2n+1}-\sqrt{2n}\right) \cdot \cos^2(n)$$
   and give a clear explanation of the answer.

   b) Determine the set of sublimits of the sequence
   $$b_n = \sin\left(\dfrac{n\pi}{2}\right) \cdot \arctan(n)$$
   and give a justification for the answer provided.

1.2. Solved Exercises

7. Consider the sequence
$$u_n = \sqrt[n]{1 + 2^{n \cdot (-1)^n}}.$$

   a) Write the subsequence of even-indexed terms and calculate its limit.

   b) Write the subsequence of odd-indexed terms and calculate its limit.

   c) Calculate $\underline{\lim} \, u_n$ and $\overline{\lim} \, u_n$.

   d) Given the previous items, what can we conclude about the convergence of the sequence?

8. Consider the sequence defined recursively:
$$\begin{cases} u_1 = \sqrt{2} \\ u_{n+1} = \sqrt{2\,u_n}, \quad \forall n \in \mathbb{N}. \end{cases}$$

   a) By induction prove that $0 < u_n < 2$, $\forall n \in \mathbb{N}$.

   b) Prove that the sequence is increasing.

   c) Prove that the sequence is convergent.

   d) Evaluate the limit of the sequence.

9. Consider the sequence
$$\begin{cases} a_1 = \sqrt{2} \\ a_{n+1} = \left(\sqrt{2}\right)^{a_n}, \quad \forall n \in \mathbb{N}. \end{cases}$$

   a) By induction show that
$$\sqrt{2} \leq a_n < 2, \quad \forall n \in \mathbb{N}.$$

   b) By induction show that the sequence (a_n) is increasing.

   c) Show that there exists $a \leq 2$ such that $a_n \to a$.

10. Let $a \in \mathbb{R}$ be a positive number. Consider the sequence of real numbers defined recursively
$$\begin{cases} x_1 = a \\ x_{n+1} = \dfrac{x_n}{2 + x_n}, \quad \forall n \in \mathbb{N}. \end{cases}$$

   a) By induction show that $x_n > 0$, $\forall n \in \mathbb{N}$.

   b) Show that the sequence is decreasing.

   c) Show that the sequence is convergent and determine its limit.

11. Let $a \in \mathbb{R}$ be a positive number. Consider the sequence of real numbers defined recursively
$$\begin{cases} x_0 = 0, \quad x_1 = a \\ x_{n+1} = x_n + x_{n-1}^2, \quad \forall n \in \mathbb{N}. \end{cases}$$

   a) Show that the sequence is increasing.

   b) Show that $x_n > 0$, $\forall n \in \mathbb{N}$.

   c) Show that if there exists $b \in \mathbb{R}$ such that $\lim x_n = b$, then $b = 0$.

   d) Given the previous items, calculate if it exists, $\lim x_n$.

12. Consider the sequence of real numbers defined recursively
$$\begin{cases} x_1 = 2 \\ x_{n+1} = \dfrac{x_n}{2} + \dfrac{1}{x_n}, \quad \forall n \in \mathbb{N}. \end{cases}$$

   a) Prove that $x_n > 0$, $\forall n \in \mathbb{N}$.

   b) Prove that $x_n > \sqrt{2}$, $\forall n \in \mathbb{N}$.

   c) Prove that (x_n) is monotonic.

   d) Show that (x_n) converges.

   e) Evaluate the limit of (x_n).

13. Consider the sequence of real numbers defined recursively
$$\begin{cases} x_1 = 3 \\ x_{n+1} = \dfrac{x_n^2 + 3}{2\,x_n}, \quad \forall n \in \mathbb{N}. \end{cases}$$

   a) Show by induction that
$$x_n - \sqrt{3} > 0, \quad \forall n \in \mathbb{N}.$$

   b) Show that the sequence (x_n) is decreasing.

   c) Show that the sequence (x_n) converges.

   d) Find the limit of the sequence (x_n).

14. Prove by definition the following limits:

    a) $\lim \dfrac{25n}{3n+1} = \dfrac{25}{3}$

    b) $\lim \dfrac{\sqrt{n}}{\sqrt{n}+1} = 1$

    c) $\lim \dfrac{2\sqrt{n}+5^{-n}}{\sqrt{n}+1} = 2$

    d) $\lim \left(1+\dfrac{\sin(n)}{n}\right) = 1$

    e) $\lim n^3 = +\infty$

    f) $\lim \dfrac{n^2}{n+1} = +\infty$

15. The Koch curve is created from a line segment of length 1. This segment is divided into three equal parts. We then remove the middle segment. Next, we replace the removed segment with two segments of length $\frac{1}{3}$. The resulting shape can be seen in the figure below.

The middle segment of each of the four segments is removed and replaced by two segments of the same length. The resulting curve has a length of $\frac{16}{9}$.

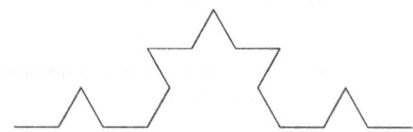

Continuing this process ad infinitum, the result is the Koch curve.

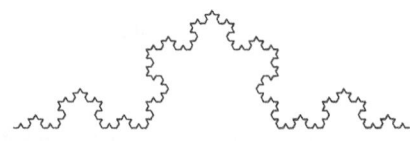

What is the length of the Koch curve?

1.2. Solved Exercises

SOLUTIONS

1. a) Let $a_n = \dfrac{n}{n^2 + 3}$. It is not possible to directly apply Theorem 1.1.5, as the indeterminate form $\frac{\infty}{\infty}$ occurs. To solve this, we can rewrite a_n as follows:

$$a_n = \frac{n}{n^2 + 3} = \frac{n^2 \cdot \dfrac{1}{n}}{n^2\left(1 + \dfrac{3}{n^2}\right)} = \frac{\dfrac{1}{n}}{1 + \dfrac{3}{n^2}}.$$

We know, from Example 1.1.16, that $\left(\dfrac{1}{n}\right)$ is a null sequence, which by Theorem 1.1.5, implies that $\dfrac{3}{n^2} \to 0$. Again by Theorem 1.1.5,

$$\lim a_n = 0.$$

Since $\lim \sqrt[n]{n} = 1$ (see Note 1 after Theorem 1.1.11), we can conclude that

$$\lim \left(\frac{n}{n^2 + 3} + \sqrt[n]{n}\right) = 1.$$

Note: The procedure described here is equivalent to dividing both terms of the fraction by the power of n with the highest exponent that appears in the expression. Alternatively, we could use Theorem 1.1.9, with $v_n = n$, keeping in mind that $p = 1$ and $q = 2$.

b) The following sequence is given by $a_n = \dfrac{n+1}{2n+5}$. Taking into account the note from the previous item, we divide both members of the fraction that defines the sequence (a_n) by n, which is the highest power in the expression.

$$a_n = \frac{n+1}{2n+5} = \frac{\dfrac{n+1}{n}}{\dfrac{2n+5}{n}} = \frac{1 + \dfrac{1}{n}}{2 + \dfrac{5}{n}}.$$

Since $\left(\dfrac{1}{n}\right)$ is a null sequence, we can apply Theorem 1.1.5 and conclude that

$$\lim a_n = \frac{1}{2}.$$

c) Let $a_n = \dfrac{n^3 + 3n^2 + 5}{n^2 + 2}$. Since n^3 is the highest power of n that appears in the fraction that defines the sequence (a_n), we can divide the numerator and denominator by n^3:

$$a_n = \frac{n^3 + 3n^2 + 5}{n^2 + 2} = \frac{\dfrac{n^3 + 3n^2 + 5}{n^3}}{\dfrac{n^2 + 2}{n^3}} = \frac{1 + \dfrac{3}{n} + \dfrac{5}{n^3}}{\dfrac{1}{n} + \dfrac{2}{n^3}}.$$

Using Theorem 1.1.5, we can conclude that

$$\lim a_n = +\infty,$$

because $\left(\dfrac{1}{n}\right)$ is a null sequence and $a_n > 0$, $\forall n \in \mathbb{N}$.

d) Let $a_n = \dfrac{35n^7 + 1}{71n^7 + 5n^6 + 1}$. If we divide both the numerator and denominator of the fraction that defines the sequence (a_n) by n raised to the highest power, n^7, we obtain

$$a_n = \frac{35n^7 + 1}{71n^7 + 5n^6 + 1} = \frac{\dfrac{35n^7 + 1}{n^7}}{\dfrac{71n^7 + 5n^6 + 1}{n^7}} = \frac{35 + \dfrac{1}{n^7}}{71 + \dfrac{5}{n} + \dfrac{1}{n^7}}.$$

Knowing that $\left(\dfrac{1}{n}\right)$ is a null sequence, we can use Theorem 1.1.5, to conclude that

$$\lim a_n = \frac{35}{71}.$$

e) Given $a_n = \dfrac{\sqrt[3]{n^2 + n} + n}{\sqrt[4]{2n^4 + 1} + \sqrt{n}}$, we can simplify the expression by dividing both the numerator and denominator by n as shown below:

$$a_n = \frac{\sqrt[3]{n^2 + n} + n}{\sqrt[4]{2n^4 + 1} + \sqrt{n}} = \frac{\dfrac{\sqrt[3]{n^2 + n} + n}{n}}{\dfrac{\sqrt[4]{2n^4 + 1} + \sqrt{n}}{n}} = \frac{\sqrt[3]{\dfrac{n^2 + n}{n^3}} + 1}{\sqrt[4]{\dfrac{2n^4 + 1}{n^4}} + \sqrt{\dfrac{n}{n^2}}} = \frac{\sqrt[3]{\dfrac{1}{n} + \dfrac{1}{n^2}} + 1}{\sqrt[4]{2 + \dfrac{1}{n^4}} + \sqrt{\dfrac{1}{n}}}.$$

Using Theorem 1.1.5, the conclusion is

$$\lim a_n = \frac{1}{\sqrt[4]{2}}.$$

f) Let $a_n = \dfrac{\sqrt{n^3 + 2n^4 + 1} - n}{-2n^2 + \sqrt[3]{n^2 + 3}}$. We can rewrite a_n using the following equalities:

$$a_n = \frac{\sqrt{n^3 + 2n^4 + 1} - n}{-2n^2 + \sqrt[3]{n^2 + 3}} = \frac{\dfrac{\sqrt{n^3 + 2n^4 + 1} - n}{n^2}}{\dfrac{-2n^2 + \sqrt[3]{n^2 + 3}}{n^2}} = \frac{\sqrt{\dfrac{n^3 + 2n^4 + 1}{n^4}} - \dfrac{1}{n}}{-2 + \sqrt[3]{\dfrac{n^2 + 3}{n^6}}} = \frac{\sqrt{\dfrac{1}{n} + 2 + \dfrac{1}{n^4}} - \dfrac{1}{n}}{-2 + \sqrt[3]{\dfrac{1}{n^4} + \dfrac{3}{n^6}}}.$$

By applying Theorem 1.1.5 to the above expression, we conclude that the limit is given by

$$\lim a_n = -\frac{\sqrt{2}}{2}.$$

g) Let $a_n = \dfrac{(1 + 2\sqrt{n})\sqrt{n + 1}}{n + \sqrt[3]{n}}$. To simplify the expression of (a_n), we can divide both the numerator and denominator of the fraction that defines it by n. This gives us

$$a_n = \frac{(1 + 2\sqrt{n})\sqrt{n + 1}}{n + \sqrt[3]{n}} = \frac{\dfrac{(1 + 2\sqrt{n})\sqrt{n + 1}}{n}}{\dfrac{n + \sqrt[3]{n}}{n}} = \frac{\dfrac{(1 + 2\sqrt{n})\sqrt{n + 1}}{\sqrt{n}\sqrt{n}}}{1 + \sqrt[3]{\dfrac{n}{n^3}}} = \frac{\left(\dfrac{1}{\sqrt{n}} + 2\right)\sqrt{1 + \dfrac{1}{n}}}{1 + \sqrt[3]{\dfrac{1}{n^2}}}.$$

By applying Theorem 1.1.5, we can determine that

$$\lim a_n = 2.$$

1.2. Solved Exercises

h) Let $a_n = \dfrac{\sqrt[3]{1-27n^3}}{1+4n}$. To find the limit of this sequence, we will manipulate the expression for a_n to simplify it.

First, we will divide both the numerator and denominator of the fraction that defines the sequence (a_n) by n, and next we simplify the expression. This gives us

$$a_n = \frac{\sqrt[3]{1-27n^3}}{1+4n} = \frac{\frac{\sqrt[3]{1-27n^3}}{n}}{\frac{1+4n}{n}} = \frac{\sqrt[3]{\frac{1-27n^3}{n^3}}}{\frac{1}{n}+4} = \frac{\sqrt[3]{\frac{1}{n^3}-27}}{\frac{1}{n}+4}.$$

Therefore, by Theorem 1.1.5,
$$\lim a_n = -\frac{3}{4}.$$

i) Consider the sequence (a_n) given by $a_n = \dfrac{n\bigl((-1)^n + \sqrt{n}\bigr)}{2+\sqrt{n^3+1}}$. In this case, by dividing both the numerator and denominator of the fraction that defines the sequence (a_n) by $n^{3/2}$, we can simplify the expression as follows:

$$a_n = \frac{n\bigl((-1)^n + \sqrt{n}\bigr)}{2+\sqrt{n^3+1}} = \frac{\frac{n\bigl((-1)^n + \sqrt{n}\bigr)}{n^{3/2}}}{\frac{2+\sqrt{n^3+1}}{n^{3/2}}} = \frac{\frac{(-1)^n + \sqrt{n}}{n^{1/2}}}{\frac{2}{\sqrt{n^3}}+\sqrt{\frac{n^3+1}{n^3}}} = \frac{\frac{(-1)^n}{\sqrt{n}}+1}{\frac{2}{\sqrt{n^3}}+\sqrt{1+\frac{1}{n^3}}}.$$

Since $((-1)^n)$ is bounded and $\left(\dfrac{1}{\sqrt{n}}\right)$ is a null sequence, we can apply Theorem 1.1.6 to conclude that $\dfrac{(-1)^n}{\sqrt{n}} \to 0$. Moreover, applying Theorem 1.1.5, we can compute the limit of the sequence and obtain

$$\lim a_n = 1.$$

j) Let $a_n = \dfrac{n\sqrt[3]{n^2+2}}{n^2+(-1)^n n}$. By dividing both the numerator and the denominator of the fraction that defines sequence (a_n) by n^2, we obtain

$$a_n = \frac{n\sqrt[3]{n^2+2}}{n^2+(-1)^n n} = \frac{\frac{n\sqrt[3]{n^2+2}}{n^2}}{\frac{n^2+(-1)^n n}{n^2}} = \frac{\frac{\sqrt[3]{n^2+2}}{n}}{1+\frac{(-1)^n}{n}} = \frac{\sqrt[3]{\frac{n^2+2}{n^3}}}{1+\frac{(-1)^n}{n}} = \frac{\sqrt[3]{\frac{1}{n}+\frac{2}{n^3}}}{1+\frac{(-1)^n}{n}}.$$

Therefore, according to Theorem 1.1.5, we arrive at the following result:

$$\lim a_n = 0.$$

k) Let $a_n = \dfrac{2n\, e^{1/n}}{(-1)^n+\sqrt{n^2+5}} = \dfrac{2n}{(-1)^n+\sqrt{n^2+5}} \cdot e^{1/n} = b_n \cdot e^{1/n}$. By dividing both the numerator and denominator of the fraction that defines the sequence (b_n) by n, we obtain

$$b_n = \frac{2n}{(-1)^n+\sqrt{n^2+5}} = \frac{2}{\frac{(-1)^n+\sqrt{n^2+5}}{n}} = \frac{2}{\frac{(-1)^n}{n}+\sqrt{\frac{n^2+5}{n^2}}} = \frac{2}{\frac{(-1)^n}{n}+\sqrt{1+\frac{5}{n^2}}}.$$

Hence,
$$\lim b_n = 2.$$

Since $\lim e^{1/n} = 1$, we can conclude that
$$\lim \frac{2n e^{1/n}}{(-1)^n + \sqrt{n^2 + 5}} = 2.$$

2. a) Consider the sequence (a_n) defined as
$$a_n = \left(\frac{n^2 - 1}{n^2}\right)^n = \left(1 - \frac{1}{n^2}\right)^n = \left[\left(1 - \frac{1}{n^2}\right)^{n^2}\right]^{1/n}, \quad \forall n \in \mathbb{N}.$$

Using Theorem 1.1.14, we can show that
$$\lim \left(1 - \frac{1}{n^2}\right)^{n^2} = e^{-1}.$$

Since $\lim \dfrac{1}{n} = 0$, we can conclude by Theorem 1.1.5 that
$$\lim a_n = (e^{-1})^0 = 1.$$

b) Denoting by a_n the general term of the sequence, we can write
$$a_n = \left(\frac{4^n - 5}{4^n + 3}\right)^{2^n} = \left[\frac{4^n\left(1 - \frac{5}{4^n}\right)}{4^n\left(1 + \frac{3}{4^n}\right)}\right]^{2^n} = \left[\frac{\left(1 - \frac{5}{4^n}\right)^{2^{2n}}}{\left(1 + \frac{3}{4^n}\right)^{2^{2n}}}\right]^{1/2^n}, \quad \forall n \in \mathbb{N}.$$

Using Theorem 1.1.14, we can find the limit of the sequence:
$$\lim \left(1 - \frac{5}{4^n}\right)^{2^{2n}} = \lim \left(1 - \frac{5}{4^n}\right)^{4^n} = e^{-5}$$
and
$$\lim \left(1 + \frac{3}{4^n}\right)^{2^{2n}} = \lim \left(1 + \frac{3}{4^n}\right)^{4^n} = e^3.$$

But by Theorem 1.1.10, $\lim \dfrac{1}{2^n} = 0$; therefore,
$$\lim a_n = \left(\frac{e^{-5}}{e^3}\right)^0 = 1.$$

c) If a_n is the general term of the sequence being discussed, we have
$$a_n = \left(\frac{n+2}{n+4}\right)^{n+1} = \left[\frac{n\left(1 + \frac{2}{n}\right)}{n\left(1 + \frac{4}{n}\right)}\right]^{n+1} = \left[\frac{1 + \frac{2}{n}}{1 + \frac{4}{n}}\right]^{n+1} = \frac{\left(1 + \frac{2}{n}\right)^n}{\left(1 + \frac{4}{n}\right)^n} \cdot \frac{1 + \frac{2}{n}}{1 + \frac{4}{n}}, \quad \forall n \in \mathbb{N}.$$

1.2. Solved Exercises

Using Theorem 1.1.13, we arrive at the following results:

$$\lim \left(1 + \frac{2}{n}\right)^n = e^2$$

and

$$\lim \left(1 + \frac{4}{n}\right)^n = e^4;$$

therefore,

$$\lim a_n = \frac{e^2}{e^4} \cdot 1 = e^{-2}.$$

d) To find the limit of the sequence (a_n), we can simplify the expression defining it by putting n in evidence in the numerator of the expression that defines (a_n) and $5n$ in the denominator. This gives us the following expression:

$$a_n = \left[\frac{n\left(1 + \frac{2}{n}\right)}{5n\left(1 + \frac{1}{n}\right)}\right]^n = \left(\frac{1}{5}\right)^n \cdot \frac{\left(1 + \frac{2}{n}\right)^n}{\left(1 + \frac{1}{n}\right)^n}, \quad \forall n \in \mathbb{N}.$$

We know from Theorem 1.1.10 that $\lim \left(\frac{1}{5}\right)^n = 0$. Additionally, by Theorem 1.1.13, we can find that

$$\lim \left(1 + \frac{2}{n}\right)^n = e^2$$

and

$$\lim \left(1 + \frac{1}{n}\right)^n = e.$$

Therefore,

$$\lim a_n = \lim \left(\frac{1}{5}\right)^n \cdot \lim \frac{\left(1 + \frac{2}{n}\right)^n}{\left(1 + \frac{1}{n}\right)^n} = 0 \cdot \frac{e^2}{e} = 0.$$

e) If we consider a_n as the general term of the sequence, we can write it as follows:

$$a_n = \left(\frac{3n+1}{3n+2}\right)^n = \left[\frac{3n\left(1 + \frac{1}{3n}\right)}{3n\left(1 + \frac{2}{3n}\right)}\right]^n = \left[\frac{1 + \frac{1}{3n}}{1 + \frac{2}{3n}}\right]^n = \frac{\left(1 + \frac{1/3}{n}\right)^n}{\left(1 + \frac{2/3}{n}\right)^n}, \quad \forall n \in \mathbb{N}.$$

According to Theorem 1.1.13, we know that

$$\lim \left(1 + \frac{1/3}{n}\right)^n = e^{1/3}$$

and

$$\lim \left(1 + \frac{2/3}{n}\right)^n = e^{2/3}.$$

Therefore,

$$\lim a_n = \frac{e^{1/3}}{e^{2/3}} = e^{-1/3}.$$

f) We have
$$a_n = \left(3 - \frac{2n+1}{n}\right)^{4n-2} = \frac{\left(3 - 2 - \frac{1}{n}\right)^{4n}}{\left(3 - 2 - \frac{1}{n}\right)^2} = \frac{\left[\left(1 - \frac{1}{n}\right)^n\right]^4}{\left(1 - \frac{1}{n}\right)^2}, \quad \forall n \in \mathbb{N}.$$

By Theorem 1.1.13, we know that
$$\lim \left(1 - \frac{1}{n}\right)^n = e^{-1}.$$

Thus, we can conclude that
$$\lim a_n = \left(e^{-1}\right)^4 = e^{-4}.$$

g) We have
$$a_n = \left(\frac{2n+5}{2n+1}\right)^{n+4} = \left[\frac{2n\left(1 + \frac{5}{2n}\right)}{2n\left(1 + \frac{1}{2n}\right)}\right]^{n+4} = \left[\left(\frac{1 + \frac{5}{2n}}{1 + \frac{1}{2n}}\right)^{2n}\right]^{1/2} \cdot \left[\frac{1 + \frac{5}{2n}}{1 + \frac{1}{2n}}\right]^4, \quad \forall n \in \mathbb{N}.$$

Using Theorem 1.1.14, we can obtain
$$\lim \left(1 + \frac{5}{2n}\right)^{2n} = e^5$$
and
$$\lim \left(1 + \frac{1}{2n}\right)^{2n} = e.$$

Therefore, we can find that
$$\lim a_n = \left(\frac{e^5}{e}\right)^{1/2} = e^2.$$

h) Knowing that $\lim_{x \to +\infty} \arctan(x) = \frac{\pi}{2}$, we can conclude[1] that $\lim e^{\arctan(n)} = e^{\pi/2}$. Now, let us consider $a_n = \left(\frac{n^2+3}{2n^2+1}\right)^n$. Putting n^2 in evidence in this expression yields

$$a_n = \left(\frac{n^2+3}{2n^2+1}\right)^n = \left[\frac{n^2\left(1 + \frac{3}{n^2}\right)}{2n^2\left(1 + \frac{1}{2n^2}\right)}\right]^n = \left(\frac{1}{2}\right)^n \cdot \frac{\left(1 + \frac{3}{n^2}\right)^n}{\left(1 + \frac{1}{2n^2}\right)^n} = \left(\frac{1}{2}\right)^n \cdot \frac{\left[\left(1 + \frac{3}{n^2}\right)^{n^2}\right]^{1/n}}{\left[\left(1 + \frac{1}{2n^2}\right)^{2n^2}\right]^{1/2n}}.$$

Applying Theorem 1.1.14, we obtain
$$\lim \left(1 + \frac{3}{n^2}\right)^{n^2} = e^3$$
and
$$\lim \left(1 + \frac{1}{2n^2}\right)^{2n^2} = e.$$

---

[1] We are using the following result: If $\lim_{x \to +\infty} f(x) = L$ and $f(n) = u_n$, $n \in \mathbb{N}$, then $\lim u_n = L$.

1.2. Solved Exercises

Therefore, as $\lim \dfrac{1}{n} = \lim \dfrac{1}{2n} = \lim \left(\dfrac{1}{2}\right)^n = 0$, we can conclude that

$$\lim a_n = 0 \cdot \dfrac{(e^3)^0}{e^0} = 0.$$

Thus, we get

$$\lim \left(\dfrac{n^2+3}{2n^2+1}\right)^n \cdot e^{\arctan(n)} = 0.$$

3. a) Let $a_n = \dfrac{3^n \sin(2^{3n}+1)}{2^{3n}+1} = \dfrac{3^n}{2^{3n}+1} \cdot \sin(2^{3n}+1)$. We know that

$$|\sin(x)| \leq 1, \quad \forall x \in \mathbb{R};$$

therefore, $\left|\sin(2^{3n}+1)\right| \leq 1$, $\forall n \in \mathbb{N}$, that is, the sequence with general term $\sin(2^{3n}+1)$ is a bounded sequence. We prove that the sequence with general term $\dfrac{3^n}{2^{3n}+1}$ is a null sequence.

$$\lim \dfrac{3^n}{2^{3n}+1} = \lim \dfrac{3^n}{8^n+1} = \lim \dfrac{\left(\dfrac{3}{8}\right)^n}{1+\left(\dfrac{1}{8}\right)^n} = 0.$$

We conclude that the given sequence is a null sequence because it is the product of a null sequence and a bounded sequence (see Theorem 1.1.6).

b) Let $a_n = \dfrac{1}{n} \cdot \cos(n+1) \cdot \log(n)$. We know that $-1 \leq \cos(x) \leq 1$, $\forall x \in \mathbb{R}$. Then

$$-1 \leq \cos(n+1) \leq 1, \quad \forall n \in \mathbb{N}.$$

Therefore,

$$-\log\!\left(\sqrt[n]{n}\right) = -\dfrac{\log(n)}{n} \leq \dfrac{1}{n} \cdot \cos(n+1) \cdot \log(n) \leq \dfrac{\log(n)}{n} = \log\!\left(\sqrt[n]{n}\right), \quad \forall n \in \mathbb{N}.$$

Because we know that $\lim \sqrt[n]{n} = 1$, we have

$$\lim \log\!\left(\sqrt[n]{n}\right) = 0.$$

It follows from the Squeeze Theorem that

$$\lim a_n = 0.$$

c) Let $a_n = \dfrac{n^2+3}{n\sqrt{n^3+2}} \cdot \cos\!\left(\sqrt{n^3+2}\right)$. For all $n \in \mathbb{N}$, we have

$$0 \leq \left|\cos\!\left(\sqrt{n^3+2}\right)\right| \leq 1.$$

Let us consider the sequence of general term $\dfrac{n^2+3}{n\sqrt{n^3+2}}$. By dividing the numerator and denominator of the fraction that defines this sequence by the highest power of n, we obtain

$$\lim \dfrac{\dfrac{n^2+3}{n^{5/2}}}{\dfrac{n\sqrt{n^3+2}}{n^{5/2}}} = \lim \dfrac{\dfrac{n^2+3}{n^{5/2}}}{\dfrac{\sqrt{n^3+2}}{n^{3/2}}} = \lim \dfrac{\sqrt{\dfrac{(n^2+3)^2}{n^5}}}{\sqrt{\dfrac{n^3+2}{n^3}}} = \lim \dfrac{\sqrt{\dfrac{1}{n}+\dfrac{6}{n^3}+\dfrac{9}{n^5}}}{\sqrt{1+\dfrac{2}{n^3}}} = 0.$$

We conclude that the sequence (a_n) is a null sequence because it is the product of a null sequence and a bounded sequence (see Theorem 1.1.6).

d) Let $a_n = \dfrac{1}{n}\sqrt[n]{n!} = \sqrt[n]{\dfrac{n!}{n^n}}$. Let $b_n = \dfrac{n!}{n^n}$. It is evident that $b_n > 0$, $\forall n \in \mathbb{N}$.

$$\lim \frac{b_{n+1}}{b_n} = \lim \frac{\dfrac{(n+1)!}{(n+1)^{n+1}}}{\dfrac{n!}{n^n}} = \lim \frac{(n+1)!\, n^n}{(n+1)^{n+1}\, n!} = \lim \frac{(n+1)\, n^n}{(n+1)^{n+1}} = \lim \left(\frac{n}{n+1}\right)^n = \frac{1}{e}.$$

By Theorem 1.1.11 we can conclude that $\lim a_n = \dfrac{1}{e}$.

e) Let $a_n = \sqrt[n]{n^2 e^{-n}} - \left(\dfrac{n^4}{n^4+1}\right)^{n^4}$ and $b_n = n^2 e^{-n}$. It is evident that $b_n > 0$, $\forall n \in \mathbb{N}$.

$$\lim \frac{b_{n+1}}{b_n} = \lim \frac{\dfrac{(n+1)^2}{e^{n+1}}}{\dfrac{n^2}{e^n}} = \lim \frac{(n+1)^2\, e^n}{e^{n+1}\, n^2} = \frac{1}{e}\lim\left(\frac{n+1}{n}\right)^2 = \frac{1}{e}.$$

By Theorem 1.1.11 we can conclude that $\lim \sqrt[n]{n^2 e^{-n}} = \dfrac{1}{e}$. Using Theorem 1.1.14 we obtain the following result:

$$\lim \left(\frac{n^4}{n^4+1}\right)^{n^4} = \lim \frac{1}{\left(\dfrac{n^4+1}{n^4}\right)^{n^4}} = \lim \frac{1}{\left(1+\dfrac{1}{n^4}\right)^{n^4}} = \frac{1}{e}.$$

We conclude that $\lim a_n = \dfrac{1}{e} - \dfrac{1}{e} = 0$.

f) Let $a_n = \dfrac{n\sin(n)}{2^n\sqrt{5n^3+1}} = \left(\dfrac{1}{2}\right)^n \cdot \dfrac{n}{\sqrt{5n^3+1}} \cdot \sin(n)$. We know that

$$|\sin(n)| \le 1, \quad \forall n \in \mathbb{N};$$

therefore, the sequence $(\sin(n))$ is a bounded sequence. We show that the sequence $\left(\dfrac{n}{\sqrt{5n^3+1}}\right)$ is a null sequence.

$$\lim \frac{n}{\sqrt{5n^3+1}} = \lim \frac{\dfrac{n}{n^{3/2}}}{\dfrac{\sqrt{5n^3+1}}{n^{3/2}}} = \lim \frac{\dfrac{1}{\sqrt{n}}}{\sqrt{\dfrac{5n^3+1}{n^3}}} = \lim \frac{\dfrac{1}{\sqrt{n}}}{\sqrt{5+\dfrac{1}{n^3}}} = 0.$$

Since $\lim \left(\dfrac{1}{2}\right)^n = 0$, we have

$$\lim \left(\frac{1}{2}\right)^n \cdot \frac{n}{\sqrt{5n^3+1}} = 0.$$

We conclude that the given sequence is a null sequence because it is the product of a null sequence by a bounded sequence (see Theorem 1.1.6).

1.2. Solved Exercises

g) It is not possible to directly apply Theorem 1.1.5, as the indeterminate form $\infty - \infty$ occurs. Dividing and multiplying the general term of the sequence by its conjugate, we obtain

$$\lim \left(\sqrt{n^2 + 2n} - n\right) = \lim \frac{\left(\sqrt{n^2 + 2n} - n\right)\left(\sqrt{n^2 + 2n} + n\right)}{\sqrt{n^2 + 2n} + n} = \lim \frac{n^2 + 2n - n^2}{\sqrt{n^2 + 2n} + n}$$

$$= \lim \frac{2n}{\sqrt{n^2 + 2n} + n} = \lim \frac{\frac{2n}{n}}{\frac{\sqrt{n^2 + 2n} + n}{n}} = \lim \frac{2}{\sqrt{\frac{n^2 + 2n}{n^2}} + 1} = \lim \frac{2}{\sqrt{1 + \frac{2}{n}} + 1} = 1.$$

h) Let $a_n = \left(\frac{(n+1)^{n+2}}{(n+2)^{n+1}} - \frac{n}{3}\right) \cdot \sin\left(\frac{1}{n}\right)$. Then

$$\lim \left(\frac{(n+1)^{n+2}}{(n+2)^{n+1}} - \frac{n}{3}\right) \cdot \sin\left(\frac{1}{n}\right) = \lim \left(\left(\frac{n+1}{n+2}\right)^{n+2} \cdot (n+2) - \frac{n}{3}\right) \cdot \sin\left(\frac{1}{n}\right)$$

$$= \lim \left(\left(1 - \frac{1}{n+2}\right)^{n+2} \cdot (n+2) \cdot \sin\left(\frac{1}{n}\right) - \frac{n}{3} \cdot \sin\left(\frac{1}{n}\right)\right)$$

$$= \lim \left(\left(1 - \frac{1}{n+2}\right)^{n+2} \cdot n \cdot \sin\left(\frac{1}{n}\right) + \left(1 - \frac{1}{n+2}\right)^{n+2} \cdot 2 \cdot \sin\left(\frac{1}{n}\right) - \frac{n}{3} \cdot \sin\left(\frac{1}{n}\right)\right)$$

$$= \lim \left(\left(1 - \frac{1}{n+2}\right)^{n+2} \cdot \frac{\sin\left(\frac{1}{n}\right)}{\frac{1}{n}} + \left(1 - \frac{1}{n+2}\right)^{n+2} \cdot 2 \cdot \sin\left(\frac{1}{n}\right) - \frac{1}{3} \cdot \frac{\sin\left(\frac{1}{n}\right)}{\frac{1}{n}}\right).$$

By Theorem 1.1.14, $\lim \left(1 - \frac{1}{n+2}\right)^{n+2} = e^{-1}$. Since $\lim \frac{1}{n} = 0$, we conclude that $\lim \sin\left(\frac{1}{n}\right) = 0$ and $\lim \frac{\sin\left(\frac{1}{n}\right)}{\frac{1}{n}} = 1$. Using Theorem 1.1.5, we arrive at the following result: $\lim a_n = e^{-1} - \frac{1}{3}$.

i) By dividing both the numerator and denominator of the fraction that defines the general term of the sequence by the power of the largest base and exponent, we obtain, by Theorem 1.1.10,

$$\lim \frac{3^n - 5}{5^n + 3} = \lim \frac{\frac{3^n - 5}{5^n}}{\frac{5^n + 3}{5^n}} = \lim \frac{\left(\frac{3}{5}\right)^n - \left(\frac{1}{5}\right)^{n-1}}{1 + \frac{3}{5^n}} = 0.$$

4. In this exercise, we apply the Squeeze Theorem. When dealing with sequences defined by summations, a useful technique is to first identify the largest and smallest terms in the summation. Then, these terms can be used to bound the sequence.

a) The given sequence has a general term a_n which is defined as the sum from $k = 1$ to $k = n - 1$ of $\frac{\sin^2(n)}{n^2 + 3k^2}$. As we show below, the largest term is $\frac{\sin^2(n)}{n^2 + 3}$ (which corresponds to $k = 1$), and the smallest is $\frac{\sin^2(n)}{n^2 + 3(n-1)^2}$ (corresponding to $k = n - 1$). Clearly,

and
$$n^2 + 3k^2 > n^2, \ \forall k, n \in \mathbb{N}$$

$$n^2 + 3k^2 \leq n^2 + 3(n-1)^2, \ \forall k, n \in \mathbb{N} \text{ such that } k \leq n-1.$$

Therefore, for all n and $1 \leq k \leq n-1$, we obtain

$$n^2 < n^2 + 3k^2 \leq n^2 + 3(n-1)^2 \Rightarrow \frac{1}{n^2} > \frac{1}{n^2 + 3k^2} \geq \frac{1}{n^2 + 3(n-1)^2}$$

$$\Rightarrow \frac{\sin^2(n)}{n^2} > \frac{\sin^2(n)}{n^2 + 3k^2} \geq \frac{\sin^2(n)}{n^2 + 3(n-1)^2}.$$

As (a_n) is defined as the sum of $n-1$ terms,

$$(n-1) \cdot \frac{\sin^2(n)}{n^2 + 3(n-1)^2} \leq \sum_{k=1}^{n-1} \frac{\sin^2(n)}{n^2 + 3k^2} < (n-1) \cdot \frac{\sin^2(n)}{n^2}, \ \forall n \in \mathbb{N}$$

$$\Leftrightarrow \frac{n-1}{n^2 + 3(n-1)^2} \cdot \sin^2(n) \leq \sum_{k=1}^{n-1} \frac{\sin^2(n)}{n^2 + 3k^2} < \frac{n-1}{n^2} \cdot \sin^2(n), \ \forall n \in \mathbb{N}$$

$$\Leftrightarrow \frac{n-1}{n^2 + 3(n-1)^2} \cdot \sin^2(n) \leq \sum_{k=1}^{n-1} \frac{\sin^2(n)}{n^2 + 3k^2} < \left(\frac{1}{n} - \frac{1}{n^2}\right) \cdot \sin^2(n), \ \forall n \in \mathbb{N}.$$

Let $b_n = \dfrac{n-1}{n^2 + 3(n-1)^2} = \dfrac{n-1}{4n^2 - 6n + 3}$. Dividing the numerator and denominator of the fraction that defines the sequence by n^2, we get

$$\lim \frac{n-1}{4n^2 - 6n + 3} = \lim \frac{\frac{1}{n} - \frac{1}{n^2}}{\frac{4n^2 - 6n + 3}{n^2}} = \lim \frac{\frac{1}{n} - \frac{1}{n^2}}{4 - \frac{6}{n} + \frac{3}{n^2}} = 0.$$

Let $c_n = \dfrac{1}{n} - \dfrac{1}{n^2}$. It is evident that $\lim c_n = 0$.

As we know that $|\sin^2(n)| \leq 1, \ \forall n \in \mathbb{N}$, the sequence $(\sin^2(n))$ is bounded. Since the product of a null sequence by a bounded sequence is a null sequence, we can assert that the sequences of general terms

$$\frac{n-1}{n^2 + 3(n-1)^2} \cdot \sin^2(n)$$

and

$$\left(\frac{1}{n} - \frac{1}{n^2}\right) \cdot \sin^2(n)$$

are null sequences. Finally, as the two limits are equal, the Squeeze Theorem allows us to conclude that

$$\lim a_n = 0.$$

b) The general term a_n of the sequence is defined as the sum from $k=1$ to $k=n$ of $\dfrac{5n}{\sqrt{n^4 + k}}$. The largest term is $\dfrac{5n}{\sqrt{n^4 + 1}}$ (which corresponds to $k=1$), and the smallest is $\dfrac{5n}{\sqrt{n^4 + n}}$ (corresponding to $k=n$), as shown below. We have

$$\sqrt{n^4 + k} > n^2, \ \forall k, n \in \mathbb{N},$$

1.2. Solved Exercises

and
$$\sqrt{n^4+k} \leq \sqrt{n^4+n}, \ \forall k,n \in \mathbb{N} \text{ such that } k \leq n.$$

Therefore, for all n and $1 \leq k \leq n$, we have
$$n^2 < \sqrt{n^4+k} \leq \sqrt{n^4+n} \Rightarrow \frac{1}{n^2} > \frac{1}{\sqrt{n^4+k}} \geq \frac{1}{\sqrt{n^4+n}}$$
$$\Rightarrow \frac{5n}{n^2} > \frac{5n}{\sqrt{n^4+k}} \geq \frac{5n}{\sqrt{n^4+n}}.$$

Because (a_n) is defined as the sum of n terms, we obtain
$$n \cdot \frac{5n}{\sqrt{n^4+n}} \leq \sum_{k=1}^{n} \frac{5n}{\sqrt{n^4+k}} < n \cdot \frac{5n}{n^2}, \ \forall n \in \mathbb{N}$$
$$\Leftrightarrow \frac{5n^2}{\sqrt{n^4+n}} \leq \sum_{k=1}^{n} \frac{5n}{\sqrt{n^4+k}} < \frac{5n^2}{n^2} = 5, \ \forall n \in \mathbb{N}.$$

Let $b_n = \frac{5n^2}{\sqrt{n^4+n}} = 5 \cdot \sqrt{\frac{n^4}{n^4+n}}$. Dividing the numerator and denominator of the radicand of this sequence by n^4, we obtain
$$\lim b_n = 5 \cdot \lim \sqrt{\frac{1}{1+\frac{1}{n^3}}} = 5.$$

Finally, as the two limits are equal, the Squeeze Theorem allows us to conclude that
$$\lim a_n = 5.$$

c) The general term a_n of the sequence is defined as the sum from $k=1$ to $k=n$ of $\frac{\sqrt[3]{2n}}{\sqrt[3]{n^4+k}}$. The largest term is $\frac{\sqrt[3]{2n}}{\sqrt[3]{n^4+1}}$ (which corresponds to $k=1$), and the smallest is $\frac{\sqrt[3]{2n}}{\sqrt[3]{n^4+n}}$ (corresponding to $k=n$), as shown below. We have
$$\sqrt[3]{n^4+k} > \sqrt[3]{n^4}, \ \forall k,n \in \mathbb{N},$$
and
$$\sqrt[3]{n^4+k} \leq \sqrt[3]{n^4+n}, \ \forall k,n \in \mathbb{N} \text{ such that } k \leq n.$$

Therefore, for all n and $1 \leq k \leq n$, we obtain
$$\sqrt[3]{n^4} < \sqrt[3]{n^4+k} \leq \sqrt[3]{n^4+n} \Rightarrow \frac{1}{\sqrt[3]{n^4}} > \frac{1}{\sqrt[3]{n^4+k}} \geq \frac{1}{\sqrt[3]{n^4+n}}$$
$$\Rightarrow \frac{\sqrt[3]{2n}}{\sqrt[3]{n^4}} > \frac{\sqrt[3]{2n}}{\sqrt[3]{n^4+k}} \geq \frac{\sqrt[3]{2n}}{\sqrt[3]{n^4+n}}.$$

Because (a_n) is defined as the sum of n terms, we obtain
$$n \cdot \frac{\sqrt[3]{2n}}{\sqrt[3]{n^4+n}} \leq \sum_{k=1}^{n} \frac{\sqrt[3]{2n}}{\sqrt[3]{n^4+k}} < n \cdot \frac{\sqrt[3]{2n}}{\sqrt[3]{n^4}}, \ \forall n \in \mathbb{N}$$
$$\Leftrightarrow \frac{\sqrt[3]{2n^4}}{\sqrt[3]{n^4+n}} \leq \sum_{k=1}^{n} \frac{\sqrt[3]{2n}}{\sqrt[3]{n^4+k}} < \frac{\sqrt[3]{2n^4}}{\sqrt[3]{n^4}} = \sqrt[3]{2}, \ \forall n \in \mathbb{N}.$$

Dividing the numerator and denominator of the fraction that defines the sequence on the left side of the inequality by $n^{4/3}$, we have

$$\lim \frac{\sqrt[3]{2n^4}}{\sqrt[3]{n^4+n}} = \lim \frac{\frac{\sqrt[3]{2n^4}}{\sqrt[3]{n^4}}}{\frac{\sqrt[3]{n^4+n}}{\sqrt[3]{n^4}}} = \lim \frac{\sqrt[3]{2}}{1+\frac{n}{\sqrt[3]{n^4}}} = \lim \frac{\sqrt[3]{2}}{1+\frac{1}{\sqrt[3]{n}}} = \sqrt[3]{2}.$$

Finally, as the two limits are equal, the Squeeze Theorem allows us to conclude that

$$\lim a_n = \sqrt[3]{2}.$$

5. Let $a_n = n + \frac{1}{2n} - \sqrt{n^2+1}$. We can simplify this expression as follows:

$$n + \frac{1}{2n} - \sqrt{n^2+1} = \frac{1}{2n} + \frac{(n-\sqrt{n^2+1})(n+\sqrt{n^2+1})}{n+\sqrt{n^2+1}} = \frac{1}{2n} + \frac{n^2-(n^2+1)}{n+\sqrt{n^2+1}} = \frac{1}{2n} - \frac{1}{n+\sqrt{n^2+1}}$$

$$= \frac{n+\sqrt{n^2+1}-2n}{2n(n+\sqrt{n^2+1})} = \frac{\sqrt{n^2+1}-n}{2n(n+\sqrt{n^2+1})} = \frac{(\sqrt{n^2+1}-n)(\sqrt{n^2+1}+n)}{2n(n+\sqrt{n^2+1})^2} = \frac{1}{2n(n+\sqrt{n^2+1})^2}.$$

By taking the limit, we can conclude that $\lim a_n = 0$ and that the sequence (a_n) is decreasing. Therefore, $\sup(a_n) = a_1 = \frac{3}{2} - \sqrt{2}$. According to Theorem 1.1.12, $\inf(a_n) = 0$.

6. a) We will use the fact that the cosine function is bounded. This means that

$$|\cos(n)| \leq 1, \quad \forall n \in \mathbb{N},$$

which implies that

$$|\cos^2(n)| \leq 1, \quad \forall n \in \mathbb{N}.$$

In addition,

$$\lim \left(\sqrt{2n+1} - \sqrt{2n}\right) = \lim \frac{\left(\sqrt{2n+1} - \sqrt{2n}\right)\left(\sqrt{2n+1} + \sqrt{2n}\right)}{\sqrt{2n+1} + \sqrt{2n}}$$

$$= \lim \frac{2n+1-2n}{\sqrt{2n+1}+\sqrt{2n}} = \lim \frac{1}{\sqrt{2n+1}+\sqrt{2n}} = 0.$$

As a result of the above, we can conclude that (a_n) is a null sequence since it is the product of a bounded sequence and a null sequence (see Theorem 1.1.6).

b) The general term of the sequence can be rewritten as follows:

$$a_n = \sin\left(\frac{n\pi}{2}\right) = \begin{cases} -1, & n = 4k-1, k \in \mathbb{N} \\ 0, & n = 2k, k \in \mathbb{N} \\ 1, & n = 4k-3, k \in \mathbb{N}. \end{cases}$$

The sequence (a_n) has three sublimits, namely, -1, 0, and 1, because it has subsequences that converge to these real numbers. The sequence $c_n = \arctan(n)$ has limit $\frac{\pi}{2}$. Therefore, we can conclude that the sublimits of the sequence $b_n = \sin\left(\frac{n\pi}{2}\right) \cdot \arctan(n)$ are $-\frac{\pi}{2}$, 0, and $\frac{\pi}{2}$.

1.2. Solved Exercises

7. a) Let $u_n = \sqrt[n]{1 + 2^{(-1)^n n}}$. The even-indexed subsequence of (u_n) is given by

$$u_{2k} = \sqrt[2k]{1 + 2^{(-1)^{2k} \, 2k}} = \sqrt[2k]{1 + 2^{2k}}, \quad k \in \mathbb{N}.$$

Let us consider the sequence $a_n = \sqrt[n]{1 + 2^n}$. By Theorem 1.1.11, as $1 + 2^n > 0$, $\forall n \in \mathbb{N}$, we can calculate the limit of (a_n) by

$$\lim \frac{1 + 2^{n+1}}{1 + 2^n} = \lim \frac{\frac{1 + 2^{n+1}}{2^{n+1}}}{\frac{1 + 2^n}{2^{n+1}}} = \lim \frac{1 + \frac{1}{2^{n+1}}}{\frac{1}{2} + \frac{1}{2^{n+1}}} = 2.$$

The limit of the sequence (a_n) is 2; therefore, all its subsequences have this limit. In particular, the even-indexed subsequence has limit 2. However, this subsequence is equal to the sequence (u_{2k}). We can affirm that $\lim u_{2k} = 2$.

b) The odd-indexed subsequence of (u_n) is given by

$$u_{2k+1} = \sqrt[2k+1]{1 + 2^{(-1)^{2k+1}(2k+1)}} = \sqrt[2k+1]{1 + 2^{-(2k+1)}} = \sqrt[2k+1]{1 + \frac{1}{2^{2k+1}}}, \quad k \in \mathbb{N},$$

and $\lim\limits_{k \to +\infty} u_{2k+1} = 1$.

c) Taking into account the results obtained in the previous items,

$$\underline{\lim} \, u_n = 1 \text{ and } \overline{\lim} \, u_n = 2.$$

d) Since $\underline{\lim} \, u_n = 1 \neq \overline{\lim} \, u_n = 2$, the sequence (u_n) is divergent.

8. a) We will use the Principle of Mathematical Induction to demonstrate that the sequence (u_n) satisfies

$$0 < u_n < 2, \quad \forall n \in \mathbb{N}.$$

For the base case $n = 1$, we have the following trivial formula:

$$0 = \sqrt{0} < \sqrt{2} = u_1 < \sqrt{4} = 2.$$

Induction hypothesis: $0 < u_n < 2$.

Induction thesis: $0 < u_{n+1} < 2$.

Proof: Using the induction hypothesis and the fact that the function $f(x) = \sqrt{2x}$ is increasing, we have

$$0 < u_n < 2 \Rightarrow 0 = \sqrt{2 \cdot 0} < \sqrt{2 u_n} = u_{n+1} < \sqrt{2 \cdot 2} = 2.$$

Therefore, the property is valid for $n + 1$.

By the Principle of Mathematical Induction, we can conclude that the inequality $0 < u_n < 2$ holds for all $n \in \mathbb{N}$.

b) We will show that the sequence (u_n) is increasing, that is,
$$u_{n+1} - u_n > 0, \quad \forall n \in \mathbb{N}.$$

For any natural number n, we have
$$u_{n+1} - u_n = \sqrt{2\,u_n} - u_n = \frac{\left(\sqrt{2\,u_n} - u_n\right)\left(\sqrt{2\,u_n} + u_n\right)}{\sqrt{2\,u_n} + u_n} = \frac{2\,u_n - u_n^2}{\sqrt{2\,u_n} + u_n} = \frac{u_n(2 - u_n)}{\sqrt{2\,u_n} + u_n} > 0$$

because in item a), we have already established that $u_n > 0$ and $2 - u_n > 0$. Therefore, the sequence is increasing.

c) Since the sequence is bounded (as shown in item a)) and increasing (as shown in item b)), we can conclude that the sequence (u_n) is convergent because by Theorem 1.1.12, every bounded monotonic sequence is convergent.

d) Let $b \in \mathbb{R}$ be the limit of the sequence. Since every subsequence of a convergent sequence is convergent to the same limit, we have
$$\lim u_{n+1} = b.$$

Hence,
$$b = \lim u_{n+1} = \lim \sqrt{2\,u_n} = \sqrt{2\,b}.$$

Therefore, b satisfies equation $b = \sqrt{2\,b}$, which means that $b^2 - 2b = 0$. Thus, $b(b - 2) = 0$, and we can exclude the solution $b = 0$ because by item b), we have
$$u_n \geq u_1 = \sqrt{2}, \quad \forall n \in \mathbb{N}.$$

Therefore, we conclude that the limit of u is $b = 2$.

9. a) Let f be the function defined by $f(x) = \left(\sqrt{2}\right)^x = e^{x \cdot \log(\sqrt{2})}$. Since $\sqrt{2} > 1$, f is continuous and increasing on $\mathbb{R}$.

For $n = 1$, the formula is trivial: $\sqrt{2} \leq a_1 = \sqrt{2} < 2$.

Induction hypothesis: $\sqrt{2} \leq a_n < 2$.

Induction thesis: $\sqrt{2} \leq a_{n+1} < 2$.

Proof: Assuming that the property is valid for n (induction hypothesis) and using the fact that f is an increasing function, we have
$$\sqrt{2} \leq a_n < 2 \Rightarrow \left(\sqrt{2}\right)^{\sqrt{2}} = f(\sqrt{2}) \leq f(a_n) = a_{n+1} < f(2) = 2.$$

Using again the monotonicity of f, we obtain
$$1 < \sqrt{2} \Rightarrow f(1) = \sqrt{2} \leq f(\sqrt{2}) = \left(\sqrt{2}\right)^{\sqrt{2}},$$

which implies that the property is valid for the order $n + 1$. Therefore, the Principle of Mathematical Induction is satisfied, and we conclude that
$$\sqrt{2} \leq a_n < 2, \quad \forall n \in \mathbb{N}.$$

1.2. Solved Exercises

b) We will show, using the Principle of Mathematical Induction, that

$$a_n < a_{n+1}, \quad \forall n \in \mathbb{N}.$$

For $n = 1$, the formula is a consequence of the calculations in item a):

$$\sqrt{2} = a_1 < a_2 = (\sqrt{2})^{\sqrt{2}}.$$

<u>Induction hypothesis:</u> $a_n < a_{n+1}$.

<u>Induction thesis:</u> $a_{n+1} < a_{n+2}$.

<u>Proof:</u> Assuming that the property is valid for $n \in \mathbb{N}$ (induction hypothesis), the validity of the property for $n+1$ is a direct consequence of the fact that f is increasing:

$$a_n < a_{n+1} \Rightarrow a_{n+1} = f(a_n) < a_{n+2} = f(a_{n+1}).$$

Therefore, we can conclude that the sequence is increasing.

c) In item a), we saw that the sequence is bounded, and in item b), we showed that it is increasing. By Theorem 1.1.12, every bounded monotonic sequence is convergent, so we can conclude that the sequence with general term a_n converges. Let $a \in \mathbb{R}$ be its limit. Since $a_n < 2$, $\forall n \in \mathbb{N}$, by Theorem 1.1.5, we come to the conclusion that $a \leq 2$.

10. a) For $n = 1$, we have $x_1 = a > 0$ by hypothesis, which implies that $x_1 > 0$.

<u>Induction hypothesis:</u> $x_n > 0$.

<u>Induction thesis:</u> $x_{n+1} > 0$.

<u>Proof:</u> We have $x_{n+1} = \dfrac{x_n}{2 + x_n}$. Since $x_n > 0$ by induction hypothesis, we know that $2 + x_n > 0$. Thus, x_{n+1} is the quotient of two positive quantities, so $x_{n+1} > 0$.

Therefore, by the Principle of Induction, we have established that $x_n > 0$, $\forall n \in \mathbb{N}$.

b) We want to prove that $x_{n+1} - x_n < 0$ for any $n \in \mathbb{N}$. We can simplify this expression as follows:

$$x_{n+1} - x_n = \frac{x_n}{2 + x_n} - x_n = \frac{x_n - 2x_n - x_n^2}{2 + x_n} = \frac{-x_n - x_n^2}{2 + x_n} = -\frac{x_n + x_n^2}{2 + x_n}.$$

Since by item a), $x_n > 0$ for all $n \in \mathbb{N}$, we have $x_n + x_n^2 > 0$, which implies that $-\dfrac{x_n + x_n^2}{2 + x_n} < 0$. Therefore, $x_{n+1} - x_n < 0$.

c) Since for all $n \in \mathbb{N}$ we have $x_n > 0$ (item a)) and (x_n) is a decreasing sequence (item b)), we have $0 < x_n \leq x_1$, that is, $0 < x_n \leq a$, for all $n \in \mathbb{N}$. In other words, the sequence of general term x_n is bounded. Because every monotone and bounded sequence is convergent, (x_n) is also convergent.

Let us assume that $\lim x_n = b$. Note that $b \geq 0$ because $x_n > 0$ for all $n \in \mathbb{N}$.

If the sequence converges to b, we also have $\lim x_{n+1} = b$. Furthermore, the sequence $\dfrac{x_n}{2 + x_n}$ is convergent, since it is the quotient of two convergent sequences where the denominator never becomes zero and has a limit different from zero.

Thus, we have

$$\lim x_{n+1} = \lim \frac{x_n}{2+x_n} \Leftrightarrow \lim x_{n+1} = \frac{\lim x_n}{\lim(2+x_n)} \Leftrightarrow b = \frac{b}{2+b}$$

$$\Leftrightarrow b - \frac{b}{2+b} = 0 \Leftrightarrow \frac{2b+b^2-b}{2+b} = 0 \Leftrightarrow \frac{b^2+b}{2+b} = 0 \Leftrightarrow b^2+b = 0 \quad (b \neq -2)$$

$$\Leftrightarrow b(b+1) = 0 \Leftrightarrow b = 0 \vee b = -1.$$

However, by Theorem 1.1.5, $b \geq 0$, which means that $\lim x_n = 0$.

11. a) We want to show that $x_{n+1} - x_n \geq 0$ for any $n \in \mathbb{N}_0$. Now, if $n = 0$, we have $x_1 - x_0 = a > 0$. If $n \geq 1$, then
$$x_{n+1} - x_n = x_n + x_{n-1}^2 - x_n = x_{n-1}^2 \geq 0.$$

b) As we saw in item a) that the sequence is increasing, we have $x_n \geq x_1 = a > 0$, $\forall n \in \mathbb{N}$.

c) Suppose there exists $b \in \mathbb{R}$ such that $b = \lim x_n$. Then, all its subsequences have limit b and
$$b = \lim x_{n+1} = \lim (x_n + x_{n-1}^2) = b + b^2$$

from where we conclude that $b = 0$.

d) If the sequence is bounded, it will converge because it is monotonic. We have shown in the previous item that if the sequence has a limit, it should be zero. This is not possible, however, since it is an increasing sequence of positive numbers. Therefore, we can conclude that it is unbounded. Since the sequence is increasing, it is bounded below by its first term. So, it has no upper bound. As a result, we can deduce that $\lim x_n = +\infty$.

12. a) For $n = 1$, $x_1 = 2 > 0$.

   Induction hypothesis: $x_n > 0$.

   Induction thesis: $x_{n+1} > 0$.

   Proof: Using the given formula $x_{n+1} = \frac{x_n}{2} + \frac{1}{x_n}$, we can see that $\frac{x_n}{2}$ and $\frac{1}{x_n}$ are both greater than zero since x_n is positive by the induction hypothesis. Therefore, the sum of these two quantities, which is x_{n+1}, is also greater than zero.

   By the Principle of Induction, we have proven that $x_n > 0$, $\forall n \in \mathbb{N}$.

   b) For $n = 1$, $x_1 = 2 > \sqrt{2}$.

   Induction hypothesis: $x_n > \sqrt{2}$.

   Induction thesis: $x_{n+1} > \sqrt{2}$.

   Proof: By definition, $x_{n+1} - \sqrt{2} = \frac{x_n}{2} + \frac{1}{x_n} - \sqrt{2} = \frac{x_n^2 + 2 - 2\sqrt{2}x_n}{2x_n} = \frac{(x_n - \sqrt{2})^2}{2x_n}$. As by induction hypothesis $x_n > \sqrt{2}$, then $(x_n - \sqrt{2})^2 > 0$, and we conclude that $x_{n+1} - \sqrt{2} > 0$.

   Then, using the Principle of Induction, we proved that $x_n > \sqrt{2}$, $\forall n \in \mathbb{N}$.

1.2. Solved Exercises

c) Let us analyze the difference $x_{n+1} - x_n$ to determine whether the sequence is increasing or decreasing:

$$x_{n+1} - x_n = \frac{x_n}{2} + \frac{1}{x_n} - x_n = -\frac{x_n}{2} + \frac{1}{x_n} = \frac{-x_n^2 + 2}{2x_n}.$$

By item b), $x_n > \sqrt{2}$, $\forall n \in \mathbb{N}$; therefore, $-x_n^2 + 2 < 0$, $\forall n \in \mathbb{N}$. Then $x_{n+1} - x_n < 0$, $\forall n \in \mathbb{N}$, thus proving that the sequence is decreasing.

d) If a sequence is decreasing, its first term is the maximum of the set of its terms; therefore, $x_1 = 2 \geq x_n$, $\forall n \in \mathbb{N}$, and (x_n) is bounded: $\sqrt{2} < x_n \leq 2$, $\forall n \in \mathbb{N}$. Thus, we can conclude that (x_n) is convergent as it is monotonic and bounded.

e) Let $a = \lim x_n$. All subsequences of (x_n) have limit a and

$$a = \lim x_{n+1} = \lim \left(\frac{x_n}{2} + \frac{1}{x_n} \right) = \frac{a}{2} + \frac{1}{a} = \frac{a^2 + 2}{2a}.$$

Solving the equation $a = \dfrac{a^2 + 2}{2a}$, we get $a = -\sqrt{2}$ or $a = \sqrt{2}$. As $x_n > \sqrt{2}$, $\forall n \in \mathbb{N}$, we conclude that $a = \sqrt{2}$.

13. a) We will show by induction that $x_n - \sqrt{3} > 0$, $\forall n \in \mathbb{N}$.

    If $n = 1$, then $x_1 = 3 > \sqrt{3}$, and the proposition is satisfied.

    <u>Induction hypothesis:</u> $x_n - \sqrt{3} > 0$.

    <u>Induction thesis:</u> $x_{n+1} - \sqrt{3} > 0$.

    Proof: By definition, $x_{n+1} - \sqrt{3} = \dfrac{x_n^2 + 3}{2\,x_n} - \sqrt{3} = \dfrac{(x_n - \sqrt{3})^2}{2x_n}$. We know that $x_n > \sqrt{3}$; therefore, $x_{n+1} - \sqrt{3} > 0$.

    Then, using the Principle of Induction, we proved that $x_n - \sqrt{3} > 0$, $\forall n \in \mathbb{N}$.

    b) We intend to show that the sequence is decreasing, that is, $x_{n+1} - x_n \leq 0$, $\forall n \in \mathbb{N}$:

    $$x_{n+1} - x_n = \frac{x_n^2 + 3}{2\,x_n} - x_n = \frac{-x_n^2 + 3}{2x_n}.$$

    By item a), $x_n > \sqrt{3}$, $\forall n \in \mathbb{N}$; therefore, $-x_n^2 + 3 < 0$, $\forall n \in \mathbb{N}$. Then $x_{n+1} - x_n < 0$, $\forall n \in \mathbb{N}$, that is to say, the sequence is decreasing.

    c) If a sequence is decreasing, its first term is the maximum of the set of terms of the sequence; therefore, $x_1 = 3 \geq x_n$, $\forall n \in \mathbb{N}$. We have (x_n) bounded: $\sqrt{3} < x_n \leq 3$, $\forall n \in \mathbb{N}$. We can conclude that (x_n) is convergent because it is monotonic and bounded.

    d) Let $a = \lim x_n$. All subsequences of (x_n) have limit a and

    $$a = \lim x_{n+1} = \lim \frac{x_n^2 + 3}{2\,x_n} = \frac{a^2 + 3}{2a}.$$

    Solving the equation $a = \dfrac{a^2 + 3}{2\,a}$, we get $a = -\sqrt{3}$ or $a = \sqrt{3}$. As $x_n > \sqrt{3}$, $\forall n \in \mathbb{N}$, we conclude that $a = \sqrt{3}$.

14. a) In order to prove by definition that $\lim \dfrac{25n}{3n+1} = \dfrac{25}{3}$, we must prove that the proposition

$$\forall \varepsilon > 0 \ \exists p \in \mathbb{N} : n > p \Rightarrow \left| \dfrac{25n}{3n+1} - \dfrac{25}{3} \right| < \varepsilon$$

is true, that is, we need to show that for each positive real ε there is at least one order after which, for every natural n,

$$\left| \dfrac{25n}{3n+1} - \dfrac{25}{3} \right| < \varepsilon.$$

Let $\varepsilon > 0$. Solving the inequality for n, we obtain

$$\left| \dfrac{25n}{3n+1} - \dfrac{25}{3} \right| < \varepsilon \Leftrightarrow \left| \dfrac{75n - 75n - 25}{3(3n+1)} \right| < \varepsilon \Leftrightarrow \dfrac{25}{3(3n+1)} < \varepsilon \Leftrightarrow n > \dfrac{25 - 3\varepsilon}{9\varepsilon}.$$

If we choose $p \in \mathbb{N}$ such that $p \geq \dfrac{25 - 3\varepsilon}{9\varepsilon}$, then, as $n > p$, we get $n > \dfrac{25 - 3\varepsilon}{9\varepsilon}$, that is,

$$\left| \dfrac{25n}{3n+1} - \dfrac{25}{3} \right| < \varepsilon.$$

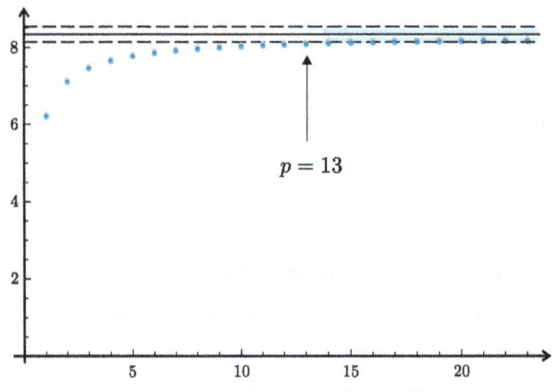

Figure 1.16: If $\varepsilon = 0.2$, then $-\varepsilon < \dfrac{25n}{3n+1} - \dfrac{25}{3} < \varepsilon$ if $n > 13$

Suppose, for instance, that we want to calculate the smallest order p from which

$$\left| \dfrac{25n}{3n+1} - \dfrac{25}{3} \right| < 0.2.$$

Substituting ε by 0.2 in the expression $\dfrac{25 - 3\varepsilon}{9\varepsilon}$, we get the value $13.\overline{5}$; therefore, as $p \in \mathbb{N}$, we can choose $p = \text{Int}(13.\overline{5}) = 13$. Thus, if $n > 13$, the desired result follows. This case is illustrated in Fig. 1.16.

b) In order to prove by definition that $\lim \dfrac{\sqrt{n}}{\sqrt{n}+1} = 1$, we should prove that the proposition

$$\forall \varepsilon > 0 \ \exists p \in \mathbb{N} : n > p \Rightarrow \left| \dfrac{\sqrt{n}}{\sqrt{n}+1} - 1 \right| < \varepsilon$$

1.2. Solved Exercises

is true, that is, we need to show that for each positive real ε there is at least one order after which, for every natural n,
$$\left|\frac{\sqrt{n}}{\sqrt{n}+1}-1\right|<\varepsilon.$$

Let $\varepsilon > 0$. Solving the inequality for n, we obtain
$$\left|\frac{\sqrt{n}}{\sqrt{n}+1}-1\right|<\varepsilon \Leftrightarrow \left|\frac{\sqrt{n}-\sqrt{n}-1}{\sqrt{n}+1}\right|<\varepsilon \Leftrightarrow \frac{1}{\sqrt{n}+1}<\varepsilon \Leftrightarrow \sqrt{n}>\frac{1-\varepsilon}{\varepsilon}.$$

If $1-\varepsilon \leq 0$, the last inequality is true regardless of the choice of $p \in \mathbb{N}$. If $1-\varepsilon > 0$, then the previous inequalities are still equivalent to
$$n > \left(\frac{1-\varepsilon}{\varepsilon}\right)^2.$$

If we choose $p \in \mathbb{N}$ such that $p \geq \left(\frac{1-\varepsilon}{\varepsilon}\right)^2$, then, as $n > p$, it follows that $n > \left(\frac{1-\varepsilon}{\varepsilon}\right)^2$, that is,
$$\left|\frac{\sqrt{n}}{\sqrt{n}+1}-1\right|<\varepsilon.$$

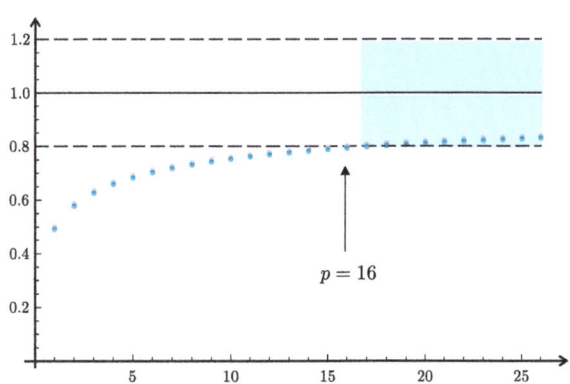

Figure 1.17: If $\varepsilon = 0.2$, then $-\varepsilon < \frac{\sqrt{n}}{\sqrt{n}+1}-1 < \varepsilon$ if $n > 16$

Suppose, for example, that we want to calculate the smallest order p from which
$$\left|\frac{\sqrt{n}}{\sqrt{n}+1}-1\right|<0.2.$$

Substituting ε by 0.2 in the expression $\left(\frac{1-\varepsilon}{\varepsilon}\right)^2$, we get the value 16; therefore, we can choose $p = 16$. Thus, if $n > 16$, the desired result follows. This case is illustrated in Fig. 1.17.

c) We intend to prove, using the definition, that $\lim \frac{2\sqrt{n}+5^{-n}}{\sqrt{n}+1} = 2$. For this, we must prove that the proposition
$$\forall \varepsilon > 0 \ \exists p \in \mathbb{N} : n > p \Rightarrow \left|\frac{2\sqrt{n}+5^{-n}}{\sqrt{n}+1}-2\right|<\varepsilon$$

is true, that is, we need to show that for each positive real ε there is at least one order after which, for every natural n,
$$\left|\frac{2\sqrt{n}+5^{-n}}{\sqrt{n}+1}-2\right|<\varepsilon.$$

Let $\varepsilon > 0$.

$$\left|\frac{2\sqrt{n}+5^{-n}}{\sqrt{n}+1}-2\right|<\varepsilon \Leftrightarrow \left|\frac{2\sqrt{n}+5^{-n}-2\sqrt{n}-2}{\sqrt{n}+1}\right|<\varepsilon \Leftrightarrow \left|\frac{5^{-n}-2}{\sqrt{n}+1}\right|<\varepsilon \Leftrightarrow \frac{2-5^{-n}}{\sqrt{n}+1}<\varepsilon.$$

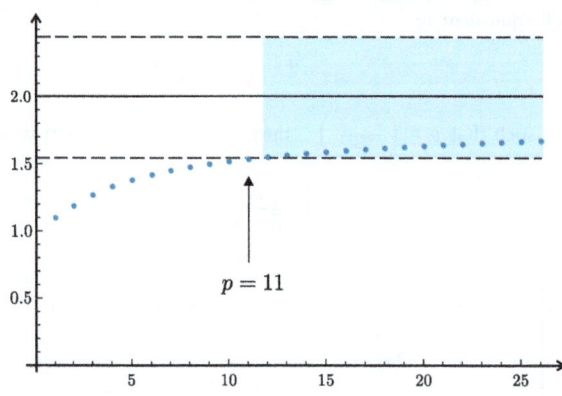

Figure 1.18: If $\varepsilon = 0.45$, then $-\varepsilon < \frac{2\sqrt{n}+5^{-n}}{\sqrt{n}+1} - 2 < \varepsilon$ if $n > 11$

However,
$$\frac{2-5^{-n}}{\sqrt{n}+1} < \frac{2}{\sqrt{n}+1}, \quad \forall n \in \mathbb{N};$$
therefore, if
$$\frac{2}{\sqrt{n}+1} < \varepsilon,$$
then
$$\frac{2-5^{-n}}{\sqrt{n}+1} < \varepsilon.$$

Solving the inequality for n, we obtain
$$\frac{2}{\sqrt{n}+1} < \varepsilon \Leftrightarrow \sqrt{n} > \frac{2-\varepsilon}{\varepsilon}.$$

If $2 - \varepsilon \leq 0$, the last inequality is true regardless of the choice of $p \in \mathbb{N}$. If $2 - \varepsilon > 0$, then the previous inequalities are still equivalent to
$$n > \left(\frac{2-\varepsilon}{\varepsilon}\right)^2.$$

If we choose $p \in \mathbb{N}$ such that $p \geq \left(\frac{2-\varepsilon}{\varepsilon}\right)^2$, then, as $n > p$, it results in $n > \left(\frac{2-\varepsilon}{\varepsilon}\right)^2$, that is,
$$\left|\frac{2\sqrt{n}+5^{-n}}{\sqrt{n}+1}-2\right|<\varepsilon.$$

1.2. Solved Exercises

Suppose, for example, that we want to calculate an order p from which

$$\left|\frac{2\sqrt{n}+5^{-n}}{\sqrt{n}+1}-2\right|<0.45.$$

Substituting ε by 0.45 in the expression $\left(\frac{2-\varepsilon}{\varepsilon}\right)^2$, we get the value $(3,\overline{4})^2 \cong 11.86$; therefore, as $p \in \mathbb{N}$, we can choose $p = \text{Int}(11.86) = 11$. Thus, if $n > 11$, the desired result is obtained. This case is illustrated in Fig. 1.18.

d) In order to prove by definition that $\lim\left(1+\frac{\sin(n)}{n}\right)=1$, we must prove that the proposition

$$\forall \varepsilon > 0 \;\; \exists p \in \mathbb{N} : n > p \Rightarrow \left|\left(1+\frac{\sin(n)}{n}\right)-1\right|<\varepsilon$$

is true, that is, we need to show that for each positive real ε there is at least one order after which, for every natural n,

$$\left|\left(1+\frac{\sin(n)}{n}\right)-1\right|<\varepsilon.$$

Let $\varepsilon > 0$.

$$\left|\left(1+\frac{\sin(n)}{n}\right)-1\right|<\varepsilon \Leftrightarrow \left|\frac{\sin(n)}{n}\right|<\varepsilon \Leftrightarrow \frac{|\sin(n)|}{n}<\varepsilon.$$

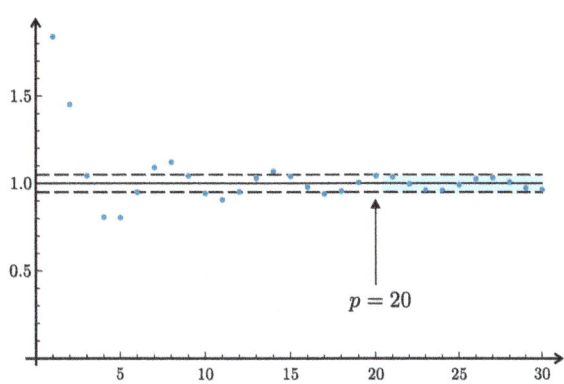

Figure 1.19: If $\varepsilon = 0.05$, then $-\varepsilon < 1+\frac{\sin(n)}{n}-1 < \varepsilon$ if $n > 20$

But

$$\frac{|\sin(n)|}{n} \leq \frac{1}{n}, \;\; \forall n \in \mathbb{N};$$

therefore, if

$$\frac{1}{n}<\varepsilon,$$

we have

$$\frac{|\sin(n)|}{n}<\varepsilon.$$

Solving the inequality for n, we obtain
$$\frac{1}{n} < \varepsilon \Leftrightarrow n > \frac{1}{\varepsilon}.$$

If we choose $p \in \mathbb{N}$ such that $p \geq \frac{1}{\varepsilon}$, then, as $n > p$, it results in $n > \frac{1}{\varepsilon}$, that is,
$$\frac{|\sin(n)|}{n} < \varepsilon.$$

Suppose, for example, we want to calculate an order p from which
$$\left|\left(1 + \frac{\sin(n)}{n}\right) - 1\right| < 0.05.$$

Substituting ε by 0.05 in the expression $\frac{1}{\varepsilon}$, we get the value 20; therefore, as $p \in \mathbb{N}$, we can choose $p = 20$. Thus, if $n > 20$, the desired result is obtained. This case is illustrated in Fig. 1.19.

e) To prove that $\lim n^3 = +\infty$ by definition, we need to show that the following proposition is true:
$$\forall L \in \mathbb{R}^+ \ \exists p \in \mathbb{N}: \ n > p \Rightarrow n^3 > L.$$

In other words, we must show that for each positive real L, there is at least one order after which, for every natural n, $n^3 > L$.

Let $L > 0$. Solving the inequality for n, we have $n^3 > L \Leftrightarrow n > \sqrt[3]{L}$. Therefore, if we choose $p \in \mathbb{N}$ such that $p > \sqrt[3]{L}$, we have proven that $\lim n^3 = +\infty$.

In Fig. 1.20 you can see that by choosing $L = 1000$, then from $n > p = \sqrt[3]{1000} = 10$, all terms of the sequence are greater than L.

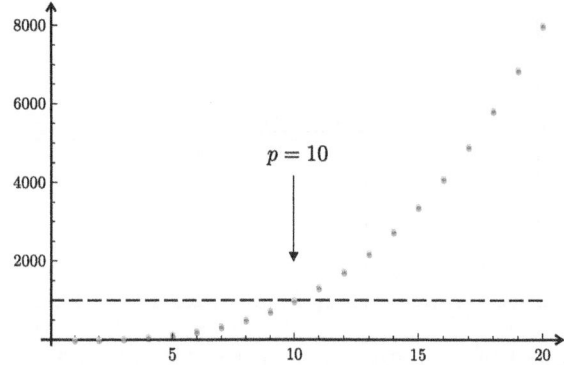

Figure 1.20: If $L = 1000$, then $n^3 > L$ if $n > 10$

f) In order to prove by definition that $\lim \frac{n^2}{n+1} = +\infty$, we must prove that the proposition
$$\forall L \in \mathbb{R}^+ \ \exists p \in \mathbb{N}: \ n > p \Rightarrow \frac{n^2}{n+1} > L$$

is true, that is, we must show that for each positive real number L, there is at least one order after which, for every natural number n,
$$\frac{n^2}{n+1} > L.$$

1.2. Solved Exercises

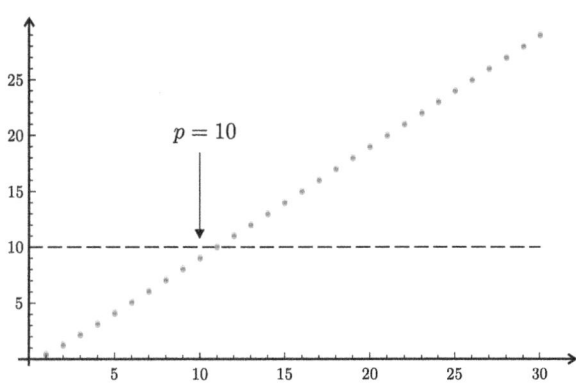

Figure 1.21: If $L = 10$, then $\frac{n^2}{n+1} > L$ if $n > 10$

Let $L > 0$. We have

$$\frac{n^2}{n+1} > L \Leftrightarrow n^2 - Ln - L > 0 \Leftrightarrow \left(n - \frac{L - \sqrt{L^2 + 4L}}{2}\right)\left(n - \frac{L + \sqrt{L^2 + 4L}}{2}\right) > 0.$$

Because $L > 0$ and $n \in \mathbb{N}$, we obtain from the previous inequality that $n > \dfrac{L + \sqrt{L^2 + 4L}}{2}$. Therefore, if we choose $p = \text{Int}\left(\dfrac{L + \sqrt{L^2 + 4L}}{2}\right)$, then for $n > p$, we get $\dfrac{n^2}{n+1} > L$.

In Fig. 1.21, you can see that if we choose $L = 10$, then from

$$n > p = \text{Int}\left(\frac{10 + \sqrt{140}}{2}\right) = 10$$

all the terms of the sequence are greater than L.

15. Let us denote N_n as the number of segments, L_n as the length of each segment, and l_n as the length of the curve in the nth iteration, that is, after repeating n times the process of constructing the curve. We will prove by induction that $N_n = 4^n$. If $n = 1$, we have four segments by construction, so $N_1 = 4^1 = 4$. Suppose (induction hypothesis) that $N_n = 4^n$. Then $N_{n+1} = N_n \times 4$ since each segment originates four new segments. Substituting the expression of N_n, we get $N_{n+1} = 4^n \times 4 = 4^{n+1}$.

In each iteration, we divide each segment by 3. During the first iteration, the length of each segment is $L_1 = \dfrac{1}{3}$. Assuming (induction hypothesis) that $L_n = \left(\dfrac{1}{3}\right)^n$, we can prove that $L_{n+1} = \left(\dfrac{1}{3}\right)^{n+1}$ as follows:

$$L_{n+1} = \frac{L_n}{3} = \frac{\left(\frac{1}{3}\right)^n}{3} = \left(\frac{1}{3}\right)^{n+1}.$$

The length of the curve in the nth iteration is $l_n = N_n \times L_n$. We know that $N_n = 4^n$, so $l_n = 4^n \left(\frac{1}{3}\right)^n = \left(\frac{4}{3}\right)^n$. The length L of the Koch curve is determined by the limit of l_n when $n \to +\infty$. Then, by Theorem 1.1.10,

$$L = \lim \left(\frac{4}{3}\right)^n = +\infty.$$

1.3 Proposed Exercises

1. Prove by definition that $\lim u_n = +\infty$ in each item listed below:

   a) $u_n = n$

   b) $u_n = n^2$

   c) $u_n = \sqrt{n}$

   d) $u_n = 2^n$

2. By definition, prove that the following sequences (u_n) are null sequences, that is, that $\lim u_n = 0$:

   a) $u_n = \dfrac{1}{n}$

   b) $u_n = \dfrac{1}{n^2}$

   c) $u_n = \dfrac{1}{\sqrt{n}}$

   d) $u_n = \dfrac{1}{2^n}$

3. Evaluate the limits of the following sequences and provide an argument for the calculations:

   a) $\dfrac{1-n}{5n+3}$

   b) $\dfrac{n^2+2}{3n+1}$

   c) $\dfrac{n^2+3n}{n+2} - \dfrac{n^2-1}{n}$

   d) $\dfrac{3n}{4n^3+1}$

   e) $\dfrac{-n^3+2}{4n^3-7}$

4. Find the limits of the following sequences and justify the calculations:

   a) $\dfrac{\sqrt{n}}{4n+1}$

   b) $\dfrac{\sqrt{n}}{\frac{1}{2}-\sqrt{n}}$

   c) $\sqrt{n^2+1} - \sqrt{n^2-1}$

   d) $(\sqrt{n+1}-\sqrt{n})\sqrt{n+\dfrac{1}{2}}$

   e) $\dfrac{1}{\sqrt{n^2+1}} + \dfrac{1}{\sqrt{n^2+2}} + \cdots + \dfrac{1}{\sqrt{n^2+n}}$

5. We say that the sequence (u_n) grows faster than the sequence (v_n) if $\dfrac{u_n}{v_n} \to +\infty$.

   a) Prove that n^n grows faster than $n!$.

   b) Prove that $n!$ grows faster than e^n.

   c) Arrange the following sequences in descending order in terms of the speed of convergence:

   $2n$, $\sqrt{10n}$, 2^n, e^n, $n!$, $\log(n)$, n^3, n^n.

6. Let (u_n) and (v_n) be two null sequences such that $v_n \neq 0$, $\forall n \in \mathbb{N}$. We say that (u_n) is of higher order than (v_n) if $\lim \dfrac{u_n}{v_n} = 0$. Order the following null sequences:

   $\dfrac{1}{2n}$, $\dfrac{1}{\sqrt{10n}}$, $\dfrac{1}{2^n}$, $\dfrac{1}{e^n}$, $\dfrac{1}{n!}$, $\dfrac{1}{\log(n)}$, $\dfrac{1}{n^3}$, $\dfrac{1}{n^n}$.

7. Calculate the limit of the sequences with general terms:

   a) $\left(\dfrac{n+3}{n+1}\right)^{2n}$

   b) $\left(\dfrac{n+5}{2n+1}\right)^{n}$

   c) $\left(\dfrac{5n+5}{2n+1}\right)^{n^2}$

   d) $\left(1-\dfrac{3}{n^2}\right)^{n}$

8. Evaluate, if it exists:

   a) $\lim \dfrac{1}{2n} \sqrt[n]{(n+1)!}$

   b) $\lim \dfrac{1}{n} \sqrt[n]{n(n+1)\cdots 2n}$

9. Determine $p \in \mathbb{R}$ such that $\lim \sqrt[n]{\dfrac{n!}{(pn)^n}} = 3$.

10. Find the limits of the following sequences and provide a clear explanation of the reasoning behind each calculation:

    a) $\big(\cos(x)\big)^n$, $x \in \mathbb{R}$

    b) $\cos^2(n) \sin\left(\dfrac{1}{n}\right)$

    c) $\dfrac{n(n-1)(n-2)(n-3)}{(n+1)(n+2)(n+3)}$

    d) $\sqrt[n]{n!\left(\dfrac{2}{n}\right)^n}$

    e) $\dfrac{1}{\sqrt{n^2+1}} + \dfrac{1}{\sqrt{n^2+2}} + \cdots + \dfrac{1}{\sqrt{n^2+2n+1}}$

1.3. Proposed Exercises

f) $\left(1-\dfrac{1}{n}\right)^n \sqrt[n]{\dfrac{n+1}{n}}$

g) $\sqrt[n]{(n+1)! - n!}$

h) $\dfrac{1}{\sqrt{n}} + \dfrac{1}{\sqrt{n+1}} + \cdots + \dfrac{1}{\sqrt{2n}}$

i) $\dfrac{1}{n^2} + \dfrac{1}{(n+1)^2} + \cdots + \dfrac{1}{(2n)^2}$

j) $\dfrac{n}{\sqrt{n^4+1}} + \dfrac{n}{\sqrt{n^4+2}} + \cdots + \dfrac{n}{\sqrt{n^4+n}}$

11. For each item, provide, when possible, examples of sequences u, v, and w such that $u_n \to +\infty$, $v_n \to -\infty$, and $w_n \to 0$.

   a) $u_n + v_n \to 1$
   b) $u_n + v_n \to -\infty$
   c) $u_n + w_n \to 1$
   d) $u_n \times w_n \to 0$
   e) $v_n \times w_n \to +\infty$
   f) $\dfrac{u_n}{w_n} \to -1$

12. Let (x_n) and (y_n) be two convergent sequences of real numbers such that $x_n \to x$ and $y_n \to y$. Show that the sequence of general term $z_n = \min\{x_n, y_n\}$ converges and that $z_n \to \min\{x, y\}$.

13. Consider the sequence (u_n) defined recursively by
$$\begin{cases} u_1 = 1, \\ u_n = \dfrac{u_{n-1} - 1}{2}, & \forall n > 1. \end{cases}$$
Explain why (u_n) is convergent and evaluate its limit.

14. Consider the sequence (u_n) defined recursively
$$\begin{cases} u_1 = 5 \\ u_{n+1} = \dfrac{5u_n - 4}{u_n} & \forall n \geq 1. \end{cases}$$

   a) Prove by induction that $\forall n \in \mathbb{N}, \ u_n > 4$.
   b) Prove that the sequence is convergent.
   c) Show that 4 is the infimum of the set of terms of the sequence.

15. Consider the sequence of real numbers defined recursively
$$\begin{cases} x_0 = 1 \\ x_{n+1} = \dfrac{1}{2}\left(x_n + \dfrac{2}{x_n}\right), & \forall n \in \mathbb{N}_0. \end{cases}$$

   a) Show by induction that
   $$\sqrt{2} \leq x_n \leq 2, \ \forall n \in \mathbb{N}.$$
   b) Show that $x_n \geq x_{n+1}, \forall n \geq 1$.
   c) Show that the sequence is convergent and calculate its limit.

16. The sequences (u_n) and (v_n) verify the following conditions:
   (i) $\forall n \in \mathbb{N} \ \ 0 < u_n < v_n$.
   (ii) (v_n) is decreasing.

   Decide whether the following statements are true or false and give a justification for the answers provided.

   a) (v_n) is convergent.
   b) (u_n) is convergent.
   c) (u_n) is decreasing.

17. Determine the upper and lower limits of the sequences with the following general terms and provide a reason for the answers:

   a) $n^{(-1)^n}$
   b) $\cos\left(\dfrac{n\pi}{3}\right)$
   c) $\sqrt{n} - (-1)^n \sqrt{n-1}$
   d) $\sin\left(\dfrac{n\pi}{4}\right)$
   e) $\sqrt{n} - (-1)^n \sqrt{n-1}$
   f) $\dfrac{1}{n^2} \cos\left(\dfrac{n\pi}{10}\right) + \left(\cos\left(\dfrac{n\pi}{2}\right)\right)^n$
   g) $\dfrac{(-1)^n n^2 + 3}{n+1}$
   h) $\sin\left(\dfrac{n\pi}{2} + a\right), \ a \in \mathbb{R}$
   i) $\left(\dfrac{1}{3}\right)^n + \dfrac{1}{2n} + 2n\left((-1)^n 3 + 3\right)$

j) $\dfrac{\left((-1)^{n+3} - (-1)^n\right) n^3 + 2}{3n+1}$

18. Show that the following are Cauchy sequences in $\mathbb{Q}$:

    a) $\left(\dfrac{1}{n^2}\right)$
    b) $\left(\dfrac{1}{2^n}\right)$

19. Show that the sequence of general term
$$1 + \dfrac{1}{2} + \cdots + \dfrac{1}{n}$$
is not a Cauchy sequence in $\mathbb{Q}$.

20. Consider the sequence of general term
$$u_n = \dfrac{n+1}{n+2}.$$
Check the sequence for convergence using the definition of a Cauchy sequence.

21. Show that the sequence
$$\begin{cases} x_1 = \dfrac{3}{2}, \\ x_{n+1} = \dfrac{x_n}{2} + \dfrac{1}{x_n}, \quad \forall n \in \mathbb{N} \end{cases}$$
is a sequence in $\mathbb{Q}$ such that $x_n^2 \to 2$. Use this result to show that (x_n) is a Cauchy sequence in $\mathbb{Q}$ that does not converge in $\mathbb{Q}$.

Hint:

(i) Show that $v_n = x_n^2 - 2$ satisfies $0 \leq v_n \leq \dfrac{1}{4^n}$.

(ii) Use the relation $x_n - x_m = \dfrac{x_n^2 - x_m^2}{x_n + x_m}$.

22. The Sierpinski carpet is a set constructed as follows: Consider a square, R, with side 1. Divide R into nine squares of equal area and remove the interior of the middle square. Repeat the process in the remaining eight squares: Divide each one into nine squares of equal area and remove the interior of the middle square. Let R_n be the region that remains after this process has been performed n times. The Sierpinski carpet, S, is the set of all points in R that are not removed by any of the previous operations, that is, S consists of the points in R that belong to R_n for any $n \geq 1$. The figure illustrates the first three steps of this process.

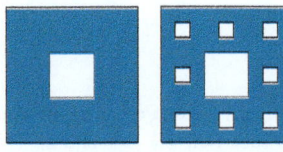
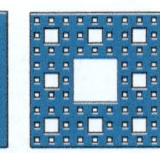

a) Calculate the area A_n of R_n, $n \geq 1$.

b) Show that $\lim A_n = 0$.

CHAPTER 2

Numerical Series

The operation of adding two numbers extends without any difficulty to a finite number of terms, even if this number is very large. However, considering an infinite number of terms raises more delicate issues addressed in this chapter.

2.1 Generalization of the Addition Operation

Addition (or sum) is the operation that corresponds to each pair of real numbers with another real number, according to specific rules. This operation satisfies certain properties and can be generalized to a finite number of terms while maintaining all properties. The definition of the sum of a finite number of terms is made recursively as follows:

$$\sum_{i=1}^{n} a_i = \begin{cases} a_1, & \text{if } n = 1 \\ \left(\sum_{i=1}^{n-1} a_i\right) + a_n, & \text{if } n > 1. \end{cases}$$

We can now consider generalizing the notion of sum to an infinite number of terms, $a_1, a_2, \ldots, a_n, \ldots$

If there is an order p from which all terms of the sequence are zero, the sum of all terms equals the sum of the first p terms:

$$\sum_{n \in \mathbb{N}} a_i = \sum_{i=1}^{p} a_i.$$

If there is a subsequence of nonzero terms, let us consider $S_n = \sum_{i=1}^{n} a_i$, the sum of the first n terms. It is natural to consider the sum of all terms as the limit of the sequence (S_n) if it exists and is finite.

© The Author(s), under exclusive license to Springer Nature Switzerland AG 2024
A. Alves de Sá, B. Louro, *Sequences and Series*,
https://doi.org/10.1007/978-3-031-67202-6_2

Zeno of Elea (490–435 B.C.) Greek philosopher and disciple of Parmenides. Most of his teaching records have come to us through secondary sources, notably through Aristotle. Zeno's great fame came from his paradoxes. These paradoxes aimed to defend his master Parmenides' theory that reality was unique, immutable, and immobile so that change, movement, time, and plurality would be nothing more than illusions. Parmenides' theory presented conclusions that contradicted what the senses conveyed; Zeno's goal was not to present evidence that reinforced the theory itself but to show that opposing theories also led to contradictions. (Source of image: Jan de Bisschop, Paradigmata graphices variorum artificum, Rijksmuseum.)

If the terms of the sequence (a_n) are all positive, it might seem at first glance that (S_n) is not convergent. In fact, assuming that the sum of an infinite number of positive terms is a real number is not an intuitive concept. In this case, intuition fails precisely because we intend to generalize to infinity a concept, that of sum, which we have intuitive for a finite number of terms. It is common for intuition to deceive us in cases of "passing" from finite to infinite.

It is true that we will not always be able to find the limit of the sequence (S_n), so, by this process, we will not be able to find the sum of an infinite number of terms if it exists. Nevertheless, we are interested in knowing how the sequence (a_n) should be so that a real number, the sum of all its terms, is associated with this sequence.

One of the historical examples illustrating these difficulties in dealing with the concept of an infinite sum of terms is one of the famous Paradoxes of Zeno, which we summarize as follows:

A walker wants to move from one place to another at a constant speed (Fig. 2.1). In T minutes, he covers half of the distance. He still has the other half to go. In $T/2$ minutes, he will cover half of what remains ($1/4$ of the total distance). Moreover, now he has $1/4$ of the distance left. He will cover half of the rest ($1/8$ of the total) in $T/4$ minutes. This argument may be repeated indefinitely. The walker spends

$$T + \frac{T}{2} + \frac{T}{4} + \frac{T}{8} + \frac{T}{16} + \ldots$$

minutes on his journey.

Given that it is the sum of infinite positive terms, we are inclined to assume that the sum is infinite. The walker will not be able to reach his goal because it will take an infinite amount of time. This reasoning contradicts intuition and practical life, which is called a paradox.

This is due to the lack of understanding of the concept of an infinite sum of terms (more precisely, of series), which came to be fully understood only in the 19th century. In this case, the walker will reach his goal in time $2T$, as intuition tells us.

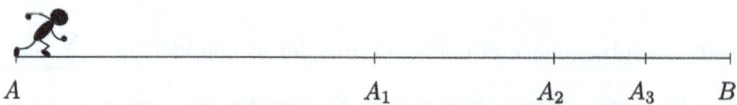

Figure 2.1: The walker's path in Zeno's Paradox

2.2 Definition of Series: Convergence – General Properties

> **Definition 2.2.1** *Let (a_n) be a sequence of real numbers. The **series generated** by a_n is the sequence (S_n) defined as follows:*
> $$\begin{aligned} S_1 &= a_1 \\ S_2 &= a_1 + a_2 \\ S_3 &= a_1 + a_2 + a_3 \\ &\vdots \\ S_n &= a_1 + a_2 + a_3 + \cdots + a_n \\ &\vdots \end{aligned}$$

To represent a series, we can use any of the following notations:
$$\sum_{n=1}^{\infty} a_n, \quad \sum a_n, \quad a_1 + a_2 + a_3 + \cdots$$

The numbers a_1, a_2, ..., are called **terms** of the series, a_n is said to be the **general term** of the series, and the sums S_1, S_2, ..., are called **partial sums**.

> **Definition 2.2.2** *The series $\sum a_n$ is **convergent** if the limit*
> $$\lim S_n = \lim \sum_{i=1}^{n} a_i$$
> *exists and is finite. If this limit does not exist or is not finite, the series is **divergent**.*

In the case of convergence, the **sum** of the series is the value S of the limit of (S_n), that is,
$$S = \lim S_n = \sum_{n=1}^{\infty} a_n.$$

Note: The identification of a series with the symbol $\sum_{n=1}^{\infty} a_n$ is a misuse of language because it is the identification of a series with its sum when it exists. However, despite being a common misuse of language, it has been found to be a valuable and harmless convention.

Example 2.2.1 An important example is the geometric series. A **geometric series** is a series generated by a geometric sequence: If (a_n) is a geometric sequence of ratio $r \neq 1$ and $a_1 \neq 0$, then

$$S_n = \sum_{i=1}^{n} a_i = \sum_{i=1}^{n} a_1 r^{i-1} = a_1 \cdot \frac{1-r^n}{1-r}.$$

We know that (S_n) is convergent if and only if $|r| < 1$, so the geometric series is convergent if and only if the absolute value of the ratio of the geometric sequence that generated it is less than 1. In this case

$$\sum_{n=1}^{\infty} a_n = \frac{a_1}{1-r}.$$

The series in Zeno's Paradox (see Sect. 2.1) is geometric with $a_1 = T$ and $r = 1/2$; it is convergent with the sum $2T$.

If $r = 1$, the general term is constant; that is,

$$\sum_{n=1}^{\infty} a_n = \sum_{n=1}^{\infty} a_1,$$

and thus, $S_n = na_1$ and, if $a_1 \neq 0$, the series will diverge.

Example 2.2.2 The series

$$\sum_{n=1}^{\infty} (-1)^n \frac{3^n}{e^{2n+1}} = \sum_{n=1}^{\infty} \frac{1}{e} \left(-\frac{3}{e^2}\right)^n$$

is geometric with ratio $r = -\dfrac{3}{e^2}$ and first term $-\dfrac{3}{e^3}$. Since $-1 < r < 1$, the series converges and its sum is

$$\frac{-\frac{3}{e^3}}{1-\left(-\frac{3}{e^2}\right)} = \frac{-3}{e\left(e^2+3\right)}.$$

Example 2.2.3 We can write the rational number $2.3\overline{25} = 2.3252525\ldots$ in the form of a fraction using a geometric series.

$$2.3\overline{25} = 2.3 + 0.025 + 0.00025 + \ldots = \frac{23}{10} + \frac{25}{10^3} + \frac{25}{10^5} + \frac{25}{10^7} + \cdots$$

2.2. Definition of Series: Convergence – General Properties

After the first term, we have a geometric series with ratio $r = \dfrac{1}{10^2}$ and first term $\dfrac{25}{10^3}$. Therefore,

$$2.3\overline{25} = \frac{23}{10} + \frac{\frac{25}{10^3}}{1 - \frac{1}{10^2}} = \frac{1151}{495}.$$

Example 2.2.4 Consider the series $\sum_{n=1}^{\infty} \dfrac{1}{\sqrt{n}}$; let us study the sequence of its partial sums and the corresponding limit:

$$S_1 = 1$$
$$S_2 = 1 + \frac{1}{\sqrt{2}}$$
$$S_3 = 1 + \frac{1}{\sqrt{2}} + \frac{1}{\sqrt{3}}$$
$$\vdots$$
$$S_n = 1 + \frac{1}{\sqrt{2}} + \frac{1}{\sqrt{3}} + \cdots + \frac{1}{\sqrt{n}}$$
$$\vdots$$

As

$$1 + \frac{1}{\sqrt{2}} + \frac{1}{\sqrt{3}} + \cdots + \frac{1}{\sqrt{n}} \geq \frac{1}{\sqrt{n}} + \frac{1}{\sqrt{n}} + \frac{1}{\sqrt{n}} + \cdots + \frac{1}{\sqrt{n}} = \frac{n}{\sqrt{n}} = \sqrt{n}$$

and $\lim \sqrt{n} = +\infty$, we conclude by Theorem 1.1.2 that the sequence (S_n) has limit $+\infty$. The series under study is divergent.

Example 2.2.5 Consider the series $\sum_{n=1}^{\infty} \dfrac{1}{n}$, commonly known as the **harmonic series**. By mathematical induction, we can prove that the subsequence (S_{2^n}) of the partial sums of this series satisfies the inequality

$$S_{2^n} \geq 1 + \frac{n}{2}, \quad \forall n \in \mathbb{N}.$$

For $n = 1$, the proposition is true:

$$S_2 = 1 + \frac{1}{2} \geq 1 + \frac{1}{2}.$$

Assuming that the proposition is valid for n, we prove that it is also valid for $n+1$.

$$S_{2^{n+1}} = S_{2^n} + \frac{1}{2^n+1} + \cdots + \frac{1}{2^{n+1}} \geq 1 + \frac{n}{2} + 2^n \frac{1}{2^{n+1}} = 1 + \frac{n+1}{2}.$$

Thus, the desired inequality is proved using the method of induction. By Theorem 1.1.2, we can conclude that the sequence (S_{2^n}) has limit $+\infty$, implying that the sequence of partial sums of the harmonic series does not have a finite limit. Therefore, the harmonic series is divergent.

Figure 2.2 shows the sequence of partial sums of the harmonic series and its general term. It illustrates how the sum of an infinite number of increasingly smaller positive quantities can have a value as large as desired.

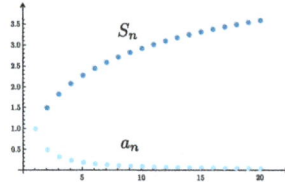

Figure 2.2: The sequences $a_n = \frac{1}{n}$ and $S_n = 1 + \cdots + \frac{1}{n}$

Example 2.2.6 Consider the series $\displaystyle\sum_{n=1}^{\infty} \frac{1}{n(n+1)}$. By observing that

$$\frac{1}{n(n+1)} = \frac{1}{n} - \frac{1}{n+1},$$

we can write the sequence of partial sums as follows (see Fig. 2.3):

$$S_1 = 1 - \frac{1}{2}$$

$$S_2 = 1 - \frac{1}{2} + \frac{1}{2} - \frac{1}{3} = 1 - \frac{1}{3}$$

$$S_3 = 1 - \frac{1}{3} + \frac{1}{3} - \frac{1}{4} = 1 - \frac{1}{4}$$

$$\vdots$$

$$S_n = 1 - \frac{1}{n+1}$$

$$\vdots$$

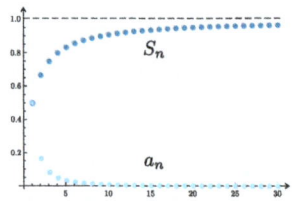

Figure 2.3: The sequences $a_n = \frac{1}{n(n+1)}$ and $S_n = 1 - \frac{1}{n+1}$

As $\lim S_n = 1$, the series is convergent and its sum is 1:

$$\sum_{n=1}^{\infty} \frac{1}{n(n+1)} = 1.$$

Example 2.2.7 The sequence of partial sums of the series

$$\sum_{n=1}^{\infty} \log\left(\frac{n}{n+1}\right) = \sum_{n=1}^{\infty} \big(\log(n) - \log(n+1)\big)$$

2.2. Definition of Series: Convergence – General Properties 73

is the sequence

$$S_1 = \log(1) - \log(2) = -\log(2)$$
$$S_2 = -\log(2) + \log(2) - \log(3) = -\log(3)$$
$$S_3 = -\log(3) + \log(3) - \log(4) = -\log(4)$$
$$\vdots$$
$$S_n = -\log(n+1)$$
$$\vdots$$

As $\lim \bigl(-\log(n+1)\bigr) = -\infty$, the series diverges.

Example 2.2.8 The general term of the series $\sum_{n=1}^{\infty} \dfrac{1}{n^2 + 3n}$ can be expressed as $\dfrac{1}{3}\left(\dfrac{1}{n} - \dfrac{1}{n+3}\right)$. The sequence of partial sums can be constructed using this expression:

$$S_1 = \frac{1}{3}\left(1 - \frac{1}{4}\right)$$

$$S_2 = \frac{1}{3}\left(1 - \frac{1}{4}\right) + \frac{1}{3}\left(\frac{1}{2} - \frac{1}{5}\right) = \frac{1}{3}\left(1 - \frac{1}{4} + \frac{1}{2} - \frac{1}{5}\right)$$

$$S_3 = \frac{1}{3}\left(1 - \frac{1}{4} + \frac{1}{2} - \frac{1}{5}\right) + \frac{1}{3}\left(\frac{1}{3} - \frac{1}{6}\right)$$
$$= \frac{1}{3}\left(1 - \frac{1}{4} + \frac{1}{2} - \frac{1}{5} + \frac{1}{3} - \frac{1}{6}\right)$$

$$S_4 = \frac{1}{3}\left(1 - \frac{1}{4} + \frac{1}{2} - \frac{1}{5} + \frac{1}{3} - \frac{1}{6}\right) + \frac{1}{3}\left(\frac{1}{4} - \frac{1}{7}\right)$$
$$= \frac{1}{3}\left(1 - \frac{1}{4} + \frac{1}{2} - \frac{1}{5} + \frac{1}{3} - \frac{1}{6} + \frac{1}{4} - \frac{1}{7}\right)$$
$$= \frac{1}{3}\left(1 + \frac{1}{2} - \frac{1}{5} + \frac{1}{3} - \frac{1}{6} - \frac{1}{7}\right)$$

$$S_5 = \frac{1}{3}\left(1 + \frac{1}{2} - \frac{1}{5} + \frac{1}{3} - \frac{1}{6} - \frac{1}{7}\right) + \frac{1}{3}\left(\frac{1}{5} - \frac{1}{8}\right)$$

$$= \frac{1}{3}\left(1 + \frac{1}{2} - \frac{1}{5} + \frac{1}{3} - \frac{1}{6} - \frac{1}{7} + \frac{1}{5} - \frac{1}{8}\right)$$

$$= \frac{1}{3}\left(1 + \frac{1}{2} + \frac{1}{3} - \frac{1}{6} - \frac{1}{7} - \frac{1}{8}\right)$$

$$\vdots$$

$$S_n = \frac{1}{3}\left(1 + \frac{1}{2} + \frac{1}{3} - \frac{1}{n+1} - \frac{1}{n+2} - \frac{1}{n+3}\right)$$

$$\vdots$$

As $\lim S_n = \frac{1}{3}\left(1 + \frac{1}{2} + \frac{1}{3}\right) = \frac{11}{18}$, the series is convergent and

$$\sum_{n=1}^{\infty} \frac{1}{n^2 + 3n} = \frac{11}{18}.$$

The last three examples are particular cases of a type of series called **telescopic series**. These are series whose general term a_n can be expressed in the form $\alpha_n - \alpha_{n+k}$, with $k \in \mathbb{N}$:

$$\sum_{n=1}^{\infty}(\alpha_n - \alpha_{n+k}), \tag{2.1}$$

allowing us to determine the expression of the sequence of partial sums by canceling consecutive terms and, consequently, in case the sequence converges to easily calculate the sum of the series. These series are convergent if and only if $\lim v_n$, where $v_n = \alpha_{n+1} + \cdots + \alpha_{n+k}$, exists and is finite. In the particular case of existing, finite, $\lim \alpha_n$, we have:

$$\sum_{n=1}^{\infty}(\alpha_n - \alpha_{n+k}) = \sum_{i=1}^{k} \alpha_i - k\,a, \tag{2.2}$$

where $a = \lim \alpha_n$. In fact, the sequence of partial sums is given by

$$S_n = \sum_{i=1}^{n}(\alpha_i - \alpha_{i+k}) = \sum_{i=1}^{n}\alpha_i - \sum_{i=1}^{n}\alpha_{i+k}$$

$$= \alpha_1 + \cdots + \alpha_k + \alpha_{k+1} + \cdots + \alpha_n$$
$$\quad -(\alpha_{k+1} + \cdots + \alpha_n + \alpha_{n+1} + \cdots + \alpha_{n+k})$$
$$= \alpha_1 + \cdots + \alpha_k - (\alpha_{n+1} + \cdots + \alpha_{n+k})$$
$$= \sum_{i=1}^{k}\alpha_i - \sum_{i=1}^{k}\alpha_{i+n}$$

2.2. Definition of Series: Convergence – General Properties

If (α_n) is convergent, then $\lim \alpha_{i+n}$ exists and $\lim \alpha_n = \lim \alpha_{i+n}$ from which it is concluded that

$$\lim S_n = \sum_{i=1}^{k} \alpha_i - \lim \left(\sum_{i=1}^{k} \alpha_{i+n} \right) = \sum_{i=1}^{k} \alpha_i - k\, a.$$

The following theorem establishes a necessary but not sufficient condition for the convergence of a series. Therefore, it is primarily used to determine whether a series is divergent. If the general term of a series is not a null sequence, the series will be divergent.

Theorem 2.2.1 *If the series $\sum a_n$ is convergent, then (a_n) is a null sequence.*

Proof: Since the series is convergent, the sequence $S_n = \sum_{i=1}^{n} a_i$ is also convergent. This is true for S_{n-1} as well, and we have $\lim S_n = \lim S_{n-1}$. Then

$$\lim a_n = \lim (S_n - S_{n-1}) = \lim S_n - \lim S_{n-1} = 0. \quad \blacksquare$$

Example 2.2.9 The series $\sum_{n=1}^{\infty} \dfrac{n}{n+1}$ is divergent because $\lim \dfrac{n}{n+1} = 1$.

Example 2.2.10 Consider the series $\sum_{n=1}^{\infty} \dfrac{1}{\sqrt{n}}$. Although

$$\lim \frac{1}{\sqrt{n}} = 0,$$

we cannot use Theorem 2.2.1 to determine if the series converges or diverges. However, in Example 2.2.4, we have already proved that this series is divergent.

Theorem 2.2.2 Let $\sum a_n$ and $\sum b_n$ be convergent series with sums A and B, respectively, and $\lambda \in \mathbb{R}$. Then:

a) The series $\sum (a_n + b_n)$ is also convergent, and its sum is $A + B$:
$$\sum_{n=1}^{\infty}(a_n + b_n) = \sum_{n=1}^{\infty} a_n + \sum_{n=1}^{\infty} b_n.$$

b) The series $\sum \lambda a_n$ is convergent, and its sum is λA:
$$\sum_{n=1}^{\infty} \lambda a_n = \lambda \sum_{n=1}^{\infty} a_n.$$

Proof:

a) Let (S_n^*) and (S_n^{**}) be the sequences of partial sums of the series $\sum a_n$ and $\sum b_n$, respectively. Because they are convergent with sums A and B, we have
$$\lim S_n^* = A \quad \text{and} \quad \lim S_n^{**} = B.$$

Let (S_n) be the sequence of partial sums of the series $\sum (a_n + b_n)$, that is,
$$S_n = \sum_{i=1}^{n}(a_i + b_i) = \sum_{i=1}^{n} a_i + \sum_{i=1}^{n} b_i = S_n^* + S_n^{**}.$$

Then
$$\lim S_n = \lim(S_n^* + S_n^{**}) = \lim S_n^* + \lim S_n^{**} = A + B,$$

that is, the series $\sum (a_n + b_n)$ is convergent and has sum $A + B$.

b) Let (S_n^*) be the sequence of partial sums of the series $\sum a_n$. By hypothesis, $\lim S_n^* = A$. Let (S_n) be the sequence of partial sums of the series $\sum \lambda a_n$. Then
$$S_n = \sum_{i=1}^{n} \lambda a_i = \lambda \sum_{i=1}^{n} a_i = \lambda S_n^*.$$

Thus,
$$\lim S_n = \lim \lambda S_n^* = \lambda \lim S_n^* = \lambda A,$$

that is, the series $\sum \lambda a_n$ is convergent and has sum λA. ∎

2.2. Definition of Series: Convergence – General Properties

Notes:

1. From the proof of item a), it is evident that the given series may diverge, and yet the series $\sum(a_n + b_n)$ can still converge. The proof also indicates that if the sequences of partial sums have infinite limits of the same sign, that is, both series are divergent, and the sequence of partial sums will be divergent. The same thing will happen if one of the series is convergent and the other divergent. If (S_n^*) and (S_n^{**}) have infinite limits with opposite signs, the series $\sum(a_n + b_n)$ may be convergent or divergent because an indeterminate form appears in the calculation of the limit.

2. Based on the proof of part b), it can be concluded that if the value of λ is not equal to 0, then the series $\sum \lambda a_n$ will converge if and only if the series $\sum a_n$ also converges. However, if the value of λ equals 0, then the series $\sum \lambda a_n$ will converge since all its terms are zero.

In order to better understand the previous theorem, we will examine an example that involves the application of our knowledge about telescopic series.

Example 2.2.11 Consider the series

$$\sum_{n=1}^{\infty} \frac{1}{n(n+3)(n+6)}.$$

We can decompose the general term as follows:

$$\frac{1}{n(n+3)(n+6)} = \frac{1}{18}\left(\frac{1}{n} - \frac{1}{n+3}\right) - \frac{1}{18}\left(\frac{1}{n+3} - \frac{1}{n+6}\right).$$

The series

$$\sum_{n=1}^{\infty}\left(\frac{1}{n} - \frac{1}{n+3}\right)$$

is telescopic with $\alpha_n = \dfrac{1}{n}$ and $k = 3$. As $\lim \alpha_n = 0$, the series is convergent, and its sum is $\alpha_1 + \alpha_2 + \alpha_3 = 1 + \dfrac{1}{2} + \dfrac{1}{3} = \dfrac{11}{6}$.

The series

$$\sum_{n=1}^{\infty}\left(\frac{1}{n+3} - \frac{1}{n+6}\right)$$

is also telescopic with $\alpha_n = \dfrac{1}{n+3}$ and $k = 3$. In this case too, $\lim \alpha_n = 0$; therefore, the series is convergent and its sum is $\dfrac{1}{4} + \dfrac{1}{5} + \dfrac{1}{6} = \dfrac{37}{60}$.
According to the previous theorem, the given series is also convergent and

$$\sum_{n=1}^{\infty} \frac{1}{n(n+3)(n+6)}$$

$$= \frac{1}{18} \sum_{n=1}^{\infty} \left(\frac{1}{n} - \frac{1}{n+3}\right) - \frac{1}{18} \sum_{n=1}^{\infty} \left(\frac{1}{n+3} - \frac{1}{n+6}\right)$$

$$= \frac{1}{18} \cdot \frac{11}{6} - \frac{1}{18} \cdot \frac{37}{60} = \frac{73}{1080}.$$

Example 2.2.12 The series

$$\sum_{n=1}^{\infty} \frac{5^{n+1}}{8^{n+2}} = \sum_{n=1}^{\infty} \frac{1}{8} \cdot \left(\frac{5}{8}\right)^{n+1} = \frac{1}{8} \sum_{n=1}^{\infty} \left(\frac{5}{8}\right)^{n+1} \quad \text{and} \quad \sum_{n=1}^{\infty} \frac{1}{n(n+1)}$$

are convergent. The first is geometric with ratio $r = \dfrac{5}{8}$ and first term $\dfrac{5^2}{8^3}$. The second is telescopic, and as we saw in Example 2.2.6, it converges and has a sum of 1. We can conclude that the series

$$\sum_{n=1}^{\infty} \left(\frac{5^{n+1}}{8^{n+2}} + \frac{1}{n(n+1)}\right)$$

is convergent and its sum is $\dfrac{25}{192} + 1 = \dfrac{217}{192}$.

Theorem 2.2.3 (Cauchy's Criterion) *A series $\sum a_n$ converges if and only if*

$$\forall \delta > 0 \ \exists p \in \mathbb{N}: \ m > n > p \Rightarrow |a_{n+1} + \cdots + a_m| < \delta.$$

<u>Proof</u>: As

$$|a_{n+1} + \cdots + a_m| = \left|\sum_{i=1}^{m} a_i - \sum_{i=1}^{n} a_i\right| = |S_m - S_n|,$$

2.2. Definition of Series: Convergence – General Properties

our aim is to prove that the series $\sum_{n=1}^{\infty} a_n$ converges if and only if

$$\forall \delta > 0 \; \exists p \in \mathbb{N}: \; m > n > p \Rightarrow |S_m - S_n| < \delta,$$

that is, we want to prove that (S_n) is a Cauchy sequence. However, by definition, the series $\sum a_n$ converges if and only if (S_n) is a convergent sequence, and in $\mathbb{R}$ a sequence is convergent if and only if it is a Cauchy sequence (see Theorem 1.1.20). ∎

Example 2.2.13 In Example 2.2.5, we showed that the harmonic series, $\sum \frac{1}{n}$, is divergent. We will prove the same result using Theorem 2.2.3. If the series is convergent, then for any given $\delta > 0$, there would exist $p \in \mathbb{N}$ such that $|a_{n+1} + \cdots + a_m| < \delta$ for all $m > n > p$. However, if we choose $m = n + n$, then

$$|a_{n+1} + \cdots + a_m| = |a_{n+1} + \cdots + a_{n+n}|$$

$$= \frac{1}{n+1} + \cdots + \frac{1}{n+n}$$

$$\geq \frac{1}{n+n} + \cdots + \frac{1}{n+n} = n \cdot \frac{1}{2n} = \frac{1}{2}.$$

Therefore, the condition of the theorem is not satisfied for $\delta < \frac{1}{2}$, which means that the harmonic series is divergent.

Corollary 1 *The convergence or divergence of a series does not depend on the first p terms, whatever $p \in \mathbb{N}$, that is, if $\sum a_n$ and $\sum b_n$ are series such that there exists $p \in \mathbb{N}$ satisfying $a_n = b_n$, $\forall n > p$, then either both series are convergent or both are divergent.*

Definition 2.2.3 *The **remainder of order** p of the series $\sum_{n=1}^{\infty} a_n$ is the series*

$$r_p = \sum_{n=1}^{\infty} a_{n+p} = \sum_{n=p+1}^{\infty} a_n.$$

According to the previous corollary, the remainder of any order will also be convergent if a series is convergent. The sum of the remainder of order p of a convergent series gives us the error made by taking the partial sum S_p as the approximate value of the sum of the series. This error is given by

$$\sum_{n=1}^{\infty} a_n - S_p = \sum_{n=1}^{\infty} a_n - \sum_{n=1}^{p} a_n = \sum_{n=1}^{\infty} a_{n+p} = r_p.$$

Corollary 2 *The convergence or divergence of a series does not change if we suppress a finite, arbitrary number of terms.*

The following theorem can be viewed as a generalization of the associative property of addition to the case of convergent series.

Theorem 2.2.4 *Let $\sum a_n$ be a convergent series and $k_1, k_2, \ldots, k_n, \ldots$ be a strictly increasing sequence of elements of $\mathbb{N}$. Let b_n be the sequence defined as follows:*

$$b_n = \begin{cases} \displaystyle\sum_{i=1}^{k_1} a_i, & \text{if } n = 1 \\ \displaystyle\sum_{i=k_{n-1}+1}^{k_n} a_i, & \text{if } n > 1. \end{cases}$$

Then, the series $\sum b_n$ is convergent and $\displaystyle\sum_{n=1}^{\infty} b_n = \sum_{n=1}^{\infty} a_n$.

<u>Proof</u>: The series $\sum a_n$ is convergent, then by Definition 2.2.2 there exists and is finite the limit

$$\lim S_n = \lim \sum_{i=1}^{n} a_i.$$

According to Theorem 1.1.15, every subsequence of (S_n) is convergent and has the same limit.

2.2. Definition of Series: Convergence – General Properties

The series $\sum_{n=1}^{\infty} b_n$ is convergent if and only if $S'_n = \sum_{i=1}^{n} b_i$ is convergent. But

$$S'_n = \sum_{i=1}^{n} b_i = \sum_{i=1}^{k_1} a_i + \sum_{i=k_1+1}^{k_2} a_i + \cdots + \sum_{i=k_{n-1}+1}^{k_n} a_i = \sum_{i=1}^{k_n} a_i = S_{k_n},$$

that is, (S'_n) is a subsequence of (S_n); therefore, it is convergent and has the same limit:

$$\sum_{n=1}^{\infty} b_n = \lim S'_n = \lim S_n = \sum_{n=1}^{\infty} a_n. \blacksquare$$

Note: If the series $\sum_{n=1}^{\infty} a_n$ is convergent, then the theorem establishes that the following "associative property" is valid:

$$a_1 + a_2 + \cdots + a_{k_1} + \cdots + a_{k_2} + \cdots = (a_1 + \cdots + a_{k_1}) + (a_{k_1+1} + \cdots + a_{k_2}) + \cdots$$

However, this property is not valid if the series is divergent. For instance, the series $\sum_{n=1}^{\infty} (-1)^n$ is divergent because its general term does not tend to zero. Nevertheless, $(-1+1) + (-1+1) + \cdots = 0$.

2.3 Alternating Series

Gottfried Wilhelm Leibniz (1646–1716) was a German philosopher and mathematician. He obtained a doctorate in Law when he was twenty years old. He pursued a career in Law and International Politics as an advisor to kings and princes. During his countless travels, Leibniz befriended the greatest intellectuals of his time. He developed differential and integral calculus simultaneously and independently of the English mathematician Isaac Newton and established the foundations of dynamics. Leibniz was one of the last intellectuals to master almost all of the knowledge of his time. (Source of image: Oil portrait by Christoph Bernhard Francke (1665–1729), Herzog Anton Ulrich-Museum.)

Definition 2.3.1 *A series is said to be **alternating** if its terms are alternatively positive and negative.*

Assuming that the first term of the series is positive, we can write

$$\sum_{n=1}^{\infty}(-1)^{n-1}a_n, \quad a_n > 0, \ \forall n \in \mathbb{N}.$$

Theorem 2.3.1 (Leibniz's Test) *If (a_n) is a decreasing sequence of positive terms and $\lim a_n = 0$, then the series $\sum (-1)^{n-1} a_n$ is convergent.*

Proof: Consider the sequence (S_n) of the partial sums of the series:

$$S_n = a_1 - a_2 + a_3 - \cdots + (-1)^{n-1}a_n.$$

We will study the subsequences of even order terms and odd order terms. Let $k \in \mathbb{N}$ be arbitrary.

$$S_{2k} = a_1 - a_2 + \cdots + a_{2k-1} - a_{2k}$$
$$S_{2k+1} = a_1 - a_2 + \cdots + a_{2k-1} - a_{2k} + a_{2k+1}.$$

We claim that the subsequence (S_{2k}) is increasing. This is because, as (a_n) is decreasing, we get:

$$S_{2k+2} - S_{2k} = a_1 - a_2 + \cdots + a_{2k-1} - a_{2k} + a_{2k+1} - a_{2k+2} -$$
$$-(a_1 - a_2 + \cdots + a_{2k-1} - a_{2k})$$
$$= a_{2k+1} - a_{2k+2} \geq 0.$$

Moreover, it is a bounded sequence since:

$$S_2 \leq S_{2k} = a_1 - [(a_2 - a_3) + (a_4 - a_5) + \cdots + (a_{2k-2} - a_{2k-1}) + a_{2k}] < a_1.$$

Hence, (S_{2k}) is monotonic and bounded, which implies that it is convergent. From $S_{2k+1} = S_{2k} + a_{2k+1}$, it follows that

$$\lim_{k \to +\infty} S_{2k+1} = \lim_{k \to +\infty} S_{2k},$$

since by hypothesis (a_n) is a null sequence.

2.3. Alternating Series

As the subsequences of even and odd terms have the same limit, (S_n) converges. Therefore, the series $\sum_{n=1}^{\infty}(-1)^{n-1}a_n$ is convergent. ∎

Note: Leibniz's test is a particular case of a more general result known as **Dirichlet's test**. The Dirichlet's test states the following:

If the sequence $(\sum_{i=1}^{n} b_i)$ is bounded, and if (a_n) is a decreasing sequence of positive terms satisfying $\lim a_n = 0$, then the series $\sum_{n=1}^{\infty} b_n a_n$ is convergent.

If we let $b_n = (-1)^n$ in Dirichlet's test, we obtain Leibniz's test.

Peter Gustave Lejeune Dirichlet (1805–1859) was a German mathematician who made valuable contributions to number theory, analysis, and mechanics. He was a professor at the universities of Breslau (1827) and Berlin (1828–1855). In 1855, he succeeded Carl Friedrich Gauss at the University of Göttingen. (Source of image: Portrait Collection of the German Museum, Munich.)

Example 2.3.1 The series $\sum_{n=1}^{\infty}(-1)^n \frac{\log(n)}{n} = \sum_{n=2}^{\infty}(-1)^n \frac{\log(n)}{n}$ is alternating. Let us check the conditions for applying Leibniz's test:

(i) $\lim \frac{\log(n)}{n} = 0$ (it is enough to note that $\lim_{x \to +\infty} \frac{\log(x)}{x}$ is an indeterminate form of type $\frac{\infty}{\infty}$; we can calculate the limit using L'Hôpital's Rule:
$$\lim_{x \to +\infty} \frac{(\log(x))'}{(x)'} = \lim_{x \to +\infty} \frac{1}{x} = 0).$$

(ii) $\frac{\log(n)}{n} > 0, \forall n > 1$.

(iii) The function $f(x) = \frac{\log(x)}{x}$ is a decreasing function on $]e, +\infty[$ because
$$f'(x) = \left(\frac{\log(x)}{x}\right)' = \frac{1 - \log(x)}{x^2} < 0, \quad \forall x > e.$$

This implies that the sequence $\left(\frac{\log(n)}{n}\right)$ is decreasing for $n \geq 3$.

Thus, we can conclude that the alternating series
$$\sum_{n=3}^{\infty}(-1)^n \frac{\log(n)}{n}$$
is convergent. As this series differs from the initial series in one term, we can affirm that the series under study is convergent.

Example 2.3.2 The series $\sum_{n=2}^{\infty} \cos(n\pi) \sin\left(\frac{\pi}{n}\right)$ is alternating because

$$\cos(n\pi) = (-1)^n \quad \text{and} \quad a_n = \sin\left(\frac{\pi}{n}\right) > 0, \quad \forall n > 1$$

(note that $0 < \frac{\pi}{n} \leq \frac{\pi}{2}$, $\forall n > 1$). We can write the series in the form $\sum_{n=2}^{\infty} (-1)^n \sin\left(\frac{\pi}{n}\right)$. Let us check the conditions of Leibniz's test:

(i) $\lim \sin\left(\frac{\pi}{n}\right) = 0$.

(ii) We have already seen that $\sin\left(\frac{\pi}{n}\right) > 0$, $\forall n > 1$.

(iii) As we also justified earlier, $0 < \frac{\pi}{n} \leq \frac{\pi}{2}$, $\forall n > 1$; the sequence $\left(\frac{\pi}{n}\right)$ is decreasing and the sine function is increasing on $\left[0, \frac{\pi}{2}\right]$, so $\left(\sin\left(\frac{\pi}{n}\right)\right)$ is decreasing for $n \geq 2$.

We can conclude that the alternating series is convergent.

Example 2.3.3 Consider the series $\sum_{n=1}^{\infty} (-1)^n \frac{1}{n^\alpha}$, $\alpha \in \mathbb{R}$.
If $\alpha \leq 0$, the series diverges because its general term does not tend to zero. If $\alpha > 0$, the series is convergent because $a_n = \frac{1}{n^\alpha}$ is a decreasing sequence with positive terms and $\lim a_n = 0$.

Not all alternating series meet the conditions of Leibniz's test. In this situation, the test is not applicable, and it is necessary to look for another approach.

Example 2.3.4 Consider the alternating series

$$\sum_{n=1}^{\infty} \frac{(-1)^n}{\sqrt{n}} \left(1 + \frac{(-1)^n}{\sqrt{n}}\right) = \sum_{n=1}^{\infty} (-1)^n \left(\frac{1}{\sqrt{n}} + \frac{(-1)^n}{n}\right).$$

Although $a_n = \frac{1}{\sqrt{n}} + \frac{(-1)^n}{n}$ is a null sequence and $a_n > 0$, $\forall n > 1$, we cannot apply Leibniz's test because (a_n) is not decreasing. However, we can

2.3. Alternating Series

show that the given series is divergent because it is the sum of a convergent series, the series $\sum (-1)^n \frac{1}{\sqrt{n}}$, with a divergent series, the series $\sum \frac{1}{n}$ (see Note 1 on page 77).

Theorem 2.3.2 *Let (a_n) be a decreasing sequence of positive terms such that $\lim a_n = 0$, and S be the sum of the series $\sum (-1)^{n-1} a_n$. Then*

$$0 \leq (-1)^n (S - S_n) \leq a_{n+1}, \quad \forall n \in \mathbb{N}.$$

Proof: We know from the proof of the previous theorem that (S_{2k}) is an increasing subsequence of (S_n), and it has the same limit, S, as the subsequence (S_{2k+1}). Similarly, (S_{2k+1}) is decreasing. This leads to the following inequalities:

$$S_{2k} \leq S \quad \text{and} \quad S \leq S_{2k+1}, \quad \forall k \in \mathbb{N}.$$

From these inequalities, we can deduce that:

$$0 \leq S_{2k-1} - S \leq S_{2k-1} - S_{2k} = a_{2k},$$
$$0 \leq S - S_{2k} \leq S_{2k+1} - S_{2k} = a_{2k+1}.$$

Therefore:
$$0 \leq S_{2k-1} - S \leq a_{2k},$$
$$0 \leq S - S_{2k} \leq a_{2k+1}.$$

This implies that
$$0 \leq (-1)^{2k-1}(S - S_{2k-1}) \leq a_{2k},$$
$$0 \leq (-1)^{2k}(S - S_{2k}) \leq a_{2k+1}.$$

From these last two inequalities, it follows that
$$0 \leq (-1)^n (S - S_n) \leq a_{n+1}. \blacksquare$$

Corollary 1 *Let (a_n) be a decreasing sequence of positive terms such that $\lim a_n = 0$ and S be the sum of the series $\sum (-1)^{n-1} a_n$. Then*

$$|S - S_n| \leq a_{n+1}, \quad \forall n \in \mathbb{N}.$$

Note: According to the previous corollary, it is evident that under the conditions of Leibniz's test, the absolute value of the error made in using a partial sum as an estimation for the sum of an alternating series is, in absolute value, always less than the absolute value of the first of the disregarded terms.

Example 2.3.5 Consider the series

$$\sum_{n=1}^{\infty}(-1)^n\frac{1}{n},$$

which is known as the **alternating harmonic series**. According to Leibniz's test, this series is convergent, as $a_n = \frac{1}{n}$ is a null decreasing sequence of positive terms. If we use the partial sum S_9 as an estimation of the sum of the series, we will make an error that in absolute value is less than or equal to $\frac{1}{10}$, which is the value of a_{10}.

2.4 Absolute Convergence

Theorem 2.4.1 *If the series $\sum |a_n|$ is convergent, then the series $\sum a_n$ is also convergent.*

Proof: By Theorem 2.2.3, the series $\sum_{n=1}^{\infty} |a_n|$ is convergent if and only if

$$\forall \delta > 0 \ \exists p \in \mathbb{N}: \ m > n > p \Rightarrow ||a_{n+1}| + \cdots + |a_m|| < \delta.$$

Since

$$|a_{n+1} + \cdots + a_m| \leq |a_{n+1}| + \cdots + |a_m|$$

and

$$||a_{n+1}| + \cdots + |a_m|| = |a_{n+1}| + \cdots + |a_m|,$$

we have

$$\forall \delta > 0 \ \exists p \in \mathbb{N}: \ m > n > p \Rightarrow |a_{n+1} + \cdots + a_m| < \delta,$$

that is, the series $\sum_{n=1}^{\infty} a_n$ is convergent. ∎

Note: It is worth noting that the converse of this theorem is not true. This means that the series $\sum a_n$ can be convergent even if the series of modules, $\sum |a_n|$, is not. For instance, consider the harmonic series, which diverges, and the alternating harmonic series, which converges. The harmonic series is the series of modules of the alternating harmonic series.

Definition 2.4.1 *A series $\sum a_n$ is **absolutely convergent** if the series $\sum |a_n|$ is convergent. A series $\sum a_n$ is **conditionally convergent** if it is convergent and the series $\sum |a_n|$ is divergent.*

Definition 2.4.2 *We say that the series $\sum b_n$ is a **rearrangement** of the series $\sum a_n$, or that it is obtained by reordering its terms, if there exists a bijection ϕ from $\mathbb{N}$ into $\mathbb{N}$ such that $b_n = a_{\phi(n)}$.*

> **Theorem 2.4.2** *If the series $\sum a_n$ is absolutely convergent, then any series obtained by rearranging its terms is absolutely convergent and has the same sum.*

This theorem generalizes the commutative property of the usual addition to absolutely convergent series. However, it should be noted that this property is not valid for conditionally convergent series. By rearranging the terms of a conditionally convergent series, we can obtain a series with a predetermined sum and even a divergent series.

> **Theorem 2.4.3** *Let $\sum a_n$ be a conditionally convergent series. Then:*
>
> a) *There are bijections $\phi : \mathbb{N} \to \mathbb{N}$ such that the series $\sum a_{\phi(n)}$ is divergent.*
>
> b) *For every real number k, there exists a bijection $\phi : \mathbb{N} \to \mathbb{N}$ such that the series $\sum a_{\phi(n)}$ is convergent and its sum is equal to k.*

Example 2.4.1 Consider the alternating harmonic series, $\sum_{n=1}^{\infty} (-1)^{n-1} \frac{1}{n}$, which we know to be conditionally convergent. Let us rearrange its terms such that each positive term is followed by two negative terms. We will obtain the following series:

$$1 - \frac{1}{2} - \frac{1}{4} + \frac{1}{3} - \frac{1}{6} - \frac{1}{8} + \frac{1}{5} - \frac{1}{10} - \frac{1}{12} + \cdots$$

For this series, we have the following partial sums:

$$S'_1 = 1$$

$$S'_2 = 1 - \frac{1}{2}$$

$$S'_3 = 1 - \frac{1}{2} - \frac{1}{4} = \frac{1}{2} - \frac{1}{4} = \frac{1}{2}\left(1 - \frac{1}{2}\right) = \frac{1}{2}S_2$$

$$S'_4 = 1 - \frac{1}{2} - \frac{1}{4} + \frac{1}{3}$$

$$S'_5 = 1 - \frac{1}{2} - \frac{1}{4} + \frac{1}{3} - \frac{1}{6}$$

2.4. Absolute Convergence

$$S'_6 = 1 - \frac{1}{2} - \frac{1}{4} + \frac{1}{3} - \frac{1}{6} - \frac{1}{8} = \left(1 - \frac{1}{2}\right) - \frac{1}{4} + \left(\frac{1}{3} - \frac{1}{6}\right) - \frac{1}{8}$$

$$= \frac{1}{2} - \frac{1}{4} + \frac{1}{6} - \frac{1}{8} = \frac{1}{2}\left(1 - \frac{1}{2} + \frac{1}{3} - \frac{1}{4}\right) = \frac{1}{2}S_4$$

$$\vdots$$

$$S'_9 = S'_6 + \frac{1}{5} - \frac{1}{10} - \frac{1}{12} = \frac{1}{2}\left(1 - \frac{1}{2} + \frac{1}{3} - \frac{1}{4}\right) + \left(\frac{1}{5} - \frac{1}{10}\right) - \frac{1}{12}$$

$$= \frac{1}{2}\left(1 - \frac{1}{2} + \frac{1}{3} - \frac{1}{4}\right) + \frac{1}{10} - \frac{1}{12} = \frac{1}{2}\left(1 - \frac{1}{2} + \frac{1}{3} - \frac{1}{4} + \frac{1}{5} - \frac{1}{6}\right)$$

$$= \frac{1}{2}S_6$$

$$\vdots$$

where $S_n = \sum_{i=1}^{n}(-1)^{i-1}\frac{1}{i}$.

Using induction, we can demonstrate that $S'_{3n} = \frac{1}{2}S_{2n}$, which implies that if we designate by S the limit of the sequence (S_n), then we have

$$\lim S'_{3n} = \frac{1}{2}S.$$

As $S'_{3n+1} = S'_{3n} + \frac{1}{2n+1}$ and $S'_{3n+2} = S'_{3n} + \frac{1}{2n+1} - \frac{1}{4n+2}$, we have

$$\lim S'_{3n+1} = \lim S'_{3n} + \lim \frac{1}{2n+1}$$

and

$$\lim S'_{3n+2} = \lim S'_{3n} + \lim \frac{1}{2n+1} - \lim \frac{1}{4n+2},$$

that is,

$$\lim S'_{3n+1} = \lim S'_{3n+2} = \lim S'_{3n} = \frac{1}{2}S.$$

Therefore, we conclude that $\lim S'_n = \frac{1}{2}S$. In other words, the series obtained by this rearrangement of the terms of the alternating harmonic series is convergent, and its sum equals half the original series' sum.

2.5 Series of Nonnegative Terms

In this section, we will establish convergence tests for a series of nonnegative terms. These tests apply to investigating the absolute convergence of series in general. We should note that convergence and absolute convergence are equivalent for a series of nonnegative terms.

Theorem 2.5.1 *Let $\sum a_n$ be a series of nonnegative terms. The series $\sum a_n$ is convergent if and only if the sequence of its partial sums is bounded.*

<u>Proof</u>: Let (S_n) be the sequence of partial sums of the series $\sum a_n$. If the series is convergent, then by definition, the sequence (S_n) converges. Consequently, it is a bounded sequence (see Theorem 1.1.7).
Suppose that (S_n) is bounded. As $a_n \geq 0$, we have
$$S_{n+1} \geq S_n, \quad \forall n \in \mathbb{N},$$
that is, (S_n) is an increasing sequence. According to Theorem 1.1.12, the two previous statements imply the convergence of (S_n), which is equivalent to saying that the series $\sum a_n$ is convergent. ∎

Theorem 2.5.2 (Cauchy's Condensation Test) *Let (a_n) be a decreasing sequence of positive numbers. The series $\sum a_n$ converges if and only if the series $\sum 2^n a_{2^n}$ converges.*

<u>Proof</u>: Consider the partial sums of the series $\sum a_n$ and $\sum 2^n a_{2^n}$:
$$S_n = \sum_{i=1}^{n} a_i \quad \text{and} \quad T_n = \sum_{i=1}^{n} 2^i a_{2^i}.$$
Note that the sequences (S_n) and (T_n) are increasing, with positive terms.

Suppose that the series $\sum 2^n a_{2^n}$ is convergent. By Theorem 2.5.1, the sequence (T_n) is bounded. Then
$$\begin{aligned} S_{2^n-1} &= a_1 + (a_2 + a_3) + (a_4 + a_5 + a_6 + a_7) + \cdots \\ &\quad + (a_{2^{n-1}} + a_{2^{n-1}+1} + \cdots + a_{2^n-1}) \\ &\leq a_1 + 2\,a_2 + 4\,a_4 + \cdots + 2^{n-1} a_{2^{n-1}} \\ &= a_1 + T_{n-1}. \end{aligned}$$

2.5. Series of Nonnegative Terms

As the sequence (T_n) is bounded, the sequence (S_{2^n-1}) is also bounded. Given that
$$S_m \leq S_{2^n-1}, \quad \forall m \leq 2^n - 1,$$
the sequence (S_n) is bounded; therefore, according to the previous theorem, the series $\sum a_n$ is convergent.

Conversely, suppose that the series $\sum a_n$ is convergent. Then, using again the previous theorem, the sequence (S_n) is bounded. We can observe that

$$\begin{aligned} S_{2^n} &= a_1 + a_2 + (a_3 + a_4) + (a_5 + a_6 + a_7 + a_8) + \cdots + \\ &\quad + (a_{2^{n-1}+1} + a_{2^{n-1}+2} + \cdots + a_{2^n}) \\ &\geq a_1 + a_2 + 2\,a_4 + 4\,a_8 + \cdots + 2^{n-1}\,a_{2^n} \\ &= a_1 + \frac{1}{2}(2\,a_2 + 4\,a_4 + 8\,a_8 + \cdots + 2^n\,a_{2^n}) \\ &= a_1 + \frac{1}{2}T_n. \end{aligned}$$

This implies that the sequence (T_n) is also bounded. Hence, the series $\sum 2^n\,a_{2^n}$ is convergent. ∎

Example 2.5.1 Consider the series $\displaystyle\sum_{n=1}^{\infty} \frac{1}{n^p}$, $p \in \mathbb{R}$, usually referred to as the p-series.

If $p \leq 0$, the series is divergent because its general term does not tend to zero.

If $p > 0$, we can apply Theorem 2.5.2 because the sequence $a_n = \dfrac{1}{n^p}$ is decreasing with positive terms. The series

$$\sum_{n=1}^{\infty} 2^n \frac{1}{(2^n)^p} = \sum_{n=1}^{\infty} \frac{2^n}{(2^p)^n} = \sum_{n=1}^{\infty} \left(\frac{1}{2^{p-1}}\right)^n$$

is geometric, with ratio $\dfrac{1}{2^{p-1}}$, and is convergent if and only if $\dfrac{1}{2^{p-1}} < 1$, that is, if and only if $p > 1$.

Conclusion: The p-series $\displaystyle\sum_{n=1}^{\infty} \frac{1}{n^p}$ converges if and only if $p > 1$.

Example 2.5.2 Consider the series $\sum_{n=2}^{\infty} \dfrac{1}{n(\log(n))^{\alpha}}$, where $\alpha \in \mathbb{R}$. The function
$$f(x) = \dfrac{1}{x(\log(x))^{\alpha}}$$
is positive and continuous on $[2, +\infty[$. We can find the derivative of f to determine whether it is increasing or decreasing. We have

$$f'(x) = 0 \Leftrightarrow -\dfrac{(\log(x))^{\alpha} + \alpha(\log(x))^{\alpha-1}}{x^2(\log(x))^{2\alpha}} = 0$$

$$\Leftrightarrow -\dfrac{(\log(x))^{\alpha-1}(\log(x) + \alpha)}{x^2(\log(x))^{2\alpha}} = 0$$

$$\Leftrightarrow \log(x) + \alpha = 0.$$

If $\alpha \geq 0$, this equation has no root in $[2, +\infty[$, which implies that $f'(x) < 0$, $\forall x \in [2, +\infty[$.

If $\alpha < 0$, then $f'(x) < 0$, $\forall x > e^{-\alpha}$, and this implies that $f'(x) < 0$, $\forall x \in]e^{-\alpha}, +\infty[$.

Then, the sequence $a_n = \dfrac{1}{n(\log(n))^{\alpha}}$ is decreasing from the order q if $q \in \mathbb{N}$ is such that $q \geq \max\{2, e^{-\alpha}\}$. The series

$$\sum_{n=q}^{\infty} \dfrac{2^n}{2^n(\log(2^n))^{\alpha}} = \sum_{n=q}^{\infty} \dfrac{1}{n^{\alpha}(\log(2))^{\alpha}} = \dfrac{1}{(\log(2))^{\alpha}} \sum_{n=q}^{\infty} \dfrac{1}{n^{\alpha}}$$

is convergent if and only if $\alpha > 1$ as we saw in the previous example. Remember that the convergence or divergence of the series does not change if we omit a finite number of terms. Hence, we can conclude that the series

$$\sum_{n=2}^{\infty} \dfrac{1}{n(\log(n))^{\alpha}}$$

is convergent if and only if $\alpha > 1$.

Note: In Theorem 2.5.2, the hypothesis that (a_n) is a decreasing sequence cannot be removed. In fact, consider the series $\sum a_n$ with

2.5. Series of Nonnegative Terms

$$a_n = \begin{cases} \dfrac{1}{2^k}, & \text{if } n = 2^k \\ 0, & \text{if } n \neq 2^k. \end{cases}$$

The series

$$\sum_{n=1}^{\infty} 2^n a_{2^n} = \sum_{n=1}^{\infty} \frac{2^n}{2^n} = \sum_{n=1}^{\infty} 1$$

is divergent. However, we can write the original series as

$$\sum_{n=1}^{\infty} a_n = \sum_{k=1}^{\infty} \frac{1}{2^k} = \sum_{k=1}^{\infty} \left(\frac{1}{2}\right)^k,$$

which is a geometric series with ratio $1/2$ and is therefore convergent.

> **Theorem 2.5.3 (Integral Test)** Let $f : [1, +\infty[\to \mathbb{R}$ be a continuous, positive, and decreasing function. For each $n \in \mathbb{N}$, let $a_n = f(n)$. Then, the series $\displaystyle\sum_{n=1}^{\infty} a_n$ and the improper integral $\displaystyle\int_1^{\infty} f(x)\,dx$ are both convergent or both divergent.

<u>Proof</u>: Let us partition the interval $[1, n]$ into $n-1$ intervals of length 1, as shown in Fig. 2.4. The total area of the rectangles inscribed in the figure is $a_2 + a_3 + a_4 + \cdots + a_n$. We have the following inequality:

$$S_n - a_1 = a_2 + a_3 + a_4 + \cdots + a_n \leq \int_1^n f(x)\,dx.$$

If we consider the area of the circumscribed rectangles (see Fig. 2.5), we have the inequality

$$S_{n-1} = a_1 + a_2 + a_3 + \cdots + a_{n-1} \geq \int_1^n f(x)\,dx.$$

From the two previous inequalities, we can conclude that

$$S_n - a_1 \leq \int_1^n f(x)\,dx \leq S_{n-1}.$$

If the integral is divergent, $\displaystyle\lim_{n \to +\infty} \int_1^n f(x)\,dx = +\infty$ since f is positive. Then, by the inequality on the right, the limit of the sequence of the partial

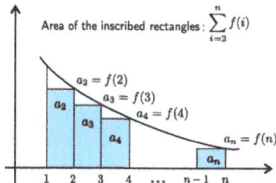

Figure 2.4: Inscribed rectangles in the graph of f

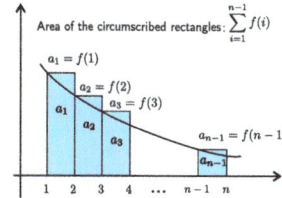

Figure 2.5: Circumscribed rectangles in the graph of f

sums of the series is also $+\infty$; that is, the series diverges. If the integral converges, then there exists and it is finite $\lim\limits_{n\to+\infty}\int_1^n f(x)\,dx$. Consequently, the sequence $S_n = \sum\limits_{i=1}^n a_i$ is bounded. As the terms of the series are positive, it is convergent by Theorem 2.5.1.

If the series is convergent, by the inequality on the right, $\int_1^n f(x)\,dx$ is bounded, so the improper integral is convergent. If the series is divergent, based on the inequality on the left, $\lim\limits_{n\to+\infty}\int_1^n f(x)\,dx = +\infty$. ∎

Note: When we use the Integral Test, it is not necessary for the series or the integral to start at $n = 1$. Additionally, the function f does not need to decrease on the interval $[1, +\infty[$, but only on some interval of the form $[N, +\infty[$, $N \in \mathbb{N}$.

Example 2.5.3 Consider the series $\sum\limits_{n=1}^{\infty} \dfrac{e^n}{1+e^{2n}}$. The real-valued function f of a real variable defined by $f(x) = \dfrac{e^x}{1+e^{2x}}$ is continuous and positive on $[1, +\infty[$. Let us study the monotonicity of f:

$$f'(x) = \frac{e^x(1+e^{2x}) - e^x\, 2e^{2x}}{(1+e^{2x})^2} = \frac{e^x(1-e^{2x})}{(1+e^{2x})^2} < 0, \quad \forall x \in [1, +\infty[.$$

Thus, the function f decreases on $[1, +\infty[$. Furthermore,

$$\lim_{t\to+\infty}\int_1^t \frac{e^x}{1+e^{2x}}\,dx = \lim_{t\to+\infty}\left(\arctan(e^t) - \arctan(e)\right) = \frac{\pi}{2} - \arctan(e).$$

Since the improper integral $\int_1^{+\infty} \dfrac{e^x}{1+e^{2x}}\,dx$ converges, then, by the Integral Test, the series $\sum\limits_{n=1}^{\infty} \dfrac{e^n}{1+e^{2n}}$ also converges.

Example 2.5.4 Consider the series $\sum\limits_{n=2}^{\infty} \dfrac{\log(n)}{n}$. The real-valued function f of a real variable defined by $f(x) = \dfrac{\log(x)}{x}$ is continuous and positive on

2.5. Series of Nonnegative Terms

$[2, +\infty[$. We can analyze the monotonicity of f by taking the derivative:

$$f'(x) = \frac{\frac{1}{x} \cdot x - \log(x)}{x^2} = \frac{1 - \log(x)}{x^2} < 0, \quad \forall x \in \,]e, +\infty[.$$

Thus, the function f decreases on $]e, +\infty[$. Now, we compute the integral:

$$\lim_{t \to +\infty} \int_1^t \frac{\log(x)}{x}\,dx = \lim_{t \to +\infty} \frac{1}{2}\left((\log(t))^2 - (\log(2))^2\right) = +\infty.$$

According to the Integral Test, the series $\sum_{n=2}^{\infty} \frac{\log(n)}{n}$ diverges since the improper integral $\int_1^{+\infty} \frac{\log(x)}{x}\,dx$ is also divergent.

Theorem 2.5.4 (General Comparison Test) Let $\sum a_n$ and $\sum b_n$ be two series of nonnegative terms such that $a_n \leq b_n$, $\forall n \in \mathbb{N}$:

a) If the series $\sum b_n$ is convergent, then the series $\sum a_n$ is convergent.

b) If the series $\sum a_n$ is divergent, then the series $\sum b_n$ is divergent.

Proof: Let (S_n) and (S_n') be the sequences of partial sums of series $\sum a_n$ and $\sum b_n$, that is,

$$S_n = \sum_{i=1}^n a_i \quad \text{and} \quad S_n' = \sum_{i=1}^n b_i.$$

Since $0 \leq a_n \leq b_n$, $\forall n \in \mathbb{N}$, we have

$$0 \leq S_n \leq S_n', \quad \forall n \in \mathbb{N}. \tag{2.3}$$

a) If the series $\sum b_n$ is convergent, then, by Theorem 2.5.1, the sequence of its partial sums, (S_n'), is bounded. By inequality (2.3), the sequence (S_n) is also bounded, that is, the series $\sum a_n$ is convergent.

b) If the series $\sum a_n$ is divergent, then by Theorem 2.5.1, the sequence (S_n) is not bounded. Inequality (2.3) implies that the sequence (S_n') is also not bounded; thus, again by Theorem 2.5.1, the series $\sum b_n$ is divergent. ∎

Note: Omitting a finite number of terms does not change the convergence or divergence of the series, as we have seen. Therefore, the previous theorem remains valid if there exists $p \in \mathbb{N}$ such that $a_n \leq b_n$, $\forall n \geq p$.

Example 2.5.5 Consider the series $\sum_{n=1}^{\infty} \frac{1}{n!}$. We know that

$$0 < \frac{1}{n!} = \frac{1}{n(n-1)(n-2)\ldots 2} \leq \frac{1}{2^{n-1}}, \quad \forall n \in \mathbb{N}.$$

Also, the series $\sum_{n=1}^{\infty} \frac{1}{2^{n-1}}$ is convergent because it is geometric with ratio $\frac{1}{2}$. By the General Comparison Test, we can conclude that the given series is convergent.

Example 2.5.6 The series $\sum_{n=1}^{\infty} \frac{\cos(n)}{n^2}$ is not of positive terms. However, we can observe that the following inequality holds:

$$0 < \left|\frac{\cos(n)}{n^2}\right| \leq \frac{1}{n^2}, \quad \forall n \in \mathbb{N}. \tag{2.4}$$

The series $\sum_{n=1}^{\infty} \frac{1}{n^2}$ is a p-series with $p = 2 > 1$ and is, therefore, convergent. By applying the General Comparison Test and taking into account inequality (2.4), we can conclude that the series $\sum_{n=1}^{\infty} \left|\frac{\cos(n)}{n^2}\right|$ is convergent. This means that the series $\sum_{n=1}^{\infty} \frac{\cos(n)}{n^2}$ is absolutely convergent.

Example 2.5.7 Consider the series $\sum_{n=1}^{\infty} (-1)^n \frac{\sin(n)}{\bigl(\log(10)\bigr)^n}$. Let us analyze the series of modules:

$$\sum_{n=1}^{\infty} \left|(-1)^n \frac{\sin(n)}{\bigl(\log(10)\bigr)^n}\right| = \sum_{n=1}^{\infty} \frac{|\sin(n)|}{\bigl(\log(10)\bigr)^n}.$$

We notice that the following inequality holds for every natural number n:

$$0 < \frac{|\sin(n)|}{\bigl(\log(10)\bigr)^n} \leq \frac{1}{\bigl(\log(10)\bigr)^n}.$$

2.5. Series of Nonnegative Terms

The series $\sum_{n=1}^{\infty} \frac{1}{\left(\log(10)\right)^n}$ is geometric with ratio $r = \frac{1}{\log(10)}$. Since

$$10 > e \Rightarrow \log(10) > 1 \Rightarrow 0 < \frac{1}{\log(10)} < 1,$$

we have $|r| < 1$. Therefore, the geometric series converges. By applying the General Comparison Test, the series $\sum_{n=1}^{\infty} \frac{|\sin(n)|}{\left(\log(10)\right)^n}$ is convergent. This implies that the given series is absolutely convergent.

Corollary 1 *Let $\sum a_n$ and $\sum b_n$ be two series of positive terms, c be a positive constant, and p be a natural number such that $a_n \leq c\, b_n$, $\forall n \geq p$:*

a) If the series $\sum b_n$ is convergent, then the series $\sum a_n$ is convergent.

b) If the series $\sum a_n$ is divergent, then the series $\sum b_n$ is divergent.

<u>Proof</u>: Let $c_n = c\, b_n$. By Theorem 2.5.4,

a) if the series $\sum c_n$ is convergent, then the series $\sum a_n$ is also convergent;

b) if the series $\sum a_n$ is divergent, then the series $\sum c_n$ is also divergent.

Since $c > 0$, the series $\sum c_n$ and $\sum b_n$ are either both convergent or both divergent. From this fact, we can conclude the desired result. ∎

Corollary 2 *Let $\sum a_n$ and $\sum b_n$ be two series such that $a_n \geq 0$ and $b_n > 0$, $\forall n \in \mathbb{N}$. If $\lim \frac{a_n}{b_n} = k \in \mathbb{R}^+$, then both series are either convergent or divergent.*

<u>Proof</u>: Suppose $\lim \frac{a_n}{b_n} = k$. By Definition 1.1.7,

$$\forall \delta > 0 \ \exists p \in \mathbb{N}: \ n > p \Rightarrow \left|\frac{a_n}{b_n} - k\right| < \delta.$$

Let $\delta = \frac{k}{2}$. From a certain order, we have

$$\left|\frac{a_n}{b_n} - k\right| < \frac{k}{2} \Leftrightarrow -\frac{k}{2} < \frac{a_n}{b_n} - k < \frac{k}{2} \Leftrightarrow \frac{k}{2} < \frac{a_n}{b_n} < \frac{3}{2}k,$$

which implies
$$a_n < \frac{3}{2} k\, b_n \quad \text{and} \quad \frac{k}{2} b_n < a_n.$$

From these inequalities and using Corollary 1, we get the intended result. ∎

Corollary 3 *Let $\sum a_n$ and $\sum b_n$ be two series such that $a_n \geq 0$ and $b_n > 0$, $\forall n \in \mathbb{N}$. If $\lim \dfrac{a_n}{b_n} = 0$, then:*

a) *If the series $\sum b_n$ is convergent, then the series $\sum a_n$ is also convergent.*

b) *If the series $\sum a_n$ is divergent, then the series $\sum b_n$ is also divergent.*

<u>Proof</u>: Let $\lim \dfrac{a_n}{b_n} = 0$. By Definition 1.1.7,

$$\forall \delta > 0 \ \exists p \in \mathbb{N}: \ n > p \Rightarrow \left|\frac{a_n}{b_n}\right| < \delta.$$

From a certain order, we have
$$\left|\frac{a_n}{b_n}\right| = \frac{a_n}{b_n} < \delta,$$

since $a_n \geq 0$ and $b_n > 0$. Consequently, $0 \leq a_n < \delta b_n$, and from Corollary 1 of Theorem 2.5.4, the result follows. ∎

Corollary 4 *Let $\sum a_n$ and $\sum b_n$ be two series such that $a_n \geq 0$ and $b_n > 0$, $\forall n \in \mathbb{N}$. If $\lim \dfrac{a_n}{b_n} = +\infty$, then:*

a) *If the series $\sum b_n$ is divergent, then the series $\sum a_n$ is also divergent.*

b) *If the series $\sum a_n$ is convergent, then the series $\sum b_n$ is also convergent.*

<u>Proof</u>: Let $\lim \dfrac{a_n}{b_n} = +\infty$. By Definition 1.1.6,

$$\forall \delta > 0 \ \exists p \in \mathbb{N}: \ n > p \Rightarrow \frac{a_n}{b_n} > \delta.$$

2.5. Series of Nonnegative Terms

From order p, we have

$$a_n > \delta b_n > 0,$$

since $b_n > 0$. The result is a consequence of Corollary 1 of Theorem 2.5.4. ∎

Corollary 5 *Let $\sum a_n$ and $\sum b_n$ be two series such that $a_n > 0$ and $b_n > 0$, $\forall n \in \mathbb{N}$. If there exists $p \in \mathbb{N}$ such that*

$$\frac{a_{n+1}}{a_n} \leq \frac{b_{n+1}}{b_n}, \quad \forall n \geq p,$$

then:

a) *If the series $\sum b_n$ is convergent, then the series $\sum a_n$ is convergent.*

b) *If the series $\sum a_n$ is divergent, then the series $\sum b_n$ is divergent.*

Proof: As $a_n > 0$ and $b_n > 0$, we can deduce that:

$$\frac{a_{n+1}}{a_n} \leq \frac{b_{n+1}}{b_n} \Leftrightarrow \frac{a_{n+1}}{b_{n+1}} \leq \frac{a_n}{b_n}, \quad \forall n \geq p.$$

This means that the sequence $\left(\dfrac{a_n}{b_n}\right)$ is decreasing after order p. Therefore, there exists a constant k (we can take $k = \dfrac{a_p}{b_p}$) such that $\dfrac{a_n}{b_n} \leq k$. In other words:

$$a_n \leq k\, b_n, \quad \forall n \geq p.$$

We can use Corollary 1 of Theorem 2.5.4 to arrive at the final result. ∎

Example 2.5.8 The terms of the series $\displaystyle\sum_{n=1}^{\infty} \frac{1 + (-1)^n}{n^2}$ are nonnegative. As

$$0 \leq \frac{1 + (-1)^n}{n^2} \leq \frac{2}{n^2}, \quad \forall n \in \mathbb{N},$$

and the series $\displaystyle\sum_{n=1}^{\infty} \frac{1}{n^2}$ is convergent, Corollary 1 of Theorem 2.5.4 allows us to conclude that the given series is convergent.

Example 2.5.9 Consider the series of positive terms $\sum_{n=1}^{\infty} \dfrac{2n^2 + n}{3n^5 + 3}$. Since

$$\lim \frac{\dfrac{2n^2 + n}{3n^5 + 3}}{\dfrac{1}{n^3}} = \lim \frac{2n^5 + n^4}{3n^5 + 3} = \frac{2}{3},$$

by Corollary 2 of Theorem 2.5.4, the series $\sum_{n=1}^{\infty} \dfrac{2n^2 + n}{3n^5 + 3}$ is convergent because $\sum_{n=1}^{\infty} \dfrac{1}{n^3}$ converges.

Example 2.5.10 The series $\sum_{n=1}^{\infty} \dfrac{\log(n)}{n^3}$ is of nonnegative terms. Since $\sum_{n=1}^{\infty} \dfrac{1}{n^2}$ is convergent and

$$\lim \frac{\dfrac{\log(n)}{n^3}}{\dfrac{1}{n^2}} = \lim \frac{\log(n)}{n} = 0,$$

by Corollary 3 of Theorem 2.5.4, the original series is also convergent.

Example 2.5.11 Consider the series

$$\sum_{n=1}^{\infty} b_n = \sum_{n=1}^{\infty} \frac{1 \times 3 \times \cdots \times (2n-1)}{2 \times 4 \times \cdots 2n} \quad \text{and} \quad \sum_{n=1}^{\infty} \frac{1}{n}.$$

Both series are of positive terms, and the second is divergent. We have

$$\frac{b_{n+1}}{b_n} = \frac{\dfrac{1 \times 3 \times \cdots \times (2n-1)(2n+1)}{2 \times 4 \times \cdots 2n(2n+2)}}{\dfrac{1 \times 3 \times \cdots \times (2n-1)}{2 \times 4 \times \cdots 2n}} = \frac{2n+1}{2n+2} \quad \text{and} \quad \frac{a_{n+1}}{a_n} = \frac{n}{n+1}.$$

It is easy to verify that $\dfrac{n}{n+1} \leq \dfrac{2n+1}{2n+2}$, which allows us to conclude, by Corollary 5 of Theorem 2.5.4, that the series $\sum_{n=1}^{\infty} b_n$ is divergent.

2.5. Series of Nonnegative Terms

Example 2.5.12 The terms of the series $\sum_{n=1}^{\infty} \dfrac{\arctan(n)+n}{n\sqrt{n}+1}$ are positive. Let us consider the series $\sum_{n=1}^{\infty} \dfrac{1}{n^{1/2}}$, which is divergent because it is a p-series with $p = \dfrac{1}{2}$. The limit

$$\lim \frac{\dfrac{\arctan(n)+n}{n\sqrt{n}+1}}{\dfrac{1}{n^{1/2}}} = \lim \frac{(\arctan(n)+n)\, n^{1/2}}{n\sqrt{n}+1} = \lim \frac{\dfrac{\arctan(n)}{n}+1}{1+\dfrac{1}{n^{3/2}}} = 1$$

is finite and different from zero. Therefore, by Corollary 2 of the General Comparison Test, the series $\sum_{n=1}^{\infty} \dfrac{\arctan(n)+n}{n\sqrt{n}+1}$ diverges.

Theorem 2.5.5 (Ratio Test) *Let $\sum a_n$ be a series of positive terms:*

a) *If there exist $r < 1$ and $p \in \mathbb{N}$ such that $\dfrac{a_{n+1}}{a_n} \leq r < 1 \ \forall n \geq p$, then the series $\sum a_n$ is convergent.*

b) *If there exists $p \in \mathbb{N}$ such that $\dfrac{a_{n+1}}{a_n} \geq 1, \ \forall n \geq p$, then the series $\sum a_n$ is divergent.*

Proof:
a) The series $\sum b_n = \sum r^n$ is geometric with ratio r and is convergent since $0 < r < 1$. Moreover, we have

$$\frac{a_{n+1}}{a_n} \leq \frac{r^{n+1}}{r^n} = r, \quad \forall n \geq p,$$

which implies by Corollary 5 of the General Comparison Test that the series $\sum a_n$ is convergent.

b) The series $\sum b_n = \sum 1$ is divergent. Additionally, we have

$$\frac{a_{n+1}}{a_n} \geq 1 = \frac{b_{n+1}}{b_n}, \quad \forall n \geq p.$$

By Corollary 5 of the General Comparison Test, we can conclude that the series $\sum a_n$ is divergent. ∎

Jean Le Rond D'Alembert (1717–1783) French mathematician and philosopher who was elected to the Academy of Sciences of Paris at the age of 24, and later to the Berlin Academy. With Diderot, he began publishing the "Encyclopedia" in 1747, writing articles on mathematics and literature. In 1772, he was appointed perpetual secretary of the French Academy. As a mathematician and scientist, D'Alembert continued the work of Newton and Leibniz and researched differential equations with partial derivatives. He was also responsible for stating the fundamental theorem of algebra, which Gauss later proved. Additionally, D'Alembert studied hydrodynamics and mechanics and formulated what is now known as "D'Alembert's principle." This principle, when applied to the Earth's movement, explains the variations of the rotation axis of the globe. (Source of image: Oil portrait by Maurice-Quentin de La Tour (1704–1788), Louvre Museum.)

Corollary 1 (D'Alembert's Test) Let $\sum a_n$ be a series of positive terms. If there exists $\lim \dfrac{a_{n+1}}{a_n} = a$ $(a \in \mathbb{R}_0^+$ or $a = +\infty)$, then:

a) If $a < 1$, the series $\sum a_n$ is convergent.

b) If $a > 1$, the series $\sum a_n$ is divergent.

Proof: We know that if $a \in \mathbb{R}$,
$$\lim \frac{a_{n+1}}{a_n} = a \Leftrightarrow \forall \delta > 0 \ \exists p \in \mathbb{N}: \ n > p \Rightarrow \left|\frac{a_{n+1}}{a_n} - a\right| < \delta.$$

a) If $a < 1$, then $1 - a > 0$. Let δ be such that $0 < \delta < 1 - a$. There exists $p \in \mathbb{N}$ such that for all $n > p$,
$$\left|\frac{a_{n+1}}{a_n} - a\right| < \delta \Leftrightarrow -\delta < \frac{a_{n+1}}{a_n} - a < \delta \Leftrightarrow a - \delta < \frac{a_{n+1}}{a_n} < a + \delta.$$

Since $\delta < 1 - a$, then $a + \delta < 1$. Item a) of the Ratio Test allows us to conclude that the series $\sum a_n$ converges.

b) If $a > 1$, let $\delta = a - 1$. There exists $p \in \mathbb{N}$ such that, for all $n > p$,
$$\left|\frac{a_{n+1}}{a_n} - a\right| < a - 1 \Leftrightarrow 1 < \frac{a_{n+1}}{a_n} < 2a - 1.$$

By item b) of the Ratio Test, the series $\sum a_n$ diverges.
If $a = +\infty$, there exists $p \in \mathbb{N}$ such that
$$\frac{a_{n+1}}{a_n} > 1, \ \forall n > p.$$

Again, by the Ratio Test, the series $\sum a_n$ diverges. ∎

Note: If $\lim \dfrac{a_{n+1}}{a_n} = 1$, D'Alembert's test is inconclusive as there are divergent and convergent series in this situation. For example, the harmonic series $\sum_{n=1}^{\infty} \dfrac{1}{n}$ is divergent and
$$\lim \frac{a_{n+1}}{a_n} = \lim \frac{n}{n+1} = 1,$$
and the series $\sum_{n=1}^{\infty} \dfrac{1}{n^2}$ is convergent and
$$\lim \frac{a_{n+1}}{a_n} = \lim \left(\frac{n}{n+1}\right)^2 = 1.$$

2.5. Series of Nonnegative Terms

However, if $\lim \dfrac{a_{n+1}}{a_n} = 1$ and the convergence is for values greater than 1, that is, there is an order $p \in \mathbb{N}$ from which $\dfrac{a_{n+1}}{a_n} \geq 1$, then, by the Ratio Test, the series $\sum\limits_{n=1}^{\infty} a_n$ diverges.

Example 2.5.13 Let $k > 0$. The series $\sum\limits_{n=1}^{\infty} \dfrac{k^n n!}{n^n}$ is of positive terms. As

$$\lim \frac{\dfrac{k^{n+1}(n+1)!}{(n+1)^{n+1}}}{\dfrac{k^n n!}{n^n}} = \lim \frac{k^{n+1}(n+1)!\, n^n}{k^n\, n!\,(n+1)^{n+1}} = \lim k \cdot \left(\frac{n}{n+1}\right)^n = k \cdot \frac{1}{e},$$

we can apply D'Alembert's test: If $\dfrac{k}{e} < 1$, the series is convergent; if $\dfrac{k}{e} > 1$, the series is divergent.

If $\dfrac{k}{e} = 1$, D'Alembert's test is inconclusive. However, as $\left(\dfrac{n+1}{n}\right)^n$ is an increasing sequence with limit e, $\left(\dfrac{n}{n+1}\right)^n$ is a decreasing sequence with limit $\dfrac{1}{e}$, which implies that $e \cdot \left(\dfrac{n}{n+1}\right)^n$ is decreasing with limit 1, that is, $\dfrac{a_{n+1}}{a_n}$ tends to 1 from above. Then, if $k = e$, the series is divergent.

Example 2.5.14 The series $\sum\limits_{n=1}^{\infty} (-1)^n \dfrac{((n+1)!)^2}{(2n)!\, 5^n}$ is alternating. Let us start by studying the series of modules, $\sum\limits_{n=1}^{\infty} \dfrac{((n+1)!)^2}{(2n)!\, 5^n}$, using D'Alembert's test. Let a_n be the general term of the series of modules:

$$\lim \frac{a_{n+1}}{a_n} = \lim \frac{\dfrac{((n+2)!)^2}{(2(n+1))!\, 5^{n+1}}}{\dfrac{((n+1)!)^2}{(2n)!\, 5^n}} = \lim \frac{((n+2)!)^2 (2n)!\, 5^n}{(2(n+1))!\, 5^{n+1}((n+1)!)^2}$$

$$= \lim \frac{((n+2)!)^2 (2n)!}{(2n+2)(2n+1)(2n)!\, 5\, ((n+1)!)^2}$$

$$= \lim \frac{(n+2)^2((n+1)!)^2}{(2n+2)(2n+1)\,5\,((n+1)!)^2}$$

$$= \lim \frac{(n+2)^2}{5\,(2n+2)(2n+1)} = \frac{1}{20}.$$

Since this value is less than 1, the series of modules is convergent; that is, the original series is absolutely convergent.

Example 2.5.15 The series $\sum_{n=1}^{\infty} \frac{(n!)^2 + n!}{(4n)! + n^4}$ is of positive terms and

$$0 < \frac{(n!)^2 + n!}{(4n)! + n^4} < \frac{2(n!)^2}{(4n)!}, \quad \forall n \in \mathbb{N}.$$

Let us study the series $\sum_{n=1}^{\infty} \frac{2(n!)^2}{(4n)!}$ by D'Alembert's test:

$$\lim \frac{\frac{2((n+1)!)^2}{(4n+4)!}}{\frac{2(n!)^2}{(4n)!}} = \lim \frac{(n+1)^2}{(4n+4)(4n+3)(4n+2)(4n+1)} = 0 < 1.$$

Therefore, the series $\sum_{n=1}^{\infty} \frac{2(n!)^2}{(4n)!}$ converges, so by the General Comparison Test, the given series converges.

Example 2.5.16 Let us consider the series of positive terms

$$\frac{1}{2} + \frac{1}{2} \cdot \frac{1}{3} + \frac{1}{2^2} \cdot \frac{1}{3} + \frac{1}{2^2} \cdot \frac{1}{3^2} + \frac{1}{2^3} \cdot \frac{1}{3^2} + \cdots,$$

that is, $a_1 = \frac{1}{2}$, $a_2 = \frac{1}{2} \cdot \frac{1}{3}$, $a_3 = \frac{1}{2^2} \cdot \frac{1}{3}$, $a_4 = \frac{1}{2^2} \cdot \frac{1}{3^2}$, ...; in general,

$$a_n = \begin{cases} \frac{1}{2^{\frac{n}{2}}} \cdot \frac{1}{3^{\frac{n}{2}}}, & \text{if } n \text{ is even} \\ \frac{1}{2^{\frac{n+1}{2}}} \cdot \frac{1}{3^{\frac{n-1}{2}}}, & \text{if } n \text{ is odd.} \end{cases}$$

2.5. Series of Nonnegative Terms

Then

$$\frac{a_{n+1}}{a_n} = \begin{cases} \dfrac{\frac{1}{2^{\frac{n+2}{2}}} \cdot \frac{1}{3^{\frac{n}{2}}}}{\frac{1}{2^{\frac{n}{2}}} \cdot \frac{1}{3^{\frac{n}{2}}}} = \dfrac{2^{\frac{n}{2}} \cdot 3^{\frac{n}{2}}}{2^{\frac{n+2}{2}} \cdot 3^{\frac{n}{2}}} = 2^{-1} = \dfrac{1}{2}, & \text{if } n \text{ is even} \\[2ex] \dfrac{\frac{1}{2^{\frac{n+1}{2}}} \cdot \frac{1}{3^{\frac{n+1}{2}}}}{\frac{1}{2^{\frac{n+1}{2}}} \cdot \frac{1}{3^{\frac{n-1}{2}}}} = \dfrac{2^{\frac{n+1}{2}} \cdot 3^{\frac{n-1}{2}}}{2^{\frac{n+1}{2}} \cdot 3^{\frac{n+1}{2}}} = 3^{-1} = \dfrac{1}{3}, & \text{if } n \text{ is odd.} \end{cases}$$

Note that we cannot use D'Alembert's test, as $\lim \dfrac{a_{n+1}}{a_n}$ does not exist. However, it follows from the Ratio Test that the series converges because $\dfrac{a_{n+1}}{a_n} \leq \dfrac{1}{2} < 1, \forall n \in \mathbb{N}$.

Note: We saw in Theorem 2.4.1 that the convergence of the series of modules implies the convergence of the series. Nevertheless, if the series of modules is divergent, nothing can be concluded about the convergence of the series. However, suppose we determine the divergence of the series of modules using D'Alembert's test. As this condition implies that the general term of the series does not converge to zero, we can conclude the divergence of the series.

Example 2.5.17 Consider the series $\sum_{n=1}^{\infty} (-1)^n \dfrac{(n!)^2}{2^n n^n}$. Let a_n be the general term of the series of modules. Using D'Alembert's test, we get

$$\lim \frac{a_{n+1}}{a_n} = \lim \frac{\frac{((n+1)!)^2}{2^{n+1}(n+1)^{n+1}}}{\frac{(n!)^2}{2^n n^n}} = \lim \frac{((n+1)!)^2 2^n n^n}{2^{n+1}(n+1)^{n+1}(n!)^2}$$

$$= \lim \frac{n+1}{2} \left(\frac{n}{n+1}\right)^n = +\infty.$$

The limit value is greater than 1, which means the series of modules is divergent. Since we used D'Alembert's test, the given series is also divergent.

> **Theorem 2.5.6 (Root Test)** Let $\sum a_n$ be a series of nonnegative terms:
>
> a) If there exist $r < 1$ and $p \in \mathbb{N}$ such that $\sqrt[n]{a_n} \leq r$, $\forall n > p$, then the series $\sum a_n$ is convergent.
>
> b) If there exist $p \in \mathbb{N}$ and a subsequence, (a_{k_n}), of (a_n) such that $\sqrt[k_n]{a_{k_n}} \geq 1$, $\forall k_n > p$, then the series $\sum a_n$ is divergent.

Proof:

a) If $\sqrt[n]{a_n} \leq r$, $\forall n > p$, then $a_n \leq r^n < 1$, $\forall n \geq p$. The series $\sum r^n$ is convergent because it is a geometric series with ratio r, where $0 < r < 1$. Therefore, the series $\sum a_n$ is convergent.

b) If $\sqrt[k_n]{a_{k_n}} \geq 1$, $\forall k_n > p$, then $a_{k_n} \geq 1$, $\forall k_n > p$, so it does not tend to zero. As a result, (a_n) is not a null sequence, which implies that the series $\sum a_n$ is divergent. ∎

> **Corollary 1** Let $\sum a_n$ be a series of nonnegative terms:
>
> a) If $\overline{\lim} \sqrt[n]{a_n} < 1$, the series $\sum a_n$ is convergent.
>
> b) If $\overline{\lim} \sqrt[n]{a_n} > 1$, the series $\sum a_n$ is divergent.

Proof: Let $a = \overline{\lim} \sqrt[n]{a_n}$.

a) Let r be such that $a < r < 1$. We can affirm that

$$\exists p \in \mathbb{N} : n > p \Rightarrow \sqrt[n]{a_n} < r,$$

which implies by item a) of the theorem that the series $\sum a_n$ converges.

b) By definition of upper limit, there is a subsequence of $\left(\sqrt[n]{a_n}\right)$ with limit $a > 1$, so this sequence has an infinity of values greater than 1. By item b) of the theorem, the series diverges. ∎

> **Corollary 2 (Cauchy's Root Test)** Let $\sum a_n$ be a series of nonnegative terms. If there exists $\lim \sqrt[n]{a_n} = a$ $(a \in \mathbb{R}_0^+$ or $a = +\infty)$, then:
>
> a) If $a < 1$, the series $\sum a_n$ is convergent.
>
> b) If $a > 1$, the series $\sum a_n$ is divergent.

2.5. Series of Nonnegative Terms

<u>Proof</u>: If there exists $\lim \sqrt[n]{a_n} = a$, then $\overline{\lim} \sqrt[n]{a_n} = \underline{\lim} \sqrt[n]{a_n} = a$. Moreover, Corollary 1 applies. ∎

Note: If $\lim \sqrt[n]{a_n} = 1$, the test is inconclusive, as there exist both convergent and divergent series in this situation. For instance, the harmonic series $\sum_{n=1}^{\infty} \frac{1}{n}$ is divergent and

$$\lim \sqrt[n]{\frac{1}{n}} = \lim \frac{1}{\sqrt[n]{n}} = 1,$$

and the series $\sum_{n=1}^{\infty} \frac{1}{n^2}$ is convergent and

$$\lim \sqrt[n]{\frac{1}{n^2}} = \lim \frac{1}{\sqrt[n]{n^2}} = 1.$$

Example 2.5.18 The series $\sum_{n=1}^{\infty} \left(\frac{n+1}{n}\right)^{n^2}$ is of positive terms. As

$$\lim \sqrt[n]{\left(\frac{n+1}{n}\right)^{n^2}} = \lim \left(\frac{n+1}{n}\right)^{n} = e > 1,$$

by Cauchy's Root Test, the series is divergent.

Example 2.5.19 Let us consider the series $\sum_{n=1}^{\infty} \frac{1}{(3+(-1)^n)^n}$.

$$\sqrt[n]{\frac{1}{(3+(-1)^n)^n}} = \begin{cases} \frac{1}{4}, & \text{if } n \text{ is even} \\ \frac{1}{2}, & \text{if } n \text{ is odd.} \end{cases}$$

Then $\sqrt[n]{a_n} \leq \frac{1}{2} < 1$, $\forall n \in \mathbb{N}$, and by the Root Test, the series is convergent.

Example 2.5.20 The series $\sum_{n=1}^{\infty} \frac{(\log(n))^{2n}}{n^n} = \sum_{n=2}^{\infty} \frac{(\log(n))^{2n}}{n^n}$ is of positive terms. We will study this series using the Cauchy's Root Test:

$$\lim \sqrt[n]{\frac{(\log(n))^{2n}}{n^n}} = \lim \frac{(\log(n))^2}{n} = 0$$

because using L'Hôpital's Rule, we have

$$\lim_{x \to +\infty} \frac{\left(\left(\log(x)\right)^2\right)'}{(x)'} = \lim_{x \to +\infty} \frac{2\log(x)}{x} = 0.$$

Since the limit is less than 1, the series is convergent.

Example 2.5.21 Let us study the convergence of the series $\sum a_n$ where

$$a_n = \begin{cases} (1 - \sqrt[n]{n})^n, & \text{if } n \text{ is odd} \\ n^2 e^{-n}, & \text{if } n \text{ is even.} \end{cases}$$

It is an alternating series. Let us study the series of modules.

$$\sqrt[n]{|a_n|} = \begin{cases} \sqrt[n]{|(1 - \sqrt[n]{n})^n|}, & \text{if } n \text{ is odd} \\ \sqrt[n]{n^2 e^{-n}}, & \text{if } n \text{ is even} \end{cases}$$

$$= \begin{cases} \sqrt[n]{n} - 1, & \text{if } n \text{ is odd} \\ e^{-1} \sqrt[n]{n^2}, & \text{if } n \text{ is even.} \end{cases}$$

Since $\lim(\sqrt[n]{n}-1) = 0$ and $\lim e^{-1}\sqrt[n]{n^2} = e^{-1}$, we obtain $\overline{\lim} \sqrt[n]{|a_n|} = e^{-1}$. As $\overline{\lim} \sqrt[n]{|a_n|} < 1$, the series $\sum_{n=1}^{\infty} |a_n|$ is convergent. Therefore, the series $\sum_{n=1}^{\infty} a_n$ is absolutely convergent.

Note: The Cauchy's Root Test is more general than D'Alembert's test. This means that if Cauchy's Root Test is inconclusive about a series, then D'Alembert's test will also be inconclusive. In fact, by Theorem 1.1.11 $\lim \frac{a_{n+1}}{a_n} = a \Rightarrow \lim \sqrt[n]{a_n} = a$. This theorem is significant because if $a = 1$, then D'Alembert's test is inconclusive, and the same happens for Cauchy's Root Test. However, it is essential to note that the reciprocal is not true. It may be possible to draw conclusions through Cauchy's Root Test, even if D'Alembert's test fails to do so.

2.5. Series of Nonnegative Terms

Example 2.5.22 Let us consider the series $\sum_{n=1}^{\infty} 2^{-n-(-1)^n}$. Using Cauchy's Root Test,

$$\lim \sqrt[n]{2^{-n-(-1)^n}} = \lim 2^{-1} 2^{-\frac{(-1)^n}{n}} = \frac{1}{2} < 1,$$

we conclude that the series is convergent, whereas D'Alembert's test fails to be applied. In fact,

$$\frac{2^{-(n+1)-(-1)^{n+1}}}{2^{-n-(-1)^n}} = 2^{-n-1-(-1)^{n+1}+n+(-1)^n} = \begin{cases} 2, & \text{if } n \text{ is even} \\ 2^{-3}, & \text{if } n \text{ is odd}. \end{cases}$$

Theorem 2.5.7 (Kummer's Test) Let $\sum a_n$ and $\sum b_n$ be two series of positive terms, with $\sum b_n$ divergent. If there exists

$$\lim \left(\frac{1}{b_n} \cdot \frac{a_n}{a_{n+1}} - \frac{1}{b_{n+1}} \right) = k, \quad k \in \overline{\mathbb{R}},$$

then:

a) If $k > 0$, the series $\sum a_n$ is convergent.

b) If $k < 0$, the series $\sum a_n$ is divergent.

Ernst Eduard Kummer (1810–1893) was a German mathematician. In 1855, he succeeded Dirichlet at the University of Berlin, where he worked with Weierstrass and Kronecker. In 1857, the Academy of Sciences of Paris awarded him a prize for his work on Fermat's Theorem. (Source of image: Photo archive of the Mathematical Research Institute Oberwolfach.)

<u>Proof</u>: If $k \in \mathbb{R}$, $\lim \left(\frac{1}{b_n} \cdot \frac{a_n}{a_{n+1}} - \frac{1}{b_{n+1}} \right) = k$ is equivalent to

$$\forall \delta > 0 \ \exists p \in \mathbb{N}: \ \forall n > p, \ \left| \frac{1}{b_n} \cdot \frac{a_n}{a_{n+1}} - \frac{1}{b_{n+1}} - k \right| < \delta.$$

But

$$\left| \frac{1}{b_n} \cdot \frac{a_n}{a_{n+1}} - \frac{1}{b_{n+1}} - k \right| < \delta \Leftrightarrow k - \delta < \frac{1}{b_n} \cdot \frac{a_n}{a_{n+1}} - \frac{1}{b_{n+1}} < k + \delta.$$

a) Let $k \in \mathbb{R}^+$ and $\delta = \dfrac{k}{2}$. There exists an order $n_0 \in \mathbb{N}$ from which we have

$$k - \frac{k}{2} < \frac{1}{b_n} \cdot \frac{a_n}{a_{n+1}} - \frac{1}{b_{n+1}} \Leftrightarrow \frac{k}{2} < \frac{1}{b_n} \cdot \frac{a_n}{a_{n+1}} - \frac{1}{b_{n+1}}$$

$$\Leftrightarrow 1 < \frac{2}{k}\left(\frac{1}{b_n} \cdot \frac{a_n}{a_{n+1}} - \frac{1}{b_{n+1}}\right) \Leftrightarrow a_{n+1} < \frac{2}{k} a_{n+1}\left(\frac{1}{b_n} \cdot \frac{a_n}{a_{n+1}} - \frac{1}{b_{n+1}}\right)$$

$$\Leftrightarrow a_{n+1} < \frac{2}{k}\left(\frac{a_n}{b_n} - \frac{a_{n+1}}{b_{n+1}}\right).$$

Adding the two members of the inequality from $n_0 + 1$ to $n + 1$, we obtain

$$\sum_{i=n_0+1}^{n+1} a_i < \sum_{i=n_0+1}^{n+1} \frac{2}{k}\left(\frac{a_{i-1}}{b_{i-1}} - \frac{a_i}{b_i}\right)$$

$$\Leftrightarrow \sum_{i=n_0+1}^{n+1} a_i < \frac{2}{k}\left(\frac{a_{n_0}}{b_{n_0}} - \frac{a_{n_0+1}}{b_{n_0+1}} + \frac{a_{n_0+1}}{b_{n_0+1}} - \frac{a_{n_0+2}}{b_{n_0+2}} + \cdots + \frac{a_n}{b_n} - \frac{a_{n+1}}{b_{n+1}}\right)$$

$$\Leftrightarrow \sum_{i=n_0+1}^{n+1} a_i < \frac{2}{k}\left(\frac{a_{n_0}}{b_{n_0}} - \frac{a_{n+1}}{b_{n+1}}\right) < \frac{2}{k}\frac{a_{n_0}}{b_{n_0}}.$$

Therefore, the sequence of partial sums of the series $\sum a_n$ is bounded because

$$0 < S_{n+1} = \sum_{i=1}^{n+1} a_i = S_{n_0} + a_{n_0+1} + \cdots + a_{n+1} \leq S_{n_0} + \frac{2}{k}\frac{a_{n_0}}{b_{n_0}},$$

and by Theorem 2.5.1, the series $\sum a_n$ converges.

If $k = +\infty$, let $\alpha > 0$ be arbitrary. There is an order $n_0 \in \mathbb{N}$ from which we have

$$\frac{1}{b_n} \cdot \frac{a_n}{a_{n+1}} - \frac{1}{b_{n+1}} > \frac{\alpha}{2},$$

and we can apply the previous reasoning.

b) Let $k \in \mathbb{R}^-$ and $\delta = -k$. There is an order $n_0 \in \mathbb{N}$ from which we have

$$\frac{1}{b_n} \cdot \frac{a_n}{a_{n+1}} - \frac{1}{b_{n+1}} < 0 \Leftrightarrow \frac{a_n}{a_{n+1}} < \frac{b_n}{b_{n+1}} \Leftrightarrow \frac{a_{n+1}}{a_n} > \frac{b_{n+1}}{b_n}.$$

As the series $\sum b_n$ is divergent, Corollary 5 of the General Comparison Test allows us to conclude that $\sum a_n$ is divergent.

2.5. Series of Nonnegative Terms

If $k = -\infty$, there is also an order $n_0 \in \mathbb{N}$ from which we have

$$\frac{1}{b_n} \cdot \frac{a_n}{a_{n+1}} - \frac{1}{b_{n+1}} < 0,$$

and we finish in the same way. ∎

Corollary 1 (Raabe's Test) *Let $\sum a_n$ be a series of positive terms. Suppose that*

$$\lim n \left(\frac{a_n}{a_{n+1}} - 1 \right) = a :$$

Then

a) *If $a < 1$, the series $\sum a_n$ is divergent.*

b) *If $a > 1$, the series $\sum a_n$ is convergent.*

Proof: In Kummer's test, consider $b_n = \dfrac{1}{n}$. The series $\sum_{n=1}^{\infty} \dfrac{1}{n}$ is divergent, and we have

$$\lim \left(\frac{1}{b_n} \cdot \frac{a_n}{a_{n+1}} - \frac{1}{b_{n+1}} \right) = \lim \left(n \cdot \frac{a_n}{a_{n+1}} - n - 1 \right)$$

$$= \lim n \left(\frac{a_n}{a_{n+1}} - 1 \right) - 1 = a - 1.$$

This demonstrates the corollary. ∎

Note: Often, cases that are inconclusive by D'Alembert's test can be solved by Raabe's test.

Joseph L. Raabe (1801–1859), was a Swiss mathematician and physicist. He was a professor at the Zurich Polytechnic Institute. His name is mainly associated with the convergence test for a series of positive terms that extends D'Alembert's Test. He also studied various aspects of planetary movements. (Source of image: Lithographie by Carl Friedrich Irminger (1813–1863).)

Example 2.5.23 Consider the series

$$\sum_{n=1}^{\infty} \frac{1 \times 3 \times \cdots \times (2n-1)}{2 \times 4 \times \cdots \times 2n} \cdot \frac{1}{n} = \sum_{n=1}^{\infty} a_n.$$

We have

$$\lim \frac{a_{n+1}}{a_n} = \lim \frac{n(2n+1)}{(n+1)(2n+2)} = 1,$$

and therefore, D'Alembert's test is inconclusive. As

$$\lim n\left(\frac{a_n}{a_{n+1}} - 1\right) = \lim n\left(\frac{(n+1)(2n+2)}{n(2n+1)} - 1\right)$$
$$= \lim \frac{(n+1)(2n+2) - n(2n+1)}{2n+1} = \lim \frac{3n+2}{2n+1} = \frac{3}{2} > 1,$$

Raabe's test shows that the series is convergent.

2.6 Products of Series

Let $\sum_{n=1}^{\infty} a_n$ and $\sum_{n=1}^{\infty} b_n$ be two convergent series with sums A and B, respectively. When thinking about the product $\left(\sum_{n=1}^{\infty} a_n\right) \times \left(\sum_{n=1}^{\infty} b_n\right)$, it will be natural to define it in such a way that the resulting series, if convergent, has sum $A \times B$. We can define, for example,

$$\left(\sum_{n=1}^{\infty} a_n\right) \times \left(\sum_{n=1}^{\infty} b_n\right) = \sum_{n=1}^{\infty} \left(a_n \sum_{k=1}^{\infty} b_k\right)$$

obtaining

$$\sum_{n=1}^{\infty} \left(a_n \sum_{k=1}^{\infty} b_k\right) = \sum_{n=1}^{\infty} a_n \cdot B = B \cdot \sum_{n=1}^{\infty} a_n = B \times A.$$

However, one may ask if it is possible to form a series whose terms are products of the form $a_n b_k$, in some order, so that the sum of this series is $A \times B$. The answer to this question is given in the following theorem:

> **Theorem 2.6.1** Let $\sum a_n$ and $\sum b_n$ be two convergent series, with sums A and B, respectively. Let ϕ be a bijective function, $\phi : \mathbb{N}^2 \to \mathbb{N}$, $\phi(i,j) = n$. To each ϕ, it corresponds to a series $\sum c_n$, with
>
> $$c_n = c_{\phi(i,j)} = a_i \times b_j.$$
>
> The series $\sum c_n$ converges, regardless of the function ϕ considered if and only if $\sum a_n$ and $\sum b_n$ are both absolutely convergent, and, in this case, we have $\sum c_n = A \times B$, and the series $\sum c_n$ is also absolutely convergent.

Note: Saying that $\sum c_n$ converges for any function ϕ considered is equivalent to state that the product series converges regardless of the order in which its terms are taken.

> **Definition 2.6.1** The **Cauchy product** of two convergent series, $\sum_{n=1}^{\infty} a_n$ and $\sum_{n=1}^{\infty} b_n$, is the series $\sum_{n=1}^{\infty} \left(\sum_{k=1}^{n} a_k b_{n-k+1}\right)$.

Note: If $n \in \mathbb{N}_0$, then the Cauchy product is written as follows:

$$\sum_{n=0}^{\infty} \left(\sum_{k=0}^{n} a_k\, b_{n-k} \right).$$

Corollary 1 *If $\sum a_n$ and $\sum b_n$ are absolutely convergent series with sums A and B, respectively, their Cauchy product is absolutely convergent and has sum $A \times B$.*

Example 2.6.1 Consider the series $\sum_{n=0}^{\infty} \dfrac{x^n}{n!}$, $x \in \mathbb{R}$. As

$$\lim \frac{\left|\dfrac{x^{n+1}}{(n+1)!}\right|}{\left|\dfrac{x^n}{n!}\right|} = \lim \frac{|x|}{n+1} = 0, \quad \forall x \in \mathbb{R},$$

the series is absolutely convergent for every $x \in \mathbb{R}$. Then, the Cauchy product of two series of this type is absolutely convergent. Let us form the product and verify that the obtained series is absolutely convergent.

$$\left(\sum_{n=0}^{\infty} \frac{x^n}{n!} \right) \times \left(\sum_{n=0}^{\infty} \frac{y^n}{n!} \right) = \sum_{n=0}^{\infty} \left(\sum_{k=0}^{n} \frac{x^k}{k!} \cdot \frac{y^{n-k}}{(n-k)!} \right)$$

$$= \sum_{n=0}^{\infty} \left(\frac{1}{n!} \sum_{k=0}^{n} \frac{n!}{k!(n-k)!} \cdot x^k y^{n-k} \right)$$

$$= \sum_{n=0}^{\infty} \frac{(x+y)^n}{n!}.$$

Note: The Cauchy product of two series that are not absolutely convergent may lead to a divergent series.

Example 2.6.2 The series $\sum_{n=0}^{\infty} \dfrac{(-1)^n}{\sqrt{n+1}}$ is conditionally convergent. Calculating the Cauchy product of the series with itself, we get

2.6. Products of Series

$$\sum_{n=0}^{\infty}\left(\sum_{k=0}^{n}\frac{(-1)^k}{\sqrt{(k+1)}}\cdot\frac{(-1)^{n-k}}{\sqrt{(n-k+1)}}\right)$$

$$=\sum_{n=0}^{\infty}\left(\sum_{k=0}^{n}\frac{(-1)^n}{\sqrt{k+1}\sqrt{n-k+1}}\right)$$

$$=\sum_{n=0}^{\infty}(-1)^n\left(\sum_{k=0}^{n}\frac{1}{\sqrt{k+1}\sqrt{n-k+1}}\right)=\sum_{n=0}^{\infty}(-1)^n a_n,$$

which is an alternating series. However, (a_n) is not a null sequence as

$$a_n=\sum_{k=0}^{n}\frac{1}{\sqrt{k+1}\sqrt{n-k+1}}\geq\sum_{k=0}^{n}\frac{1}{\sqrt{n+1}\sqrt{n+1}}=1;$$

thus, the product series is divergent.

Theorem 2.6.2 (Mertens) *If at least one of the convergent series, $\sum a_n$ and $\sum b_n$, is absolutely convergent, then their Cauchy product is convergent, and its sum is the product of the sums of the given series.*

Theorem 2.6.3 *If the series $\sum a_n$ and $\sum b_n$ are convergent, with sums A and B, respectively, then if their Cauchy product is convergent, its sum is $A \times B$.*

Example 2.6.3 The series $\sum_{n=1}^{\infty}\frac{(-1)^n}{n}$ is conditionally convergent, and the series $\sum_{n=1}^{\infty}\frac{(-1)^n}{n^2}$ is absolutely convergent. By Mertens's Theorem, the Cauchy product of the series, which is alternating, is convergent:

$$\left(\sum_{n=1}^{\infty}\frac{(-1)^n}{n}\right)\times\left(\sum_{n=1}^{\infty}\frac{(-1)^n}{n^2}\right)=\sum_{n=1}^{\infty}\left(\sum_{k=1}^{n}\frac{(-1)^k}{k}\cdot\frac{(-1)^{n-k+1}}{(n-k+1)^2}\right)$$

$$=\sum_{n=1}^{\infty}\left((-1)^{n+1}\sum_{k=1}^{n}\frac{1}{k(n-k+1)^2}\right).$$

2.7 Solved Exercises

1. Show that the following series are convergent and find their sums:

   a) $\sum_{n=1}^{\infty} \dfrac{1}{n^2 + 3n + 2}$

   b) $\sum_{n=1}^{\infty} \left(\arctan(n+3) - \arctan(n) \right)$

   c) $\sum_{n=0}^{\infty} \left(\dfrac{1}{e^n} - \dfrac{1}{e^{n+1}} \right)$

   d) $\sum_{n=1}^{\infty} \left(\dfrac{1}{\arctan(n)} - \dfrac{1}{\arctan(n+1)} \right)$

   e) $\sum_{n=1}^{\infty} \left(\dfrac{1}{2^{2n-1}} - \dfrac{1}{2^{3n-2}} \right)$

   f) $\sum_{n=0}^{\infty} \dfrac{3^n + 7^n}{3^n \cdot 7^n}$

   g) $\sum_{n=0}^{\infty} \dfrac{1}{4^{n-1}} \cdot \dfrac{2}{5^n} \cdot \dfrac{3}{6^{n+1}}$

   h) $\sum_{n=2}^{\infty} \dfrac{\log(\frac{n+2}{n})}{\log(n) \log(n+2)}$

2. Show that the series
$$\sum_{n=2}^{\infty} \left(\dfrac{1}{(n+1)^2} + \dfrac{1}{\pi^n} \right)$$
is convergent and find its sum, knowing that $\sum_{n=1}^{\infty} \dfrac{1}{n^2} = \dfrac{\pi^2}{6}$.

3. Write the following repeating decimals as fractions:

   a) $0,\overline{5}$

   b) $0,\overline{34}$

   c) $1,\overline{345}$

   d) $0,324\overline{101}$

4. Test the convergence of the following series using a comparison test:

   a) $\sum_{n=1}^{\infty} \dfrac{2n^2 - 1}{3n^5 + 2n + 1}$

   b) $\sum_{n=1}^{\infty} \left(\sin\left(\dfrac{1}{n}\right) \right)^2$

   c) $\sum_{n=1}^{\infty} \dfrac{\log(n^7 + 1)}{n^2}$

   d) $\sum_{n=1}^{\infty} \left(\sqrt{n + \dfrac{1}{n}} - \sqrt{n} \right)$

   e) $\sum_{n=1}^{\infty} \dfrac{\sqrt{n+1} - \sqrt{n}}{\sqrt{n+1} + \sqrt{n}}$

   f) $\sum_{n=1}^{\infty} \dfrac{\cos(n)}{n^2 + 4}$

   g) $\sum_{n=1}^{\infty} \dfrac{1}{n^3 + 10 \cos(n)}$

   h) $\sum_{n=1}^{\infty} \dfrac{1 - (-1)^n}{\sqrt{n}}$

5. Investigate the convergence of the following series:

   a) $\sum_{n=1}^{\infty} (-1)^n \dfrac{\log(n)}{n}$

   b) $\sum_{n=1}^{\infty} (-1)^{n+1} \dfrac{2}{e^n + e^{-n}}$

   c) $\sum_{n=1}^{\infty} (-1)^n \dfrac{1}{\log(3n)}$

   d) $\sum_{n=1}^{\infty} (-1)^n \dfrac{\sin(\frac{1}{n})}{n}$

   e) $\sum_{n=1}^{\infty} (-1)^n \dfrac{3^n}{e^{2n+1}}$

   f) $\sum_{n=1}^{\infty} \dfrac{\sin(\frac{n\pi}{2})}{\sqrt{n+1}}$

6. Consider the series $\sum_{n=1}^{\infty} \dfrac{(-1)^{n+1}}{n}$:

   a) Study it for convergence.

   b) Indicate a partial sum S_n that approximates the sum of the series with an error less than $\dfrac{1}{1000}$.

2.7. Solved Exercises

c) Indicate an upper bound of the error committed when S_5 is taken as the sum of the series.

7. Use the Ratio Test or D'Alembert's Test to determine the convergence or divergence of the following series:

a) $\sum_{n=1}^{\infty} \dfrac{n^n}{\pi^n\, n!}$

b) $\sum_{n=0}^{\infty} \dfrac{n\, 2^n}{4\, n^3 + 1}$

c) $\sum_{n=1}^{\infty} \dfrac{(n+1)! - n!}{n!\,(n+1)!}$

d) $\sum_{n=1}^{\infty} \dfrac{2 \cdot 4 \cdot 6 \cdots (2n+2)}{1 \cdot 4 \cdot 7 \cdots (3n+1)}$

e) $\sum_{n=1}^{\infty} \dfrac{(n+1)!}{n^n\, \sqrt{3n+2}}$

f) $\sum_{n=1}^{\infty} \dfrac{3^n + n!}{n! + n^n}$

g) $\sum_{n=1}^{\infty} \dfrac{n!}{(2n)! + 2^n}$

h) $\sum_{n=0}^{\infty} \dfrac{n! + 3n}{((n+1)!)^2}$

8. Use the Root Test or Cauchy's Root Test to determine the convergence or divergence of the following series:

a) $\sum_{n=3}^{\infty} \left(1 - \dfrac{2}{n}\right)^{n^2}$

b) $\sum_{n=0}^{\infty} e^{-n^2}$

c) $\sum_{n=1}^{\infty} \left(\sqrt[n]{2} - 1\right)^n$

d) $\sum_{n=1}^{\infty} \left(\cos\left(\dfrac{\pi}{6} + \dfrac{1}{n}\right)\right)^n$

e) $\sum_{n=0}^{\infty} \left(\dfrac{1}{2} + (-1)^n \dfrac{n}{4n+1}\right)^n$

f) $\sum_{n=1}^{\infty} \left(\dfrac{n^2 - 2}{3n^2}\right)^n$

9. Test for convergence or divergence the following series. If convergence occurs, indicate whether it is conditional or absolute:

a) $\sum_{n=2}^{\infty} \dfrac{(-1)^n}{\log(n^n)}$

b) $\sum_{n=1}^{\infty} \left(\dfrac{n^2}{\sqrt{n^5+1}} + \left(1 + \dfrac{2}{n}\right)^n\right)$

c) $\sum_{n=2}^{\infty} \log\left(1 - \dfrac{1}{n^2}\right)$

d) $\sum_{n=1}^{\infty} \dfrac{\sin(n) + 2^n}{n + 5^n}$

e) $\sum_{n=1}^{\infty} (-1)^n \dfrac{e^{n+1}}{n^n}$

f) $\sum_{n=1}^{\infty} \dfrac{(-1)^n + \dfrac{1}{n}}{n}$

g) $\sum_{n=1}^{\infty} \log\left(1 + \dfrac{1}{2^n}\right)$

h) $\sum_{n=1}^{\infty} \left(1 - \sqrt[n]{n}\right)^n$

i) $\sum_{n=1}^{\infty} \dfrac{(-1)^n}{1 - (-1)^n\, n^2}$

j) $\sum_{n=1}^{\infty} \left(\dfrac{(-1)^n + 4}{n + 4^n} + \left(\dfrac{3}{n+2}\right)^{2n}\right)$

k) $\sum_{n=1}^{\infty} \left(1 - \dfrac{1}{n}\right)^n$

l) $\sum_{n=2}^{\infty} \dfrac{1}{n\,(\log(n))^3}$

m) $\sum_{n=1}^{\infty} \int_n^{n+2} \dfrac{1}{x^2}\, dx$

n) $\sum_{n=1}^{\infty} \dfrac{(-1)^n}{\sqrt{n+1} + \sqrt[3]{n+1}}$

o) $\sum_{n=1}^{\infty} \dfrac{1}{n(\sqrt{n+1} + \sqrt{n})}$

p) $\sum_{n=1}^{\infty} (-1)^n \left(1 - \cos\left(\dfrac{1}{n}\right)\right)$

q) $\sum_{n=1}^{\infty} (-1)^n\, 2^n \sin\left(\dfrac{1}{3^n}\right)$

r) $\sum_{n=1}^{\infty} (-1)^n \dfrac{n!\sqrt{n}}{n^n \sqrt{n+1}}$

s) $\sum_{n=1}^{\infty} (\sqrt{n}+1) \sin\left(\dfrac{1}{n^2}\right)$

t) $\sum_{n=2}^{\infty} \dfrac{1}{\log(n!)}$
Hint: Note that $n! < n^n$

10. Assess the convergence or divergence of the given series. If the series converges, specify whether the convergence is absolute or conditional:

a) $\sum_{n=0}^{\infty} \dfrac{n \cos(n\pi)}{\sqrt{n^3+1}}$

b) $\sum_{n=1}^{\infty} \dfrac{n!}{3 \times 5 \times \cdots \times (2n+1)}$

c) $\sum_{n=1}^{\infty} \dfrac{1+\cos^2(n)}{\sqrt{n}}$

d) $\sum_{n=0}^{\infty} \dfrac{3^n (n+1)!}{(n+1)^n}$

e) $\sum_{n=1}^{\infty} \sin\left(\dfrac{(-1)^n}{n}\right)$

f) $\sum_{n=1}^{\infty} \dfrac{\sqrt{n}}{n^2+\cos(n)}$

g) $\sum_{n=1}^{\infty} \dfrac{3^n}{2n+n^n}$

h) $\sum_{n=1}^{\infty} \dfrac{1+n(-1)^n}{1+2n^3}$

i) $\sum_{n=1}^{\infty} \dfrac{2^n}{n!+1}$

j) $\sum_{n=0}^{\infty} \left(\dfrac{n!}{(2n)!} - \dfrac{1}{2^n}\right)$

k) $\sum_{n=1}^{\infty} \dfrac{n(2n)!}{4^n (n!)^2}$

l) $\sum_{n=2}^{\infty} \dfrac{\sin(n)}{n^3 \log(n)}$

m) $\sum_{n=1}^{\infty} (-1)^n \left(1-\dfrac{1}{2n}\right)^n$

n) $\sum_{n=1}^{\infty} \dfrac{1}{2n+5} \arcsin\left(\dfrac{1}{n}\right)$

11. Determine whether the following series are convergent or divergent. If convergence occurs, indicate if it is conditional or absolute:

a) $\sum_{n=1}^{\infty} \dfrac{1}{n} \sin\left(\dfrac{1}{2^n}\right)$

b) $\sum_{n=1}^{\infty} \dfrac{\sqrt[3]{n^2}}{n\sqrt{n}+2n^2}$

c) $\sum_{n=2}^{\infty} \dfrac{\cos(\pi n)}{\sqrt{n^2-1}}$

d) $\sum_{n=2}^{\infty} \dfrac{(-1)^n}{\sqrt{n}\log(n)}$

e) $\sum_{n=1}^{\infty} \dfrac{4+\sin(n)}{\sqrt[3]{n}+1}$

f) $\sum_{n=1}^{\infty} \dfrac{n^{2n} \sin(n^3)}{(3n^2+5)^n}$

g) $\sum_{n=3}^{\infty} (-1)^n \arcsin\left(\dfrac{n-1}{n^2+1}\right)$

h) $\sum_{n=1}^{\infty} \dfrac{4 \times 7 \times \cdots \times (3n+1)}{8 \times 11 \times \cdots \times (3n+5)}$

i) $\sum_{n=1}^{\infty} \left(\dfrac{\arctan(n^3)}{\sqrt{n}+n^2} + \dfrac{2^n (2n)!}{3^n (2n+1)!}\right)$

j) $\sum_{n=1}^{\infty} \dfrac{\sin\left(\frac{1}{\sqrt{n}}\right)}{n+\sqrt{n}}$

k) $\sum_{n=1}^{\infty} \left(n \sin\left(\dfrac{1}{n}\right) - (n+2)\sin\left(\dfrac{1}{n+2}\right)\right)$

l) $\sum_{n=1}^{\infty} \dfrac{2^n+3}{(n+1)!}$

m) $\sum_{n=1}^{\infty} \dfrac{\arctan(n+1) - \arctan(n)}{n^2}$

n) $\sum_{n=1}^{\infty} \dfrac{\log(n)}{n \sin\left(\frac{1}{n}\right)}$

o) $\sum_{n=1}^{\infty} (\log(2))^n \log\left(\dfrac{n+1}{n}\right)$

2.7. Solved Exercises

p) $\sum_{n=3}^{\infty} \dfrac{1}{n \log(n) \left(\log\left(\log(n) \right) \right)^p}$, $p \in \mathbb{R}$

12. a) Test for convergence or divergence the series
$$\sum_{n=1}^{\infty} \dfrac{(n+1)^n}{3^n n!}.$$
b) Based on the previous item, find the limit of the sequence with general term $\dfrac{(n+1)^n}{3^n n!}$.

13. Test for convergence or divergence the series $\sum a_n$ where
$$a_n = \begin{cases} \left(\dfrac{n}{n+1}\right)^{n^2}, & \text{if } n = 3k,\ k \in \mathbb{N} \\ \dfrac{1}{n!}, & \text{if } n = 3k+1,\ k \in \mathbb{N}_0 \\ \dfrac{(-1)^n}{(n+1)^n}, & \text{if } n = 3k+2,\ k \in \mathbb{N}_0. \end{cases}$$

If convergence occurs, indicate whether it is conditional or absolute.

14. Investigate, by two distinct processes, the convergence or divergence of the series $\sum a_n$ where
$$a_n = \begin{cases} \left(\dfrac{n}{2}\right)^2, & \text{if } n \text{ is even} \\ \dfrac{(n+1)(n-1)}{2^2}, & \text{if } n \text{ is odd.} \end{cases}$$

If convergence occurs, indicate whether it is conditional or absolute.

15. Let $(a_n)_{n \in \mathbb{N}}$ be a sequence of positive terms such that the series $\sum n a_n$ is convergent. Prove that the series $\sum (a_n)^2$ is convergent.

16. Show that, if $a_n > 0$, $\forall n \in \mathbb{N}$, and the series $\sum a_n$ converges, then the series $\sum \log(1 + a_n)$ converges.

17. Consider the series $\sum_{n=1}^{\infty} (-1)^{n-1} a_n$ where
$$a_n = \begin{cases} \dfrac{1}{n}, & \text{if } n \text{ is odd} \\ \dfrac{1}{n^2}, & \text{if } n \text{ is even.} \end{cases}$$

a) Show that Leibniz's test cannot be applied.
b) Show that the series is divergent.

18. Let (a_n) be a sequence of real numbers. Consider the sequence (b_n) defined by
$$b_n = \begin{cases} a_{\frac{n+1}{2}}, & \text{if } n \text{ is odd} \\ -a_{\frac{n}{2}}, & \text{if } n \text{ is even.} \end{cases}$$

Show that:

a) The series $\sum b_n$ is convergent if and only if $\lim a_n = 0$.
b) The series $\sum b_n$ is absolutely convergent if and only if the series $\sum a_n$ is absolutely convergent.

19. Show that if (a_n) is a decreasing sequence and $\sum a_n$ converges, then $\lim(n\, a_n) = 0$.

Hint: Relate the convergence of the series to the fact that the sequence of partial sums is a Cauchy sequence (choose m and n conveniently.)

20. Let $\sum a_n$ be a convergent series of positive terms. Show that the series $\sum \dfrac{\sqrt{a_n}}{n^p}$ converges if $p > \dfrac{1}{2}$.

Hint: Consider the inequality $ab \leq \dfrac{a^2 + b^2}{2}$, $\forall a, b \in \mathbb{R}$.

21. Let a_n be a sequence of positive terms such that $\dfrac{a_{n+1}}{a_n} \geq 1 - \dfrac{1}{n}$, $\forall n \in \mathbb{N}$. Show that the series $\sum a_n$ is divergent.

22. Let $g: \mathbb{R} \to \mathbb{R}$ be a function of class C^2. Suppose there exists $p \in \mathbb{N}$ such that g, g', $(g')^2 - g\, g''$ are all positive functions on $[p, +\infty[$. Show that the series
$$\sum_{n=p}^{\infty} \dfrac{g'(n)}{g(n)}$$
is convergent if and only if $\lim_{x \to +\infty} g(x)$ exists and is finite.

23. Test for convergence or divergence the series
$$\sum_{n=2}^{\infty} \arctan(v_n)$$
where $v_2 = k > 0$ and $v_{n+1} = v_n \sin\left(\dfrac{\pi}{n}\right)$, $n \geq 2$.

24. The Sierpinski triangle is a geometric figure constructed as follows: A square is divided into four equal squares, and the upper right square is removed. Each of the remaining three squares is divided in the same way, and the upper right square of each is removed.

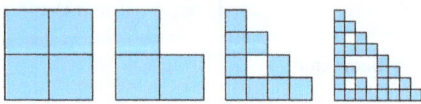

Continuing this process to infinity, we obtain a figure called the Sierpinski triangle.

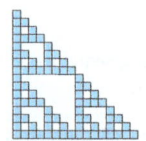  

Suppose the length of the side of the initial square is equal to 1. What is the total area of the squares removed from the initial square? What can we conclude about the area of the Sierpinski triangle?

25. Consider an equilateral triangle with side length 1. Construct a sequence of circles approaching the vertices of the triangle such that each circle is tangent to the sides of the triangle and to the circles that precede and follow it in the sequence. Find the sum of the areas of the circles.

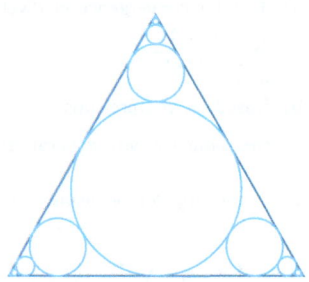

26. Consider a square inscribed in a circle of radius 1. In this square, inscribe a circle and in it a new square, and so on, as suggested by the figure. Find the area of the colored region.

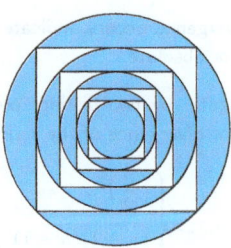

2.7. Solved Exercises

SOLUTIONS

1. a) Let a_n be the general term of the series $\sum_{n=1}^{\infty} \dfrac{1}{n^2 + 3n + 2}$. Since $n^2 + 3n + 2 = (n+1)(n+2)$, we get

$$a_n = \frac{1}{n^2 + 3n + 2} = \frac{A}{n+1} + \frac{B}{n+2}$$

$$= \frac{A(n+2) + B(n+1)}{(n+1)(n+2)}$$

$$= \frac{(A+B)n + 2A + B}{(n+1)(n+2)},$$

from which

$$\begin{cases} A + B = 0 \\ 2A + B = 1 \end{cases} \Leftrightarrow \begin{cases} A = 1 \\ B = -1. \end{cases}$$

Therefore,

$$a_n = \frac{1}{n+1} - \frac{1}{n+2}.$$

The series is telescopic with $\alpha_n = \dfrac{1}{n+1}$ and $k = 1$, using the notation of formula (2.1) on page 74. In this case, $\lim \alpha_n = 0$, so we can conclude that the series converges and, using formula (2.2) on the page mentioned above, its sum is

$$\sum_{n=1}^{\infty} \left(\frac{1}{n+1} - \frac{1}{n+2} \right) = \alpha_1 = \frac{1}{2}.$$

Alternatively, we can write the sequence of partial sums and determine its limit as follows:

$$S_1 = \frac{1}{2} - \frac{1}{3}$$

$$S_2 = \frac{1}{2} - \frac{1}{3} + \frac{1}{3} - \frac{1}{4} = \frac{1}{2} - \frac{1}{4}$$

$$S_3 = \frac{1}{2} - \frac{1}{4} + \frac{1}{4} - \frac{1}{5} = \frac{1}{2} - \frac{1}{5}$$

$$\vdots$$

$$S_n = \frac{1}{2} - \frac{1}{n+2}$$

$$\vdots$$

As $\lim S_n = \dfrac{1}{2}$, the series is convergent and its sum is $\dfrac{1}{2}$:

$$\sum_{n=1}^{\infty} \frac{1}{n^2 + 3n + 2} = \frac{1}{2}.$$

b) Let us consider the series $\sum_{n=1}^{\infty} \left(\arctan(n+3) - \arctan(n) \right) = -\sum_{n=1}^{+\infty} \left(\arctan(n) - \arctan(n+3) \right)$. It is telescopic with $\alpha_n = \arctan(n)$ and $k = 3$, using the notation of formula (2.1) on page 74. In this case, $\lim \alpha_n = \frac{\pi}{2}$, therefore, we can conclude that the series converges, and, using the formula (2.2), its sum is

$$-\sum_{n=1}^{\infty} (\arctan(n) - \arctan(n+3)) = -\left(\alpha_1 + \alpha_2 + \alpha_3 - 3 \cdot \frac{\pi}{2} \right) = \frac{5\pi}{4} - \arctan(2) - \arctan(3).$$

An alternative path to solve this problem is to write the sequence of partial sums and find its limit:

$$S_1 = \arctan(4) - \arctan(1) = \arctan(4) - \frac{\pi}{4}$$
$$S_2 = \arctan(4) - \frac{\pi}{4} + \arctan(5) - \arctan(2)$$
$$S_3 = \arctan(4) - \frac{\pi}{4} + \arctan(5) - \arctan(2) + \arctan(6) - \arctan(3)$$
$$S_4 = \arctan(4) - \frac{\pi}{4} + \arctan(5) - \arctan(2) + \arctan(6) - \arctan(3) + \arctan(7) - \arctan(4)$$
$$= -\frac{\pi}{4} + \arctan(5) - \arctan(2) + \arctan(6) - \arctan(3) + \arctan(7)$$
$$S_5 = -\frac{\pi}{4} + \arctan(5) - \arctan(2) + \arctan(6) - \arctan(3) + \arctan(7) + \arctan(8) - \arctan(5)$$
$$= -\frac{\pi}{4} - \arctan(2) + \arctan(6) - \arctan(3) + \arctan(7) + \arctan(8)$$
$$= -\frac{\pi}{4} - \arctan(2) - \arctan(3) + \arctan(6) + \arctan(7) + \arctan(8)$$
$$\vdots$$
$$S_n = -\frac{\pi}{4} - \arctan(2) - \arctan(3) + \arctan(n+1) + \arctan(n+2) + \arctan(n+3)$$
$$\vdots$$

As $\lim S_n = -\frac{\pi}{4} - \arctan(2) - \arctan(3) + \frac{3\pi}{2} = \frac{5\pi}{4} - \arctan(2) - \arctan(3)$, the series is convergent and

$$\sum_{n=1}^{+\infty} \left(\arctan(n+3) - \arctan(n) \right) = \frac{5\pi}{4} - \arctan(2) - \arctan(3).$$

c) The series $\sum_{n=0}^{\infty} \left(\frac{1}{e^n} - \frac{1}{e^{n+1}} \right)$ is telescopic, with $\alpha_n = \frac{1}{e^n}$ and $k = 1$, using the notation of formula (2.1) on page 74. As $\lim \frac{1}{e^n} = 0$, the series is convergent. The sum of the series is

$$\sum_{n=0}^{\infty} \left(\frac{1}{e^n} - \frac{1}{e^{n+1}} \right) = \frac{1}{e^0} - \lim \frac{1}{e^n} = 1.$$

Another process to find the sum is to write the general term in the form $\frac{1}{e^n}\left(1 - \frac{1}{e}\right)$, highlighting the fact that this series is geometric with ratio $r = \frac{1}{e}$. As $0 < \frac{1}{e} < 1$, the series converges and:

$$\sum_{n=0}^{\infty} \left(\frac{1}{e^n} - \frac{1}{e^{n+1}} \right) = \sum_{n=0}^{\infty} \frac{1}{e^n}\left(1 - \frac{1}{e}\right) = \left(1 - \frac{1}{e}\right) \sum_{n=0}^{\infty} \frac{1}{e^n} = \left(1 - \frac{1}{e}\right) \frac{1}{1 - \frac{1}{e}} = 1.$$

2.7. Solved Exercises

d) The series $\sum_{n=1}^{\infty}\left(\dfrac{1}{\arctan(n)} - \dfrac{1}{\arctan(n+1)}\right)$ is telescopic, with $a_n = \dfrac{1}{\arctan(n)}$ and $k = 1$, using the notation of formula (2.1) on page 74. In this case, $\lim a_n = \dfrac{2}{\pi}$. Therefore, we can conclude that the series converges and, using formula (2.2), has sum

$$\sum_{n=1}^{\infty}\left(\dfrac{1}{\arctan(n)} - \dfrac{1}{\arctan(n+1)}\right) = \dfrac{1}{\arctan(1)} - \dfrac{2}{\pi} = \dfrac{4}{\pi} - \dfrac{2}{\pi} = \dfrac{2}{\pi}.$$

Alternatively, we can write the sequence of partial sums and find its limit:

$$S_1 = \dfrac{1}{\arctan(1)} - \dfrac{1}{\arctan(2)} = \dfrac{4}{\pi} - \dfrac{1}{\arctan(2)}$$

$$S_2 = \dfrac{4}{\pi} - \dfrac{1}{\arctan(2)} + \dfrac{1}{\arctan(2)} - \dfrac{1}{\arctan(3)} = \dfrac{4}{\pi} - \dfrac{1}{\arctan(3)}$$

$$S_3 = \dfrac{4}{\pi} - \dfrac{1}{\arctan(3)} + \dfrac{1}{\arctan(3)} - \dfrac{1}{\arctan(4)} = \dfrac{4}{\pi} - \dfrac{1}{\arctan(4)}$$

$$\vdots$$

$$S_n = \dfrac{4}{\pi} - \dfrac{1}{\arctan(n+1)}$$

$$\vdots$$

As

$$\lim S_n = \lim\left(\dfrac{4}{\pi} - \dfrac{1}{\arctan(n+1)}\right) = \dfrac{4}{\pi} - \dfrac{2}{\pi} = \dfrac{2}{\pi},$$

the series is convergent, and its sum is

$$\sum_{n=1}^{\infty}\left(\dfrac{1}{\arctan(n)} - \dfrac{1}{\arctan(n+1)}\right) = \dfrac{2}{\pi}.$$

e) The following two series $\sum_{n=1}^{\infty} \dfrac{1}{2^{2n-1}}$ and $\sum_{n=1}^{\infty} \dfrac{1}{2^{3n-2}}$ are geometric series. To see this, observe that:

$$\sum_{n=1}^{\infty} \dfrac{1}{2^{2n-1}} = \sum_{n=1}^{\infty} \dfrac{1}{2^{2n} \cdot 2^{-1}} = 2\sum_{n=1}^{\infty} \dfrac{1}{2^{2n}} = 2\sum_{n=1}^{\infty} \dfrac{1}{4^n}$$

and

$$\sum_{n=1}^{\infty} \dfrac{1}{2^{3n-2}} = \sum_{n=1}^{\infty} \dfrac{1}{2^{3n} \cdot 2^{-2}} = 4\sum_{n=1}^{\infty} \dfrac{1}{2^{3n}} = 4\sum_{n=1}^{\infty} \dfrac{1}{8^n}.$$

The first series has ratio $r = \dfrac{1}{4}$, and the second has ratio $r = \dfrac{1}{8}$. The series are convergent as in both cases $|r| < 1$. The sum of the series can be calculated as follows:

$$\sum_{n=1}^{\infty} \dfrac{1}{2^{2n-1}} = 2 \cdot \dfrac{\frac{1}{4}}{1 - \frac{1}{4}} = \dfrac{2}{3} \quad \text{and} \quad \sum_{n=1}^{\infty} \dfrac{1}{2^{3n-2}} = 4 \cdot \dfrac{\frac{1}{8}}{1 - \frac{1}{8}} = \dfrac{4}{7}.$$

By Theorem 2.2.2 the series $\sum_{n=1}^{\infty}\left(\dfrac{1}{2^{2n-1}} - \dfrac{1}{2^{3n-2}}\right)$ is convergent, and its sum is equal to $\dfrac{2}{3} - \dfrac{4}{7} = \dfrac{2}{21}$.

f) The series $\sum_{n=0}^{\infty} \frac{3^n + 7^n}{3^n \cdot 7^n}$ can be written in the form $\sum_{n=0}^{\infty} \left(\frac{1}{7^n} + \frac{1}{3^n}\right)$. Both $\sum_{n=0}^{\infty} \frac{1}{7^n}$ and $\sum_{n=0}^{\infty} \frac{1}{3^n}$ are geometric series with ratios $r = \frac{1}{7}$ and $r = \frac{1}{3}$, respectively.

The series are convergent as in both cases $|r| < 1$. Using the formula for the sum of a convergent geometric series, we get

$$\sum_{n=0}^{\infty} \frac{1}{7^n} = \frac{1}{1 - \frac{1}{7}} = \frac{7}{6} \quad \text{and} \quad \sum_{n=0}^{\infty} \frac{1}{3^n} = \frac{1}{1 - \frac{1}{3}} = \frac{3}{2}.$$

By applying Theorem 2.2.2, we can conclude that the given series $\sum_{n=0}^{\infty} \frac{3^n + 7^n}{3^n \cdot 7^n}$ is convergent, and its sum is equal to $\frac{7}{6} + \frac{3}{2} = \frac{8}{3}$.

g) The series $\sum_{n=0}^{\infty} \frac{1}{4^{n-1}} \cdot \frac{2}{5^n} \cdot \frac{3}{6^{n+1}}$ is geometric with ratio $r = \frac{1}{120}$ because

$$\frac{1}{4^{n-1}} \cdot \frac{2}{5^n} \cdot \frac{3}{6^{n+1}} = \frac{6 \cdot 4}{4^n \cdot 5^n \cdot 6^n \cdot 6} = \frac{4}{120^n}.$$

Thus, we can calculate the sum as follows:

$$\sum_{n=0}^{\infty} \frac{1}{4^{n-1}} \cdot \frac{2}{5^n} \cdot \frac{3}{6^{n+1}} = 4 \sum_{n=0}^{\infty} \frac{1}{120^n} = 4 \cdot \frac{1}{1 - \frac{1}{120}} = \frac{480}{119}.$$

h) Using the properties of logarithms, we can obtain the following equality:

$$\sum_{n=2}^{\infty} \frac{\log(\frac{n+2}{n})}{\log(n) \log(n+2)} = \sum_{n=2}^{\infty} \left(\frac{1}{\log(n)} - \frac{1}{\log(n+2)}\right).$$

Thus the series is telescopic, with $\alpha_n = \frac{1}{\log(n)}$ and $k = 2$, using the notation of formula (2.1) on page 74. The series is convergent because $\lim \frac{1}{\log(n)} = 0$, and

$$\sum_{n=2}^{\infty} \left(\frac{1}{\log(n)} - \frac{1}{\log(n+2)}\right) = \frac{1}{\log(2)} + \frac{1}{\log(3)}.$$

2. By reindexing the series, we can write $\sum_{n=2}^{\infty} \frac{1}{(n+1)^2} = \sum_{n=3}^{\infty} \frac{1}{n^2}$. Additionally,

$$\sum_{n=1}^{\infty} \frac{1}{n^2} = 1 + \frac{1}{4} + \sum_{n=3}^{\infty} \frac{1}{n^2}.$$

Knowing that $\sum_{n=1}^{\infty} \frac{1}{n^2} = \frac{\pi^2}{6}$, then $\sum_{n=3}^{\infty} \frac{1}{n^2} = \frac{\pi^2}{6} - 1 - \frac{1}{4} = \frac{\pi^2}{6} - \frac{5}{4}$, that is,

$$\sum_{n=2}^{\infty} \frac{1}{(n+1)^2} = \frac{\pi^2}{6} - \frac{5}{4}.$$

2.7. Solved Exercises

The series $\sum_{n=2}^{\infty} \frac{1}{\pi^n}$ is convergent because it is geometric with ratio $\frac{1}{\pi}$. By Theorem 2.2.2, the sum of these two series is convergent and

$$\sum_{n=2}^{\infty} \left(\frac{1}{(n+1)^2} + \frac{1}{\pi^n} \right) = \sum_{n=2}^{\infty} \frac{1}{(n+1)^2} + \sum_{n=2}^{\infty} \frac{1}{\pi^n} = \frac{\pi^2}{6} - \frac{5}{4} + \frac{\frac{1}{\pi^2}}{1 - \frac{1}{\pi}} = \frac{\pi^2}{6} - \frac{5}{4} + \frac{1}{\pi(\pi-1)}:$$

3. a) The rational number $0,\overline{5} = 0,55555\ldots$ can be written in the form of a fraction using geometric series. By writing

$$0,\overline{5} = 0,5 + 0,05 + \ldots = \frac{5}{10} + \frac{5}{10^2} + \frac{5}{10^3} + \frac{5}{10^4} + \cdots$$

it is clear that we have a geometric series with ratio $r = \frac{1}{10}$ and first term $\frac{5}{10}$. Therefore,

$$0,\overline{5} = \frac{\frac{5}{10}}{1 - \frac{1}{10}} = \frac{5}{9}.$$

b) As in the previous item, we can use geometric series to write the rational number $0,\overline{34} = 0,343434\ldots$ in the form of a fraction. By writing

$$0,\overline{34} = 0,34 + 0,0034 + \ldots = \frac{34}{10^2} + \frac{34}{10^4} + \frac{34}{10^6} + \frac{34}{10^8} + \cdots$$

it is evident that we have a geometric series with ratio $r = \frac{1}{10^2}$ and first term $\frac{34}{10^2}$. Therefore,

$$0,\overline{34} = \frac{\frac{34}{10^2}}{1 - \frac{1}{10^2}} = \frac{34}{99}.$$

c) We can write the rational number $1,\overline{345}$ in the following form:

$$1,\overline{345} = 1 + 0,345 + 0,000345 + \ldots = 1 + \frac{345}{10^3} + \frac{345}{10^6} + \frac{345}{10^9} + \frac{345}{10^{12}} + \cdots$$

After the first term, we have a geometric series with ratio $r = \frac{1}{10^3}$ and first term $\frac{345}{10^3}$. Therefore,

$$1,\overline{345} = 1 + \frac{\frac{345}{10^3}}{1 - \frac{1}{10^3}} = 1 + \frac{345}{999} = \frac{448}{333}.$$

d) By writing the rational number $0,324\overline{101}$ in the form

$$0,324\overline{101} = 0,324 + 0,000101 + 0,000000101 + \ldots = \frac{324}{10^3} + \frac{101}{10^6} + \frac{101}{10^9} + \frac{101}{10^{12}} + \cdots$$

we see that after the first term, we have a geometric series with ratio $r = \frac{1}{10^3}$ and first term $\frac{101}{10^6}$. Therefore,

$$0,324\overline{101} = \frac{324}{10^3} + \frac{\frac{101}{10^6}}{1 - \frac{1}{10^3}} = \frac{324}{10^3} + \frac{101}{999 \times 10^3} = \frac{323777}{999000}.$$

4. a) The series $\sum_{n=1}^{\infty} \dfrac{2n^2-1}{3n^5+2n+1}$ is of positive terms. To study this series, we will compare it with the series $\sum_{n=1}^{\infty} \dfrac{1}{n^3}$, which is convergent because it is a p-series with $p=3$:

$$\lim \frac{\frac{2n^2-1}{3n^5+2n+1}}{\frac{1}{n^3}} = \lim \frac{(2n^2-1)n^3}{3n^5+2n+1} = \lim \frac{2n^5-n^3}{3n^5+2n+1} = \frac{2}{3}.$$

Since $\dfrac{2}{3} \in \mathbb{R}^+$, according to Corollary 2 of the General Comparison Test, we can conclude that the given series $\sum_{n=1}^{\infty} \dfrac{2n^2-1}{3n^5+2n+1}$ is convergent.

b) The series $\sum_{n=1}^{\infty} \left(\sin\left(\dfrac{1}{n}\right)\right)^2$ is of positive terms because $\dfrac{1}{n} \in \,]0,1] \subset \,]0, \dfrac{\pi}{2}[$, $\forall n \in \mathbb{N}$. To determine its convergence, we compare it with the series $\sum_{n=1}^{\infty} \dfrac{1}{n^2}$, which we know is convergent as it is a p-series with $p=2$. We calculate the limit

$$\lim \frac{\left(\sin\left(\frac{1}{n}\right)\right)^2}{\frac{1}{n^2}} = \lim \left[\frac{\sin\left(\frac{1}{n}\right)}{\frac{1}{n}}\right]^2 = 1,$$

which belongs to $\mathbb{R}^+$. Therefore, by Corollary 2 of the General Comparison Test, the given series is convergent.

c) The series $\sum_{n=1}^{\infty} \dfrac{\log(n^7+1)}{n^2}$ is a series of positive terms. We can analyze this series by comparing it with the series $\sum_{n=1}^{\infty} \dfrac{1}{n^{3/2}}$, which is convergent as it is a p-series with $p = \dfrac{3}{2}$:

$$\lim \frac{\frac{\log(n^7+1)}{n^2}}{\frac{1}{n^{3/2}}} = \lim \frac{n^{3/2}\log(n^7+1)}{n^2} = \lim \frac{\log(n^7+1)}{n^{1/2}} = 0.$$

Since the series $\sum_{n=1}^{\infty} \dfrac{1}{n^{3/2}}$ is convergent, the fact that this limit is zero allows us to conclude, by Corollary 3 of the General Comparison Test, that the series $\sum_{n=1}^{\infty} \dfrac{\log(n^7+1)}{n^2}$ is also convergent.

Note: For calculating the limit $\lim \dfrac{\log(n^7+1)}{n^{1/2}}$, we used L'Hôpital's Rule applied to the calculation of $\lim_{x \to +\infty} \dfrac{\log(x^7+1)}{x^{1/2}}$, where the indeterminate form of type $\dfrac{\infty}{\infty}$ arises:

2.7. Solved Exercises

$$\lim_{x \to +\infty} \frac{\left(\log(x^7+1)\right)'}{(x^{1/2})'} = \lim_{x \to +\infty} \frac{\frac{7x^6}{x^7+1}}{\frac{1}{2\sqrt{x}}} = \lim_{x \to +\infty} \frac{14x^6\sqrt{x}}{x^7+1} = 0.$$

d) Let us begin by rewriting the general term of the series:

$$a_n = \sqrt{n + \frac{1}{n}} - \sqrt{n} = \frac{\left(\sqrt{n+\frac{1}{n}} - \sqrt{n}\right)\left(\sqrt{n+\frac{1}{n}} + \sqrt{n}\right)}{\sqrt{n+\frac{1}{n}} + \sqrt{n}} = \frac{\frac{1}{n}}{\sqrt{n+\frac{1}{n}} + \sqrt{n}}$$

$$= \frac{1}{n\left(\sqrt{n+\frac{1}{n}} + \sqrt{n}\right)} = \frac{1}{n\left(\sqrt{\frac{n^2+1}{n}} + \sqrt{n}\right)} = \frac{1}{\sqrt{n^3+n} + \sqrt{n^3}}.$$

As $a_n > 0$, $\forall n \in \mathbb{N}$, the series is of positive terms. Let us compare it with the series $\sum_{n=1}^{\infty} \frac{1}{n^{3/2}}$, which is convergent because it is a p-series with $p = \frac{3}{2}$. The limit

$$\lim \frac{\frac{1}{\sqrt{n^3+n}+\sqrt{n^3}}}{\frac{1}{n^{3/2}}} = \lim \frac{\sqrt{n^3}}{\sqrt{n^3+n}+\sqrt{n^3}} = \frac{1}{2}$$

is finite and different from zero; therefore, by Corollary 2 of the General Comparison Test, the series $\sum_{n=1}^{\infty} \left(\sqrt{n+\frac{1}{n}} - \sqrt{n}\right)$ is convergent.

e) The series $\sum_{n=1}^{\infty} \frac{\sqrt{n+1}-\sqrt{n}}{\sqrt{n+1}+\sqrt{n}}$ is of positive terms. Considering that

$$\frac{\sqrt{n+1}-\sqrt{n}}{\sqrt{n+1}+\sqrt{n}} = \frac{(\sqrt{n+1}-\sqrt{n})(\sqrt{n+1}+\sqrt{n})}{(\sqrt{n+1}+\sqrt{n})^2} = \frac{1}{(\sqrt{n+1}+\sqrt{n})^2},$$

we can write the series in the form $\sum_{n=1}^{\infty} \frac{1}{(\sqrt{n+1}+\sqrt{n})^2}$. We will analyze this series by comparing it with the harmonic series:

$$\lim \frac{\frac{1}{(\sqrt{n+1}+\sqrt{n})^2}}{\frac{1}{n}} = \lim \frac{n}{(\sqrt{n+1}+\sqrt{n})^2} = \frac{1}{4}.$$

Since $\frac{1}{4} \in \mathbb{R}^+$, we can conclude, by Corollary 2 of the General Comparison Test, that the given series $\sum_{n=1}^{\infty} \frac{\sqrt{n+1}-\sqrt{n}}{\sqrt{n+1}+\sqrt{n}}$ is divergent.

f) The series $\sum_{n=1}^{\infty} \dfrac{\cos(n)}{n^2+4}$ is not of positive terms. However, by bounding the modulus of the general term, we can obtain the following inequality:

$$0 < \left|\dfrac{\cos(n)}{n^2+4}\right| \leq \dfrac{1}{n^2+4} \leq \dfrac{1}{n^2}, \quad \forall n \in \mathbb{N}. \tag{2.5}$$

It is worth noting that the series $\sum_{n=1}^{\infty} \dfrac{1}{n^2}$ is convergent since it is a p-series with $p=2$. Using the inequality (2.5), we can apply the General Comparison Test to show that the series $\sum_{n=1}^{\infty} \left|\dfrac{\cos(n)}{n^2+4}\right|$ converges. Therefore, the original series $\sum_{n=1}^{\infty} \dfrac{\cos(n)}{n^2+4}$ is absolutely convergent.

g) Recall that $-1 \leq \cos(n) \leq 1$, $\forall n \in \mathbb{N}$. Using this fact, we obtain $n^3 + 10\cos(n) \geq n^3 - 10$, $\forall n \in \mathbb{N}$. Thus, we have that

$$n^3 + 10\cos(n) \geq n^3 - 10 > 0, \quad \forall n \geq 3.$$

This inequality implies that

$$0 < \dfrac{1}{n^3 + 10\cos(n)} \leq \dfrac{1}{n^3 - 10}, \quad \forall n \geq 3. \tag{2.6}$$

The series $\sum_{n=3}^{\infty} \dfrac{1}{n^3-10}$ is of positive terms, and we can study its convergence by comparing it with the series $\sum_{n=3}^{\infty} \dfrac{1}{n^3}$, which we know to be convergent. Since

$$\lim \dfrac{\dfrac{1}{n^3-10}}{\dfrac{1}{n^3}} = \lim \dfrac{n^3}{n^3-10} = 1$$

is a positive real number, the series $\sum_{n=3}^{\infty} \dfrac{1}{n^3-10}$ is also convergent. Using inequality (2.6) and the General Comparison Test, we can conclude that the series $\sum_{n=3}^{\infty} \dfrac{1}{n^3+10\cos(n)}$ is convergent. This series differs from the given series only in a finite number of terms; therefore, the series $\sum_{n=1}^{\infty} \dfrac{1}{n^3+10\cos(n)}$ converges.

h) Consider the series $\sum_{n=1}^{\infty} \dfrac{1-(-1)^n}{\sqrt{n}}$, where a_n is its general term. We can express a_n as follows:

$$a_n = \begin{cases} 0, & \text{if } n \text{ is even} \\ \dfrac{2}{\sqrt{n}}, & \text{if } n \text{ is odd.} \end{cases}$$

2.7. Solved Exercises

Hence, $\sum_{n=1}^{\infty} \frac{1-(-1)^n}{\sqrt{n}} = \sum_{n=1}^{\infty} \frac{2}{\sqrt{2n-1}}$. This is a series of positive terms. Let us now consider the series $\sum_{n=1}^{\infty} \frac{1}{\sqrt{n}}$, which is divergent because it is a p-series with $p = \frac{1}{2}$. The limit

$$\lim \frac{\frac{2}{\sqrt{2n-1}}}{\frac{1}{\sqrt{n}}} = \lim \frac{2\sqrt{n}}{\sqrt{2n-1}} = \lim \sqrt{\frac{4n}{2n-1}} = \sqrt{2}$$

belongs to $\mathbb{R}^+$; therefore, by Corollary 2 of the General Comparison Test, the given series is divergent.

5. a) The series $\sum_{n=1}^{\infty} (-1)^n \frac{\log(n)}{n} = \sum_{n=2}^{\infty} (-1)^n \frac{\log(n)}{n}$ is alternating. To begin, let us study the series of modules $\sum_{n=2}^{\infty} \frac{\log(n)}{n}$, comparing it with the harmonic series. We can see that

$$\lim \frac{\frac{\log(n)}{n}}{\frac{1}{n}} = \lim \log(n) = +\infty.$$

This result allows us to conclude, by Corollary 3 of the General Comparison Test, that the series of modules is divergent. Since the initial series is alternating, let us check if we can apply Leibniz's test:

(i) $\lim \frac{\log(n)}{n} = 0$ (notice that $\lim_{x \to +\infty} \frac{\log(x)}{x}$ is an indeterminate form of type $\frac{\infty}{\infty}$ that can be solved using L'Hôpital's Rule: $\lim_{x \to +\infty} \frac{(\log(x))'}{(x)'} = \lim_{x \to +\infty} \frac{1}{x} = 0$).

(ii) $\frac{\log(n)}{n} > 0$, $\forall n > 1$.

(iii) The function $f(x) = \frac{\log(x)}{x}$ is decreasing on $]e, +\infty[$. Indeed, $f'(x) = \frac{1 - \log(x)}{x^2} < 0 \Leftrightarrow x > e$, which implies that $\left(\frac{\log(n)}{n}\right)$ is decreasing for $n \geq 3$.

Therefore, we can state that the alternating series is convergent. As the series of modules is divergent, the given series is conditionally convergent, taking into account Definition 2.4.1.

Note: The series $\sum_{n=2}^{\infty} \frac{\log(n)}{n}$ and $\sum_{n=3}^{\infty} \frac{\log(n)}{n}$ are both divergent since they differ only in a finite number of terms.

b) The series $\sum_{n=1}^{\infty} (-1)^{n+1} \frac{2}{e^n + e^{-n}}$ is alternating, given that $\frac{2}{e^n + e^{-n}} > 0$, $\forall n \in \mathbb{N}$. To begin with, let us study the series of modules:

$$\sum_{n=1}^{\infty} \left| (-1)^{n+1} \frac{2}{e^n + e^{-n}} \right| = \sum_{n=1}^{\infty} \frac{2}{e^n + e^{-n}}.$$

We have

$$0 < \frac{2}{e^n + e^{-n}} < \frac{2}{e^n}, \quad \forall n \in \mathbb{N}. \tag{2.7}$$

The series $\sum_{n=1}^{\infty} \frac{1}{e^n} = \sum_{n=1}^{\infty} \left(\frac{1}{e}\right)^n$ is geometric with ratio $r = \frac{1}{e}$. It is convergent because $|r| = \frac{1}{e} < 1$. By Theorem 2.2.2, the series $\sum_{n=1}^{\infty} \frac{2}{e^n}$ is also convergent. Using the General Comparison Test and considering inequality (2.7), the series $\sum_{n=1}^{\infty} \frac{2}{e^n + e^{-n}}$ is convergent. As this is the series of modules of the given series, by Definition 2.4.1, we can conclude that the initial series is absolutely convergent.

c) The series $\sum_{n=1}^{\infty} (-1)^n \frac{1}{\log(3n)}$ is alternating because $a_n = \frac{1}{\log(3n)} > 0, \forall n \in \mathbb{N}$. To study the convergence of the series, let us first consider the series of modules, $\sum_{n=1}^{\infty} \frac{1}{\log(3n)}$. We can compare this series with the harmonic series:

$$\lim \frac{\frac{1}{\log(3n)}}{\frac{1}{n}} = \lim \frac{n}{\log(3n)} = +\infty.$$

It follows by Corollary 4 of the General Comparison Test that the series $\sum_{n=1}^{\infty} \frac{1}{\log(3n)}$ is divergent. Since the initial series is alternating, we can check the conditions of Leibniz's test to see if it converges:

(i) $\lim \frac{1}{\log(3n)} = 0$.

(ii) $\frac{1}{\log(3n)} > 0, \forall n \geq 1$.

(iii) The sequence $\big(\log(3n)\big)$ is increasing; therefore, $\left(\frac{1}{\log(3n)}\right)$ is a decreasing sequence.

According to Leibniz's test, the alternating series is convergent. However, as the series of its modules is divergent, the given series is conditionally convergent, considering Definition 2.4.1.

d) The series $\sum_{n=1}^{\infty} (-1)^n \frac{\sin\left(\frac{1}{n}\right)}{n}$ is alternating, because $a_n = \frac{\sin\left(\frac{1}{n}\right)}{n} > 0, \forall n \geq 1$ (recall that $0 < \frac{1}{n} < \frac{\pi}{2}$, $\forall n \geq 1$). The series of modules is $\sum_{n=1}^{\infty} \frac{\sin\left(\frac{1}{n}\right)}{n}$. Consider the convergent series $\sum_{n=1}^{\infty} \frac{1}{n^2}$, which is a p-series with $p = 2$. The limit

$$\lim \frac{\frac{\sin\left(\frac{1}{n}\right)}{n}}{\frac{1}{n^2}} = \lim \frac{\sin\left(\frac{1}{n}\right)}{\frac{1}{n}} = 1$$

is finite and different from zero; therefore, by Corollary 2 of the General Comparison Test, the series of modules is convergent; thus, considering Definition 2.4.1, we can conclude the given series is absolutely convergent.

2.7. Solved Exercises

e) The series $\sum_{n=1}^{\infty}(-1)^n \dfrac{3^n}{e^{2n+1}}$ is alternating because $\dfrac{3^n}{e^{2n+1}} > 0$, $\forall n \in \mathbb{N}$. To analyze the absolute convergence of the series, we consider the series of modules:

$$\sum_{n=1}^{\infty}\left|(-1)^n\dfrac{3^n}{e^{2n+1}}\right| = \sum_{n=1}^{\infty}\dfrac{3^n}{e^{2n+1}} = \sum_{n=1}^{\infty}\dfrac{1}{e}\left(\dfrac{3}{e^2}\right)^n = \dfrac{1}{e}\sum_{n=1}^{\infty}\left(\dfrac{3}{e^2}\right)^n.$$

The series $\sum_{n=1}^{\infty}\left(\dfrac{3}{e^2}\right)^n$ is geometric with ratio $r = \dfrac{3}{e^2}$. Since $|r| < 1$, the series is convergent. We know by Theorem 2.2.2 that the product of a constant by a convergent series is convergent; therefore, the series of modules is convergent. Then, the given series is absolutely convergent, again considering Definition 2.4.1.

f) Considering that

$$\sin\left(\dfrac{n\pi}{2}\right) = \begin{cases} 0, & \text{if } n \text{ is even} \\ (-1)^{\frac{n-1}{2}}, & \text{if } n \text{ is odd}, \end{cases}$$

we can write the series $\sum_{n=1}^{\infty} \dfrac{\sin\left(\frac{n\pi}{2}\right)}{\sqrt{n+1}}$ as

$$\sum_{n=1}^{\infty}\dfrac{(-1)^{n-1}}{\sqrt{2n}} = \dfrac{1}{\sqrt{2}}\sum_{n=1}^{\infty}\dfrac{(-1)^{n-1}}{\sqrt{n}}.$$

This series is alternating, and its series of modules is $\sum_{n=1}^{\infty}\dfrac{1}{\sqrt{n}}$, which is divergent. We will now apply Leibniz's test to determine the convergence of our given series:

(i) $\lim \dfrac{1}{\sqrt{n}} = 0$.

(ii) $\dfrac{1}{\sqrt{n}} > 0$, $\forall n \in \mathbb{N}$.

(iii) $\dfrac{1}{\sqrt{n}} - \dfrac{1}{\sqrt{n+1}} > 0$, $\forall n \in \mathbb{N}$, and therefore, the sequence $\left(\dfrac{1}{\sqrt{n}}\right)$ is decreasing.

Thus, the alternating series is convergent. However, since the series of absolute values is divergent, we know the given series is conditionally convergent.

6. a) The series $\sum_{n=1}^{\infty}(-1)^{n+1}\dfrac{1}{n}$ is alternating. The series of modules $\sum_{n=1}^{\infty}\dfrac{1}{n}$ is the harmonic series, which is divergent.

To determine the convergence of the alternating series, let us apply Leibniz's test:

(i) $\lim \dfrac{1}{n} = 0$.

(ii) $\dfrac{1}{n} > 0$, $\forall n \in \mathbb{N}$.

(iii) $\dfrac{1}{n} - \dfrac{1}{n+1} = \dfrac{1}{n(n+1)} > 0$, $\forall n \in \mathbb{N}$, and therefore, the sequence is decreasing.

Thus, the alternating series is convergent. As its series of modules is divergent, the given series is conditionally convergent.

b) From the Corollary of Theorem 2.3.2, we know that $|S - S_n| \leq a_{n+1}$. It is sufficient to require that $a_{n+1} < \frac{1}{1000}$, that is, $n > 999$. Therefore, the intended sum is S_{1000}.

c) By the corollary referred to in the previous item, $|S - S_5| \leq a_6 = \frac{1}{6}$, and this value is an upper bound of the error.

7. a) The series $\sum_{n=1}^{\infty} \frac{n^n}{\pi^n \, n!}$ is of positive terms. Let a_n be the general term of the series. We can find the limit of the ratio of successive terms as follows:

$$\lim \frac{a_{n+1}}{a_n} = \lim \frac{\frac{(n+1)^{n+1}}{\pi^{n+1}(n+1)!}}{\frac{n^n}{\pi^n \, n!}} = \lim \frac{(n+1)^{n+1} \pi^n \, n!}{n^n \, \pi^{n+1} (n+1)!}$$

$$= \lim \frac{(n+1)(n+1)^n \, n!}{n^n \, \pi \, (n+1) \, n!} = \lim \frac{1}{\pi}\left(\frac{n+1}{n}\right)^n = \frac{e}{\pi}.$$

Since this value is less than 1, D'Alembert's test guarantees that the series is convergent.

b) The series $\sum_{n=0}^{\infty} \frac{n \, 2^n}{4n^3 + 1}$ is of positive terms. Let a_n be the general term of the series. We will take the limit of the ratio of successive terms of the series:

$$\lim \frac{a_{n+1}}{a_n} = \lim \frac{\frac{(n+1)2^{n+1}}{4(n+1)^3 + 1}}{\frac{n \, 2^n}{4n^3 + 1}} = \lim \frac{(n+1)\, 2^{n+1} \, (4n^3 + 1)}{n \, 2^n \, (4(n+1)^3 + 1)}$$

$$= \lim \frac{2(n+1)(4n^3 + 1)}{n(4(n+1)^3 + 1)} = \lim \frac{2(n+1)(4n^3 + 1)}{n(4n^3 + 12n^2 + 12n + 5)} = 2.$$

As this limit has a value greater than 1, the series is divergent by D'Alembert's test.

c) Noting that

$$\sum_{n=1}^{\infty} \frac{(n+1)! - n!}{n!\,(n+1)!} = \sum_{n=1}^{\infty} \frac{(n+1)n! - n!}{n!\,(n+1)!} = \sum_{n=1}^{\infty} \frac{n}{(n+1)!},$$

let a_n be the general term of the series. As

$$\lim \frac{a_{n+1}}{a_n} = \lim \frac{\frac{n+1}{(n+2)!}}{\frac{n}{(n+1)!}} = \lim \frac{(n+1)(n+1)!}{n(n+2)!} = \lim \frac{n+1}{n(n+2)} = 0 < 1,$$

it follows from D'Alembert's test that the series is convergent.

d) The series $\sum_{n=1}^{\infty} \frac{2 \cdot 4 \cdot 6 \cdots (2n+2)}{1 \cdot 4 \cdot 7 \cdots (3n+1)}$ is of positive terms. Let a_n be the general term of the series. We have

$$\lim \frac{a_{n+1}}{a_n} = \lim \frac{\frac{2 \cdot 4 \cdot 6 \cdots (2n+2)(2n+4)}{1 \cdot 4 \cdot 7 \cdots (3n+1)(3n+4)}}{\frac{2 \cdot 4 \cdot 6 \cdots (2n+2)}{1 \cdot 4 \cdot 7 \cdots (3n+1)}} = \lim \frac{2n+4}{3n+4} = \frac{2}{3}.$$

2.7. Solved Exercises

As this value is less than 1, we can conclude by D'Alembert's test that the series is convergent.

e) The series $\sum_{n=1}^{\infty} \dfrac{(n+1)!}{n^n \sqrt{3n+2}}$ is of positive terms. Let a_n be the general term of the series. Evaluating $\lim\limits_{n \to \infty} \dfrac{a_{n+1}}{a_n}$, we obtain

$$\lim \frac{a_{n+1}}{a_n} = \lim \frac{\dfrac{(n+2)!}{(n+1)^{n+1}\sqrt{3n+5}}}{\dfrac{(n+1)!}{n^n\sqrt{3n+2}}} = \lim \frac{(n+2)!\, n^n \sqrt{3n+2}}{(n+1)!\,(n+1)^{n+1}\sqrt{3n+5}}$$

$$= \lim \frac{(n+2)\, n^n \sqrt{3n+2}}{(n+1)^{n+1}\sqrt{3n+5}} = \lim \frac{n+2}{n+1}\sqrt{\frac{3n+2}{3n+5}}\left(\frac{n}{n+1}\right)^n = \frac{1}{e}.$$

Since this value is less than 1, the series is convergent according to D'Alembert's test.

f) The series $\sum_{n=1}^{\infty} \dfrac{3^n + n!}{n! + n^n}$ is of positive terms. It is easy to prove by induction that $3^n < n!$, $\forall n \geq 7$, which implies that

$$\frac{3^n + n!}{n! + n^n} \leq \frac{2\, n!}{n^n}, \quad \forall n \geq 7.$$

Let a_n be the general term of the series $\sum_{n=1}^{\infty} \dfrac{2\, n!}{n^n}$. D'Alembert's test confirms that the series converges since

$$\lim \frac{a_{n+1}}{a_n} = \lim \frac{\dfrac{2(n+1)!}{(n+1)^{n+1}}}{\dfrac{2\, n!}{n^n}} = \lim \frac{(n+1)!\, n^n}{n!\,(n+1)^{n+1}} = \lim \left(\frac{n}{n+1}\right)^n = \frac{1}{e} < 1.$$

By the General Comparison Test, we can conclude that the series $\sum_{n=1}^{\infty} \dfrac{3^n + n!}{n! + n^n}$ also converges.

Note: Two series that differ only in a finite number of terms are both convergent or both divergent.

g) The series $\sum_{n=1}^{\infty} \dfrac{n!}{(2\,n)! + 2^n}$ is of positive terms. We can establish the inequality

$$\frac{n!}{(2\,n)! + 2^n} \leq \frac{n!}{(2\,n)!}, \quad \forall n \in \mathbb{N}.$$

Let us investigate the convergence of the series $\sum_{n=1}^{\infty} \dfrac{n!}{(2\,n)!}$. If a_n denotes the general term of this series, then

$$\lim \frac{a_{n+1}}{a_n} = \lim \frac{\dfrac{(n+1)!}{(2(n+1))!}}{\dfrac{n!}{(2\,n)!}} = \lim \frac{(n+1)!\,(2n)!}{n!\,(2(n+1))!} = \lim \frac{n+1}{(2n+2)(2n+1)} = 0.$$

By D'Alembert's test, the series is convergent since this limit is less than 1. Applying the General Comparison Test, the original series $\sum_{n=1}^{\infty} \dfrac{n!}{(2\,n)! + 2^n}$ is convergent.

h) The series $\sum_{n=0}^{\infty} \dfrac{n!+3n}{((n+1)!)^2}$ is of positive terms. Let us consider the series $\sum_{n=0}^{\infty} a_n = \sum_{n=0}^{\infty} \dfrac{n!}{((n+1)!)^2}$ and $\sum_{n=0}^{\infty} b_n = \sum_{n=0}^{\infty} \dfrac{3n}{((n+1)!)^2}$. If both series are convergent, it follows from Theorem 2.2.2 that the given series is convergent. Let us begin by finding the limit of $\dfrac{a_{n+1}}{a_n}$:

$$\lim \dfrac{a_{n+1}}{a_n} = \lim \dfrac{\dfrac{(n+1)!}{((n+2)!)^2}}{\dfrac{n!}{((n+1)!)^2}} = \lim \dfrac{(n+1)!((n+1)!)^2}{n!((n+2)!)^2} = \lim \dfrac{n+1}{(n+2)^2} = 0 < 1.$$

As the limit is less than 1, we can use D'Alembert's test to conclude that the series $\sum_{n=0}^{\infty} a_n$ is convergent.

Similarly, the series $\sum_{n=0}^{\infty} b_n$ is convergent:

$$\lim \dfrac{b_{n+1}}{b_n} = \lim \dfrac{\dfrac{3(n+1)}{((n+2)!)^2}}{\dfrac{3n}{((n+1)!)^2}} = \lim \dfrac{(n+1)((n+1)!)^2}{n\,((n+2)!)^2} = \lim \dfrac{n+1}{n(n+2)^2} = 0 < 1.$$

Since both series are convergent, we can conclude that the given series is also convergent as it is the sum of two convergent series.

8. a) Consider the series of positive terms $\sum_{n=3}^{\infty} \left(1 - \dfrac{2}{n}\right)^{n^2}$ and let a_n represent its general term. Since

$$\lim \sqrt[n]{a_n} = \lim \sqrt[n]{\left(1 - \dfrac{2}{n}\right)^{n^2}} = \lim \left(1 - \dfrac{2}{n}\right)^n = e^{-2} = \dfrac{1}{e^2} < 1,$$

the series converges by Cauchy's Root Test.

b) The series $\sum_{n=0}^{\infty} e^{-n^2} = \sum_{n=0}^{\infty} \left(\dfrac{1}{e}\right)^{n^2}$ is of positive terms. We have

$$\lim \sqrt[n]{\left(\dfrac{1}{e}\right)^{n^2}} = \lim \left(\dfrac{1}{e}\right)^n = 0.$$

As this value is less than 1, Cauchy's Root Test ensures the convergence of the series.

c) The series $\sum_{n=1}^{\infty} \left(\sqrt[n]{2} - 1\right)^n$ is of positive terms. We find that:

$$\lim \sqrt[n]{\left(\sqrt[n]{2} - 1\right)^n} = \lim \left(\sqrt[n]{2} - 1\right) = 0,$$

since $\lim \sqrt[n]{2} = 1$. As this value is less than 1, Cauchy's Root Test allows us to conclude that the series is convergent.

2.7. Solved Exercises

d) The series $\sum_{n=1}^{\infty} \left(\cos\left(\frac{\pi}{6} + \frac{1}{n} \right) \right)^n$ is of positive terms because $0 < \frac{\pi}{6} + \frac{1}{n} < \frac{\pi}{2}$, $\forall n \in \mathbb{N}$. Applying Cauchy's Root Test, we conclude that the series is convergent. In fact:

$$\lim \sqrt[n]{\left(\cos\left(\frac{\pi}{6} + \frac{1}{n} \right) \right)^n} = \lim \cos\left(\frac{\pi}{6} + \frac{1}{n} \right) = \cos\left(\frac{\pi}{6} \right) = \frac{\sqrt{3}}{2} < 1.$$

e) The series $\sum_{n=0}^{\infty} \left(\frac{1}{2} + (-1)^n \frac{n}{4n+1} \right)^n$ is of positive terms. The sequence of general term

$$\sqrt[n]{\left(\frac{1}{2} + (-1)^n \frac{n}{4n+1} \right)^n} = \frac{1}{2} + (-1)^n \frac{n}{4n+1}$$

has two sublimits: $\frac{3}{4}$ and $\frac{1}{4}$. Since the upper limit is less than 1, we can apply Corollary 1 of the Root Test to conclude that the series is convergent.

f) We can write $\sum_{n=1}^{\infty} \left(\frac{n^2-2}{3n^2} \right)^n = -\frac{1}{3} + \sum_{n=2}^{\infty} \left(\frac{n^2-2}{3n^2} \right)^n$. The series $\sum_{n=2}^{\infty} \left(\frac{n^2-2}{3n^2} \right)^n$ is of positive terms and differs from the given series in only one term. We can state that the two series are either both convergent or both divergent. Using Cauchy's Root Test, we can show that the series converges because

$$\lim \sqrt[n]{\left(\frac{n^2-2}{3n^2} \right)^n} = \lim \frac{n^2-2}{3n^2} = \frac{1}{3} < 1.$$

9. a) The series $\sum_{n=2}^{\infty} \frac{(-1)^n}{\log(n^n)}$ is alternating because $a_n = \frac{1}{\log(n^n)} > 0$, $\forall n \geq 2$; its general term can be written in the form $\frac{(-1)^n}{n \log(n)}$. Let us start by studying the series of modules, $\sum_{n=2}^{\infty} \left| \frac{(-1)^n}{n \log(n)} \right| = \sum_{n=2}^{\infty} \frac{1}{n \log(n)}$, applying the Cauchy's Condensation Test. The sequence $\left(\frac{1}{n \log(n)} \right)_{n \in \mathbb{N} \setminus \{1\}}$ is obviously decreasing. The series

$$\sum_{n=2}^{\infty} \frac{2^n}{2^n \log(2^n)} = \sum_{n=2}^{\infty} \frac{1}{n \log(2)} = \frac{1}{\log(2)} \sum_{n=2}^{\infty} \frac{1}{n}$$

is divergent as it is the product of a constant by the harmonic series. Then, the series of modules is divergent by Cauchy's Condensation Test. As a result, if convergent, the series under study is conditionally convergent. Since the series is alternating and satisfies the following conditions:

(i) $\frac{1}{n \log(n)} > 0$

(ii) $\lim \frac{1}{n \log(n)} = 0$

(iii) As mentioned earlier, the sequence $\left(\frac{1}{n \log(n)} \right)_{n \in \mathbb{N} \setminus \{1\}}$ is decreasing

it follows from the Leibniz's test that the alternating series is convergent. From the study of the series of modules, we can state that the series is conditionally convergent, considering Definition 2.4.1.

b) Let us evaluate the limit of the general term of the series:

$$\lim \left(\frac{n^2}{\sqrt{n^5+1}} + \left(1+\frac{2}{n}\right)^n \right) = \lim \frac{n^2}{\sqrt{n^5+1}} + \lim \left(1+\frac{2}{n}\right)^n = \lim \sqrt{\frac{n^4}{n^5+1}} + e^2 = e^2.$$

As the general term of the series does not converge to zero, the series is divergent (refer to Theorem 2.2.1).

c) First, we note that $0 < 1 - \frac{1}{n^2} < 1$, $\forall n \geq 2$, which implies that $\log\left(1 - \frac{1}{n^2}\right) < 0$, $\forall n \geq 2$. Thus, the series is of negative terms. Next, we consider the series of modules

$$\sum_{n=2}^{\infty} \left| \log\left(1 - \frac{1}{n^2}\right) \right| = \sum_{n=2}^{\infty} \left(-\log\left(1 - \frac{1}{n^2}\right) \right).$$

We can compare this series with the series $\sum_{n=2}^{\infty} \frac{1}{n^2}$, which is convergent because it is a p-series with $p=2$. The limit

$$\lim \frac{-\log\left(1 - \frac{1}{n^2}\right)}{\frac{1}{n^2}} = -\lim n^2 \log\left(1 - \frac{1}{n^2}\right) = -\lim \log\left(1 - \frac{1}{n^2}\right)^{n^2} = -\log(e^{-1}) = 1$$

is finite and different from zero. Therefore, the series of modules is convergent by Corollary 2 of the General Comparison Test. Then, by Theorem 2.4.1, the given series is absolutely convergent, considering Definition 2.4.1.

d) The series $\sum_{n=1}^{\infty} \frac{\sin(n) + 2^n}{n + 5^n}$ is of positive terms because $-1 \leq \sin(n) \leq 1$ and $2^n > 1$, $\forall n \in \mathbb{N}$. The previous inequalities allow us to obtain the following estimate:

$$0 < \frac{\sin(n) + 2^n}{n + 5^n} < \frac{1 + 2^n}{5^n} = \frac{1}{5^n} + \frac{2^n}{5^n} = \left(\frac{1}{5}\right)^n + \left(\frac{2}{5}\right)^n, \quad \forall n \in \mathbb{N}.$$

The series $\sum_{n=1}^{\infty} \left(\frac{1}{5}\right)^n$ is convergent because it is geometric with ratio $r = \frac{1}{5}$, and the series $\sum_{n=1}^{\infty} \left(\frac{2}{5}\right)^n$ is convergent because it is geometric with ratio $r = \frac{2}{5}$. Then, the series

$$\sum_{n=1}^{\infty} \left(\left(\frac{1}{5}\right)^n + \left(\frac{2}{5}\right)^n \right)$$

is convergent by Theorem 2.2.2. By the General Comparison Test, the original series is convergent. As it is a series of positive terms, it is absolutely convergent.

e) The series $\sum_{n=1}^{\infty} (-1)^n \frac{e^{n+1}}{n^n}$ is alternating because $a_n = \frac{e^{n+1}}{n^n} > 0$, $\forall n \in \mathbb{N}$. To determine the convergence of the series, we first examine the series of modules, $\sum_{n=1}^{\infty} \frac{e^{n+1}}{n^n}$, using D'Alembert's test:

2.7. Solved Exercises

$$\lim \frac{a_{n+1}}{a_n} = \lim \frac{\frac{e^{n+2}}{(n+1)^{n+1}}}{\frac{e^{n+1}}{n^n}} = \lim \frac{e^{n+2}n^n}{e^{n+1}(n+1)^{n+1}} = \lim \frac{e}{n+1}\left(\frac{n}{n+1}\right)^n = 0 \times \frac{1}{e} = 0.$$

As this value is less than 1, the series of modules is convergent. Therefore, the given series is absolutely convergent by Theorem 2.4.1, considering Definition 2.4.1.

f) To study the convergence of the series $\sum_{n=1}^{\infty} \frac{(-1)^n + \frac{1}{n}}{n} = \sum_{n=2}^{\infty} \frac{(-1)^n + \frac{1}{n}}{n}$, we will start by rewriting its general term:

$$\frac{(-1)^n + \frac{1}{n}}{n} = (-1)^n \cdot \frac{n + (-1)^n}{n^2}.$$

Let $a_n = \frac{n + (-1)^n}{n^2}$. It is evident that $a_n > 0$, $\forall n \in \mathbb{N} \setminus \{1\}$, and therefore the series under study is alternating. Let us study the series of modules, $\sum_{n=1}^{\infty} \frac{n + (-1)^n}{n^2}$, by comparison with the harmonic series. The limit

$$\lim \frac{\frac{n + (-1)^n}{n^2}}{\frac{1}{n}} = \lim \frac{n(n + (-1)^n)}{n^2} = \lim \frac{n + (-1)^n}{n} = 1$$

belongs to $\mathbb{R}^+$. As the harmonic series is divergent, the series of modules is also divergent, by Corollary 2 of the General Comparison Test.

Let us consider the series $\sum_{n=1}^{\infty} \frac{(-1)^n}{n}$ and $\sum_{n=1}^{\infty} \frac{1}{n^2}$. Taking into account that $\frac{(-1)^n + \frac{1}{n}}{n} = \frac{(-1)^n}{n} + \frac{1}{n^2}$ and that the two previous series are convergent (the first one is the alternating harmonic series, and the second one is a p-series with $p = 2$), the series $\sum_{n=1}^{\infty} \frac{(-1)^n + \frac{1}{n}}{n}$ is convergent by Theorem 2.2.2. From the study of the series of modules, we can affirm that the series is conditionally convergent, according to Definition 2.4.1.

Note: The Leibniz's Test is not applicable because the sequence $\left(\frac{n + (-1)^n}{n^2}\right)$ is not monotonic (see Fig. 2.6).

g) Bearing in mind that $1 + \frac{1}{2^n} > 1$, $\forall n \geq 1$, we conclude that $\log\left(1 + \frac{1}{2^n}\right) > 0$, $\forall n \geq 1$. Thus, the series $\sum_{n=1}^{\infty} \log\left(1 + \frac{1}{2^n}\right)$ is of positive terms. Let us compare this series with the series $\sum_{n=1}^{\infty} \frac{1}{2^n}$, which is convergent as it is a geometric series with ratio $r = \frac{1}{2}$. The limit

$$\lim \frac{\log\left(1 + \frac{1}{2^n}\right)}{\frac{1}{2^n}} = \lim 2^n \log\left(1 + \frac{1}{2^n}\right) = \lim \log\left(1 + \frac{1}{2^n}\right)^{2^n} = \log(e) = 1$$

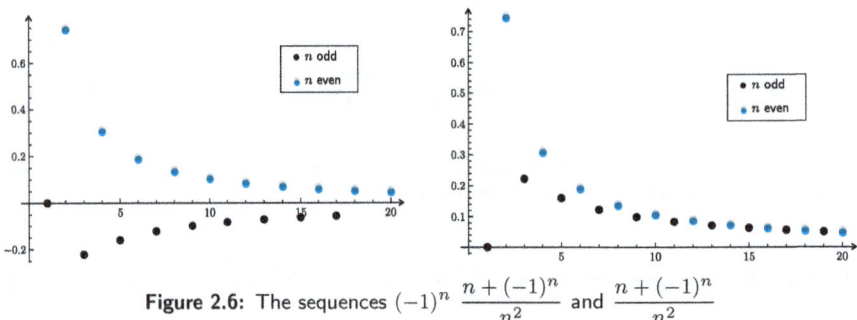

Figure 2.6: The sequences $(-1)^n \dfrac{n+(-1)^n}{n^2}$ and $\dfrac{n+(-1)^n}{n^2}$

is finite and different from zero; therefore, the series is convergent by Corollary 2 of the General Comparison Test. It is absolutely convergent since it is a series of positive terms.

h) The series $\sum_{n=1}^{\infty} \left(1 - \sqrt[n]{n}\right)^n = \sum_{n=1}^{\infty} (-1)^n \left(\sqrt[n]{n} - 1\right)^n$ is an alternating series. Let us begin by studying the series of modules using Cauchy's Root Test:

$$\lim \sqrt[n]{\left(\sqrt[n]{n} - 1\right)^n} = \lim \left(\sqrt[n]{n} - 1\right) = 0$$

since $\lim \sqrt[n]{n} = 1$. As this value is less than 1, the series of modules is convergent. It follows by Theorem 2.4.1 that the given series is convergent, and by Definition 2.4.1, it is absolutely convergent.

i) To determine the convergence of the series $\sum_{n=1}^{\infty} \dfrac{(-1)^n}{1 - (-1)^n \, n^2}$, we first rewrite its general term as

$$\frac{(-1)^n}{1 - (-1)^n \, n^2} = \frac{1}{(-1)^n - n^2} = \frac{-1}{n^2 + (-1)^{n+1}}.$$

We can simplify the series as follows:

$$\sum_{n=1}^{\infty} \frac{(-1)^n}{1 - (-1)^n \, n^2} = \sum_{n=1}^{\infty} \frac{-1}{n^2 + (-1)^{n+1}} = -\sum_{n=1}^{\infty} \frac{1}{n^2 + (-1)^{n+1}}.$$

Note that the general term of this last series is positive, whatever $n \in \mathbb{N}$. We can compare the series with the series $\sum_{n=1}^{\infty} \dfrac{1}{n^2}$, which we know is convergent because it is a p-series with $p = 2$. The limit

$$\lim \frac{\dfrac{1}{n^2 + (-1)^{n+1}}}{\dfrac{1}{n^2}} = \lim \frac{n^2}{n^2 + (-1)^{n+1}} = \lim \frac{1}{1 + \dfrac{(-1)^{n+1}}{n^2}} = 1$$

belongs to $\mathbb{R}^+$. Therefore, by Corollary 2 of the General Comparison Test, the series is convergent, and the convergence is absolute.

2.7. Solved Exercises

j) Consider the series $\sum_{n=1}^{\infty} a_n = \sum_{n=1}^{\infty} \frac{(-1)^n + 4}{n + 4^n}$ and $\sum_{n=1}^{\infty} b_n = \sum_{n=1}^{\infty} \left(\frac{3}{n+2}\right)^{2n}$. The series $\sum_{n=1}^{\infty} a_n$ is of positive terms; in fact, the expression for a_n is as follows:

$$a_n = \begin{cases} \dfrac{5}{n+4^n} & \text{if } n \text{ is even} \\ \dfrac{3}{n+4^n} & \text{if } n \text{ is odd}. \end{cases}$$

Consequently, $a_n \leq \frac{5}{4^n}$, $\forall n \in \mathbb{N}$. The series $\sum_{n=1}^{\infty} \frac{5}{4^n}$ is a convergent geometric series because the ratio is $r = \frac{1}{4}$. Since $a_n \leq \frac{5}{4^n}$, $\forall n \in \mathbb{N}$, the General Comparison Test implies that $\sum_{n=1}^{\infty} a_n$ is convergent.

We can use the Cauchy's Root Test to study the series of positive terms, $\sum_{n=1}^{\infty} b_n$:

$$\lim \sqrt[n]{\left(\frac{3}{n+2}\right)^{2n}} = \lim \left(\frac{3}{n+2}\right)^2 = 0 < 1.$$

This implies that the series $\sum_{n=1}^{\infty} b_n$ is convergent.

Since both $\sum_{n=1}^{\infty} a_n$ and $\sum_{n=1}^{\infty} b_n$ are convergent, we can conclude that the given series is convergent by Theorem 2.2.2. Since it is a series of positive terms, it is absolutely convergent.

k) The series $\sum_{n=1}^{\infty} \left(1 - \frac{1}{n}\right)^n$ has a general term that is not a null sequence. Specifically,

$$\lim \left(1 - \frac{1}{n}\right)^n = e^{-1}.$$

Theorem 2.2.1 can be applied to establish that this series is divergent.

l) We will use the Cauchy Condensation Test to determine the convergence of the series $\sum_{n=2}^{\infty} \frac{1}{n \left(\log(n)\right)^3}$. It is clear that the sequence $\left(\frac{1}{n \left(\log(n)\right)^3}\right)$ is decreasing with positive terms. We can apply the Cauchy Condensation Test by considering the series

$$\sum_{n=2}^{\infty} \frac{2^n}{2^n \left(\log(2^n)\right)^3} = \sum_{n=2}^{\infty} \frac{1}{n^3 \left(\log(2)\right)^3} = \frac{1}{\left(\log(2)\right)^3} \sum_{n=2}^{\infty} \frac{1}{n^3},$$

which is convergent as it is the product of a constant by a convergent p-series, $\sum_{n=2}^{\infty} \frac{1}{n^3}$. Therefore, by the Cauchy Condensation Test, the series $\sum_{n=2}^{\infty} \frac{1}{n \left(\log(n)\right)^3}$ is convergent. Furthermore, since it is a series of positive terms, it is absolutely convergent.

m) If $\dfrac{1}{x^2} > 0$ for all $x \neq 0$, we can say that the integral is positive, which implies that the series consists of positive terms. As

$$\int_n^{n+2} \frac{1}{x^2}\, dx = \left[-\frac{1}{x}\right]_n^{n+2} = \frac{1}{n} - \frac{1}{n+2},$$

we can write the series as

$$\sum_{n=1}^{\infty} \int_n^{n+2} \frac{1}{x^2}\, dx = \sum_{n=1}^{\infty} \left(\frac{1}{n} - \frac{1}{n+2}\right).$$

This series is telescopic with $\alpha_n = \dfrac{1}{n}$ and $k = 2$, using the notation of formula (2.1) on page 74. The telescopic series converges if (α_n) is a convergent sequence. As $\alpha_n \to 0$, the series $\sum_{n=1}^{\infty} \left(\dfrac{1}{n} - \dfrac{1}{n+2}\right)$ is convergent. Furthermore, it is absolutely convergent since it consists of positive terms.

n) The series $\sum_{n=1}^{\infty} \dfrac{(-1)^n}{\sqrt{n+1} + \sqrt[3]{n+1}}$ is alternating. The series of modules is the series $\sum_{n=1}^{\infty} \dfrac{1}{\sqrt{n+1} + \sqrt[3]{n+1}}$, which we can compare with the series $\sum_{n=1}^{\infty} \dfrac{1}{n^{1/2}}$, which is divergent because it is a p-series with $p = \dfrac{1}{2}$. As the value of the limit

$$\lim \frac{\dfrac{1}{\sqrt{n+1} + \sqrt[3]{n+1}}}{\dfrac{1}{n^{1/2}}} = \lim \frac{\sqrt{n}}{\sqrt{n+1} + \sqrt[3]{n+1}} = 1$$

belongs to $\mathbb{R}^+$, Corollary 2 of the General Comparison Test guarantees that the series of modules is divergent. Let us study the alternating series by applying the Leibniz's test:

(i) $\lim \dfrac{1}{\sqrt{n+1} + \sqrt[3]{n+1}} = 0$.

(ii) $\dfrac{1}{\sqrt{n+1} + \sqrt[3]{n+1}} > 0,\ \forall n \in \mathbb{N}$.

(iii) The sequence $\left(\dfrac{1}{\sqrt{n+1} + \sqrt[3]{n+1}}\right)$ is decreasing, because $\left(\sqrt{n+1} + \sqrt[3]{n+1}\right)$ is obviously increasing.

The test conditions are satisfied; therefore, the series is convergent. As the series of modules is divergent, we conclude that the alternating series is conditionally convergent.

o) The series $\sum_{n=1}^{\infty} \dfrac{1}{n(\sqrt{n+1} + \sqrt{n})}$ is of positive terms. Let us study this series by comparing it with the series $\sum_{n=1}^{\infty} \dfrac{1}{n^{3/2}}$, which we know to be convergent, as it is a p-series with $p = \dfrac{3}{2}$. The limit

$$\lim \frac{\dfrac{1}{n(\sqrt{n+1} + \sqrt{n})}}{\dfrac{1}{n^{3/2}}} = \lim \frac{n^{3/2}}{n(\sqrt{n+1} + \sqrt{n})} = \frac{1}{2}$$

belongs to $\mathbb{R}^+$. It follows from Corollary 2 of the General Comparison Test that the given series is convergent. It is absolutely convergent since it is a series of positive terms.

2.7. Solved Exercises

p) The series $\sum_{n=1}^{\infty} (-1)^n \left(1 - \cos\left(\frac{1}{n}\right)\right)$ is alternating, because $1 - \cos\left(\frac{1}{n}\right) > 0$, $\forall n \geq 1$. The series of modules is the series $\sum_{n=1}^{\infty} \left(1 - \cos\left(\frac{1}{n}\right)\right)$. As the value of the limit

$$\lim \frac{1 - \cos\left(\frac{1}{n}\right)}{\frac{1}{n^2}} = \frac{1}{2}$$

belongs to $\mathbb{R}^+$. As the series $\sum_{n=1}^{\infty} \frac{1}{n^2}$ is convergent because it is a p-series with $p = 2$, then by Corollary 2 of the General Comparison Test, the series of modules is convergent. Therefore, by Theorem 2.4.1, the given series is convergent, and it is absolutely convergent (see Definition 2.4.1).

q) The series $\sum_{n=1}^{\infty} (-1)^n \, 2^n \sin\left(\frac{1}{3^n}\right)$ is alternating, because $a_n = 2^n \sin\left(\frac{1}{3^n}\right) > 0$, $\forall n \geq 1$ (note that $0 < \frac{1}{3^n} < \frac{\pi}{2}$, $\forall n \geq 1$). The series of modules is $\sum_{n=1}^{\infty} 2^n \sin\left(\frac{1}{3^n}\right)$. Considering that if $x > 0$, then $\sin(x) \leq x$, we get

$$2^n \sin\left(\frac{1}{3^n}\right) \leq 2^n \left(\frac{1}{3}\right)^n = \left(\frac{2}{3}\right)^n, \quad \forall n \in \mathbb{N}.$$

The series $\sum_{n=1}^{\infty} \left(\frac{2}{3}\right)^n$ is geometric with ratio $r = \frac{2}{3}$; therefore, it is convergent. According to the General Comparison Test, the series of modules is convergent, so the given series is absolutely convergent.

r) The series $\sum_{n=1}^{\infty} (-1)^n \frac{n! \sqrt{n}}{n^n \sqrt{n+1}}$ is alternating because $\frac{n! \sqrt{n}}{n^n \sqrt{n+1}} > 0$, $\forall n \in \mathbb{N}$. Let us study the series of modules, $\sum_{n=1}^{\infty} \frac{n! \sqrt{n}}{n^n \sqrt{n+1}}$, using D'Alembert's test. Taking $a_n = \frac{n! \sqrt{n}}{n^n \sqrt{n+1}}$:

$$\lim \frac{a_{n+1}}{a_n} = \lim \frac{\frac{(n+1)! \sqrt{n+1}}{(n+1)^{n+1} \sqrt{n+2}}}{\frac{n! \sqrt{n}}{n^n \sqrt{n+1}}} = \lim \frac{(n+1)! \, n^n \, (n+1)}{n! \, (n+1)^{n+1} \sqrt{n(n+2)}}$$

$$= \lim \frac{(n+1)^2 \, n^n}{(n+1)^{n+1} \sqrt{n(n+2)}} = \lim \frac{n+1}{\sqrt{n(n+2)}} \left(\frac{n}{n+1}\right)^n = \frac{1}{e}.$$

Since this value is less than 1, the series is convergent. As it is the series of modules of the initial series, we can affirm that the series $\sum_{n=1}^{\infty} (-1)^n \frac{n! \sqrt{n}}{n^n \sqrt{n+1}}$ is absolutely convergent.

s) The series $\sum_{n=1}^{\infty} (\sqrt{n} + 1) \sin\left(\frac{1}{n^2}\right)$ is of positive terms because $0 < \frac{1}{n^2} < \frac{\pi}{2}$. Let us consider the series

$\sum_{n=1}^{\infty} \dfrac{1}{n^{3/2}}$, which we know to be convergent because it is a p-series with $p = \dfrac{3}{2}$. The limit

$$\lim \frac{(\sqrt{n}+1)\sin\left(\dfrac{1}{n^2}\right)}{\dfrac{1}{n^{3/2}}} = \lim \frac{\sin\left(\dfrac{1}{n^2}\right)}{\dfrac{1}{n^2}} \cdot \frac{\sqrt{n}+1}{\sqrt{n}} = 1$$

belongs to $\mathbb{R}^+$, and therefore, by Corollary 2 of the General Comparison Test, the series is convergent. As it is of positive terms, it is absolutely convergent.

t) The series $\sum_{n=2}^{\infty} \dfrac{1}{\log(n!)}$ is of positive terms. Knowing that $n! < n^n$ and that the logarithmic function is increasing, we have

$$0 < \frac{1}{\log(n^n)} = \frac{1}{n \log(n)} < \frac{1}{\log(n!)}, \quad \forall n \geq 2.$$

We proved in exercise 9a) that the series $\sum_{n=2}^{\infty} \dfrac{1}{n \log(n)}$ is divergent. By the General Comparison Test, the series $\sum_{n=2}^{\infty} \dfrac{1}{\log(n!)}$ is divergent.

10. a) The series $\sum_{n=0}^{\infty} \dfrac{n \cos(n\pi)}{\sqrt{n^3+1}} = \sum_{n=1}^{\infty} \dfrac{n \cos(n\pi)}{\sqrt{n^3+1}}$ is alternating because $\cos(n\pi) = (-1)^n$ and $\dfrac{n}{\sqrt{n^3+1}} > 0$, $\forall n \in \mathbb{N}$. First, we will examine the series of modules, $\sum_{n=1}^{\infty} \dfrac{n}{\sqrt{n^3+1}}$, by comparing it with the series $\sum_{n=1}^{\infty} \dfrac{1}{n^{1/2}}$, which is divergent because it is a p-series with $p = \dfrac{1}{2}$. The limit

$$\lim \frac{\dfrac{n}{\sqrt{n^3+1}}}{\dfrac{1}{n^{1/2}}} = \lim \frac{\sqrt{n^3}}{\sqrt{n^3+1}} = 1$$

is finite and different from zero. Therefore, by Corollary 2 of the General Comparison Test, the series $\sum_{n=1}^{\infty} \dfrac{n}{\sqrt{n^3+1}}$ is divergent.

Next, we will apply Leibniz's test to the alternating series:

(i) $\lim \dfrac{n}{\sqrt{n^3+1}} = 0$.

(ii) $\dfrac{n}{\sqrt{n^3+1}} > 0$, $\forall n \in \mathbb{N}$.

(iii) To verify that it is a decreasing sequence, consider the real function of real variable $f(x) = \dfrac{x}{\sqrt{x^3+1}}$. The derivative of f is $f'(x) = \dfrac{-x^3+2}{2(x^3+1)\sqrt{x^3+1}}$, and therefore, $f'(x) < 0$, $\forall x > \sqrt[3]{2}$. The sequence then decreases for $n \geq 2$.

Since the test conditions are satisfied, we can conclude that the alternating series is convergent. As the series of modules is divergent, the given series is said to be conditionally convergent.

2.7. Solved Exercises 143

b) The series $\sum_{n=1}^{\infty} \dfrac{n!}{3 \times 5 \times \cdots \times (2n+1)}$ is of positive terms. Let us study it using D'Alembert's test. Let a_n be the general term of the series:

$$\lim \dfrac{a_{n+1}}{a_n} = \lim \dfrac{\dfrac{(n+1)!}{3 \times 5 \times \cdots \times (2n+1)(2n+3)}}{\dfrac{n!}{3 \times 5 \times \cdots \times (2n+1)}} = \lim \dfrac{n+1}{2n+3} = \dfrac{1}{2}.$$

Since this limit is less than 1, it follows from D'Alembert's test that the series is convergent. The series is absolutely convergent, as it is of positive terms.

c) The series $\sum_{n=1}^{\infty} \dfrac{1 + \cos^2(n)}{\sqrt{n}}$ is of positive terms. We have the following inequality

$$\dfrac{1}{\sqrt{n}} \leq \dfrac{1 + \cos^2(n)}{\sqrt{n}}, \quad \forall n \in \mathbb{N}.$$

Since the series $\sum_{n=1}^{\infty} \dfrac{1}{\sqrt{n}}$ is divergent as it is a p-series with $p = \dfrac{1}{2}$, the General Comparison Test allows us to conclude that the given series is divergent.

d) The series $\sum_{n=0}^{\infty} \dfrac{3^n (n+1)!}{(n+1)^n}$ is of positive terms. Let us study it using D'Alembert's test. Let a_n be the general term of the series:

$$\lim \dfrac{a_{n+1}}{a_n} = \lim \dfrac{\dfrac{3^{n+1} (n+2)!}{(n+2)^{n+1}}}{\dfrac{3^n (n+1)!}{(n+1)^n}} = 3 \lim \left(\dfrac{n+1}{n+2}\right)^n = \dfrac{3}{e}.$$

This limit value is greater than 1; therefore, by D'Alembert's test, the series is divergent.

e) We know that the function $\sin(x)$ is odd; therefore, the series $\sum_{n=1}^{\infty} \sin\left(\dfrac{(-1)^n}{n}\right)$ is alternating and can be written in the form $\sum_{n=1}^{\infty} (-1)^n \sin\left(\dfrac{1}{n}\right)$. Let us begin by studying the series of modules, $\sum_{n=1}^{\infty} \sin\left(\dfrac{1}{n}\right)$ by comparing it with the harmonic series. The limit

$$\lim \dfrac{\sin\left(\dfrac{1}{n}\right)}{\dfrac{1}{n}} = 1$$

belongs to $\mathbb{R}^+$. It follows from Corollary 2 of the General Comparison Test that the series of modules is divergent.

Let us study the alternating series by applying Leibniz's test:

(i) $\lim \sin\left(\dfrac{1}{n}\right) = 0$.

(ii) $\sin\left(\dfrac{1}{n}\right) > 0, \forall n \in \mathbb{N}$.

(iii) The sequence $\left(\dfrac{1}{n}\right)$ is decreasing, the sine function is increasing on $[0, \frac{\pi}{2}]$, and therefore, the sequence $\left(\sin\left(\dfrac{1}{n}\right)\right)$ is decreasing.

We conclude that the alternating series is convergent. Since the series of modules is divergent, the alternating series is conditionally convergent.

f) The series $\sum_{n=1}^{\infty} \dfrac{\sqrt{n}}{n^2 + \cos(n)}$ is of positive terms. The limit

$$\lim \frac{\dfrac{\sqrt{n}}{n^2 + \cos(n)}}{\dfrac{1}{n^{3/2}}} = \lim \frac{n^2}{n^2 + \cos(n)} = 1$$

is finite and different from zero. The series $\sum_{n=1}^{\infty} \dfrac{1}{n^{3/2}}$ is convergent, since it is a p-series with $p = \dfrac{3}{2}$. The General Comparison Test ensures that the original series is convergent. As it is of positive terms, it converges absolutely.

g) The series $\sum_{n=1}^{\infty} \dfrac{3^n}{2n + n^n}$ is of positive terms, and the inequality $\dfrac{3^n}{2n + n^n} \leq \dfrac{3^n}{n^n}$ holds in $\mathbb{N}$. Let us study the series $\sum_{n=1}^{\infty} \dfrac{3^n}{n^n}$ using the Cauchy's Root Test. Let a_n be the general term of the series:

$$\lim \sqrt[n]{a_n} = \lim \frac{3}{n} = 0.$$

Since this limit is less than 1, the series is convergent. Using the General Comparison Test, we can affirm that the series under study is convergent and it is absolutely convergent since it is of positive terms.

h) The series $\sum_{n=1}^{\infty} \dfrac{1 + n(-1)^n}{1 + 2n^3}$ is alternating, since we can write its general term as $(-1)^n \dfrac{n + (-1)^n}{1 + 2n^3}$. Let us start by studying the series of modules, $\sum_{n=1}^{\infty} \dfrac{n + (-1)^n}{1 + 2n^3}$. The limit

$$\lim \frac{\dfrac{n + (-1)^n}{1 + 2n^3}}{\dfrac{1}{n^2}} = \lim \frac{n^3 + (-1^n)n^2}{2n^3 + 1} = \frac{1}{2}$$

is finite and different from zero. Since the series $\sum_{n=1}^{\infty} \dfrac{1}{n^2}$ is a p-series with $p = 2$, it converges. Then, Corollary 2 of the General Comparison Test allows us to conclude that the series $\sum_{n=1}^{\infty} \dfrac{n + (-1)^n}{1 + 2n^3}$ is convergent. Therefore, by Theorem 2.4.1, the alternating series is convergent, and it converges absolutely (see Definition 2.4.1).

2.7. Solved Exercises

i) The series $\sum_{n=1}^{\infty} \dfrac{2^n}{n!+1}$ is of positive terms, and the inequality $\dfrac{2^n}{n!+1} \leq \dfrac{2^n}{n!}$ holds in $\mathbb{N}$. Let us study the series $\sum_{n=1}^{\infty} \dfrac{2^n}{n!}$ using D'Alembert's test. Let a_n be the general term of the series:

$$\lim \frac{a_{n+1}}{a_n} = \lim \frac{\dfrac{2^{n+1}}{(n+1)!}}{\dfrac{2^n}{n!}} = \lim \frac{2}{n+1} = 0.$$

This limit has a value less than 1; therefore, according to D'Alembert's test, the series is convergent. By the General Comparison Test, the initial series is convergent and absolutely convergent as it is of positive terms.

j) Let us consider the series $\sum_{n=0}^{\infty} \dfrac{n!}{(2n)!}$ and $\sum_{n=0}^{\infty} \dfrac{1}{2^n}$. The first one, which is of positive terms, can be studied using D'Alembert's test. Let a_n be the general term of the series:

$$\lim \frac{a_{n+1}}{a_n} = \lim \frac{\dfrac{(n+1)!}{(2n+2)!}}{\dfrac{n!}{(2n)!}} = \lim \frac{(n+1)!(2n)!}{n!(2n+2)!} = \lim \frac{n+1}{(2n+2)(2n+1)} = 0.$$

This limit is less than 1; therefore, by D'Alembert's test, the series is convergent. As it is of positive terms, it is absolutely convergent.

The second series is geometric with ratio $r = \dfrac{1}{2}$; therefore, it is convergent. Again, as it is of positive terms, it converges absolutely.

The original series is the sum of the series $\sum_{n=0}^{\infty} \dfrac{n!}{(2n)!}$ and $\sum_{n=0}^{\infty} -\dfrac{1}{2^n}$, so it is convergent and absolutely convergent (consider the inequality: $|a-b| \leq |a|+|b|, \forall a, b \in \mathbb{R}$).

k) The series $\sum_{n=1}^{\infty} \dfrac{n(2n)!}{4^n (n!)^2}$ is of positive terms. Let a_n be the general term of the series,

$$\lim \frac{a_{n+1}}{a_n} = \lim \frac{\dfrac{(n+1)(2n+2)!}{4^{n+1}((n+1)!)^2}}{\dfrac{n(2n)!}{4^n (n!)^2}} = \lim \frac{(n+1)(2n+2)!\, 4^n (n!)^2}{n(2n)!\, 4^{n+1}((n+1)!)^2}$$

$$= \lim \frac{(n+1)(2n+2)(2n+1)}{4\, n(n+1)^2} = \lim \frac{2(n+1)^2(2n+1)}{4\, n(n+1)^2}$$

$$= \lim \frac{2n+1}{2n} = 1,$$

and therefore, D'Alembert's test is inconclusive. The sequence $\left(\dfrac{2n+1}{2n}\right)$ tends to 1 by values greater than 1, from which we conclude, by the Ratio Test, that the series is divergent.

Note: We can also, alternatively, study the series using Raabe's test (a common procedure when D'Alembert's test is inconclusive). In this case,

$$\lim\ n\left(\frac{a_n}{a_{n+1}} - 1\right) = \lim\ n\left(\frac{2n}{2n+1} - 1\right) = \lim \frac{-n}{2n+1} = -\frac{1}{2}.$$

As this value is less than 1, the series is divergent, as we already saw.

l) The modulus of the general term of the given series is bounded by the general term of the series $\sum_{n=2}^{\infty} \frac{1}{n^3 \log(n)}$, since

$$\left|\frac{\sin(n)}{n^3 \log(n)}\right| \leq \frac{1}{n^3 \log(n)}, \quad \forall n \in \mathbb{N} \setminus \{1\}.$$

As the series $\sum_{n=1}^{\infty} \frac{1}{n^3}$ is convergent as it is a p-series with $p = 3$ and

$$\lim \frac{\frac{1}{n^3 \log(n)}}{\frac{1}{n^3}} = \lim \frac{1}{\log(n)} = 0,$$

Corollary 3 of the General Comparison Test allows us to conclude that the series $\sum_{n=2}^{\infty} \frac{1}{n^3 \log(n)}$ is convergent. Consequently, the series of modules is convergent, making the given series absolutely convergent.

m) The general term of the series $\sum_{n=1}^{\infty} (-1)^n \left(1 - \frac{1}{2n}\right)^n$ is not a null sequence. In fact,

$$\lim \left(1 - \frac{1}{2n}\right)^n = \lim \left(\left(1 - \frac{1}{2n}\right)^{2n}\right)^{\frac{1}{2}} = e^{-1/2},$$

which implies that the sequence of general term $(-1)^n \left(1 - \frac{1}{2n}\right)^n$ does not have a limit. From Theorem 2.2.1, it follows that the series is divergent.

n) The series $\sum_{n=1}^{\infty} \frac{1}{2n+5} \cdot \arcsin\left(\frac{1}{n}\right)$ is of positive terms. Consider the series $\sum_{n=1}^{\infty} \frac{1}{n^2}$, which we know to be convergent as it is a p-series with $p = 2$. The limit

$$\lim \frac{\frac{1}{2n+5} \cdot \arcsin\left(\frac{1}{n}\right)}{\frac{1}{n^2}} = \lim \frac{1}{2n+5} \cdot \frac{\arcsin\left(\frac{1}{n}\right)}{\frac{1}{n}} \cdot \frac{1}{\frac{1}{n}} = \frac{1}{2}$$

belongs to $\mathbb{R}^+$. By Corollary 2 of the General Comparison Test, the series $\sum_{n=1}^{\infty} \frac{1}{2n+5} \cdot \arcsin\left(\frac{1}{n}\right)$ is convergent. As it is of positive terms, it is absolutely convergent.

2.7. Solved Exercises

11. a) The series $\sum_{n=1}^{\infty} \frac{1}{n} \sin\left(\frac{1}{2^n}\right)$ consists of positive terms because $0 < \frac{1}{2^n} < \frac{\pi}{2}$, $\forall n \in \mathbb{N}$. Consider the series $\sum_{n=1}^{\infty} \frac{1}{2^n}$. This series is convergent since it is a geometric series with $r = \frac{1}{2}$. To show that the first series is also convergent, we can use Corollary 3 of the General Comparison Test. Specifically, we can take the limit

$$\lim \frac{\frac{1}{n} \sin\left(\frac{1}{2^n}\right)}{\frac{1}{2^n}} = \lim \frac{1}{n} \cdot \frac{\sin\left(\frac{1}{2^n}\right)}{\frac{1}{2^n}} = 0 \times 1 = 0.$$

Since the limit value equals zero, we can conclude that the series under consideration is convergent. Moreover, as it is of positive terms, it is absolutely convergent.

b) The series $\sum_{n=1}^{\infty} \frac{\sqrt[3]{n^2}}{n\sqrt{n} + 2n^2}$ is of positive terms. Consider the series $\sum_{n=1}^{\infty} \frac{1}{n^{4/3}}$, which is convergent as it is a p-series with $p = \frac{4}{3}$. The limit

$$\lim \frac{\frac{\sqrt[3]{n^2}}{n\sqrt{n} + 2n^2}}{\frac{1}{n^{4/3}}} = \lim \frac{n^2}{n\sqrt{n} + 2n^2} = \frac{1}{2}$$

is finite and different from zero; therefore, by Corollary 2 of the General Comparison Test, the series $\sum_{n=1}^{\infty} \frac{\sqrt[3]{n^2}}{n\sqrt{n} + 2n^2}$ is convergent. As it is a series of positive terms, it is absolutely convergent.

c) The series $\sum_{n=2}^{\infty} \frac{\cos(\pi n)}{\sqrt{n^2 - 1}}$ is alternating because $\cos(n\pi) = (-1)^n$ and $\frac{1}{\sqrt{n^2 - 1}} > 0$. We will study the series of modules, $\sum_{n=2}^{\infty} \frac{1}{\sqrt{n^2 - 1}}$, by comparing it with the harmonic series. The limit

$$\lim \frac{\frac{1}{\sqrt{n^2 - 1}}}{\frac{1}{n}} = \lim \frac{\sqrt{n^2}}{\sqrt{n^2 - 1}} = 1$$

is finite and different from zero; therefore, by Corollary 2 of the General Comparison Test, the series $\sum_{n=2}^{+\infty} \frac{1}{\sqrt{n^2 - 1}}$ is divergent.

Let us apply Leibniz's test to study the alternating series:

(i) $\lim \frac{1}{\sqrt{n^2 - 1}} = 0$.

(ii) $\frac{1}{\sqrt{n^2 - 1}} > 0$, $\forall n \in \mathbb{N} \setminus \{1\}$.

(iii) The sequence $(\sqrt{n^2 - 1})$ is increasing so $\left(\frac{1}{\sqrt{n^2 - 1}}\right)$ is decreasing for $n \geq 2$.

The conditions for applying the test are satisfied; therefore, the alternating series is convergent. As the series of modules is divergent, the given series is conditionally convergent.

d) The series $\sum_{n=2}^{\infty} \frac{(-1)^n}{\sqrt{n}\log(n)}$ is alternating because $a_n = \frac{1}{\sqrt{n}\log(n)} > 0$, $\forall n \geq 2$. Let us start by studying the series of modules

$$\sum_{n=2}^{\infty} \left| \frac{(-1)^n}{\sqrt{n}\log(n)} \right| = \sum_{n=2}^{+\infty} \frac{1}{\sqrt{n}\log(n)}.$$

Given the limit

$$\lim \frac{\frac{1}{\sqrt{n}\log(n)}}{\frac{1}{n^{3/4}}} = \lim \frac{n^{3/4}}{\sqrt{n}\log(n)} = \lim \frac{n^{1/4}}{\log(n)} = +\infty,$$

and the fact that the series $\sum_{n=2}^{\infty} \frac{1}{n^{3/4}}$ is divergent because it is a p-series with $p = \frac{3}{4}$, by Corollary 4 of the General Comparison Test, the series of modules is divergent. Consequently, if convergent, the original series will be conditionally convergent. Since the series is alternating, let us check if we are in the conditions of Leibniz's test:

(i) $\frac{1}{\sqrt{n}\log(n)} > 0$.

(ii) $\lim \frac{1}{\sqrt{n}\log(n)} = 0$.

(iii) Since $(\sqrt{n}\log(n))$ is clearly increasing, the sequence $\left(\frac{1}{\sqrt{n}\log(n)}\right)_{n\in\mathbb{N}\setminus\{1\}}$ is decreasing.

As we are in the conditions of Leibniz's test, we can conclude that the alternating series is convergent. From the study on the series of modules, we can affirm that the series is conditionally convergent.

e) The series $\sum_{n=1}^{\infty} \frac{4+\sin(n)}{\sqrt[3]{n}+1}$ is of positive terms. We can observe that $0 < \frac{3}{\sqrt[3]{n}+1} \leq \frac{4+\sin(n)}{\sqrt[3]{n}+1}$, $\forall n \in \mathbb{N}$. As the series $\sum_{n=1}^{\infty} \frac{1}{\sqrt[3]{n}}$ is a p-series with $p = \frac{1}{3}$, it is divergent. Since the limit

$$\lim \frac{\frac{3}{\sqrt[3]{n}+1}}{\frac{1}{\sqrt[3]{n}}} = \lim \frac{3\sqrt[3]{n}}{\sqrt[3]{n}+1} = 3$$

is finite and different from zero, Corollary 2 of the General Comparison Test allows us to conclude that the series $\sum_{n=1}^{\infty} \frac{3}{\sqrt[3]{n}+1}$ is divergent. Hence, the General Comparison Test ensures that the series $\sum_{n=1}^{\infty} \frac{4+\sin(n)}{\sqrt[3]{n}+1}$ is divergent.

f) We begin by analyzing the series of modules, as the series $\sum_{n=1}^{\infty} \frac{n^{2n}\sin(n^3)}{(3n^2+5)^n}$ is not of positive terms. The inequality

$$0 \leq \left| \frac{n^{2n}\sin(n^3)}{(3n^2+5)^n} \right| \leq \frac{n^{2n}}{(3n^2+5)^n}$$

2.7. Solved Exercises

holds true in $\mathbb{N}$.

We evaluate the limit

$$\lim \sqrt[n]{\frac{n^{2n}}{(3n^2+5)^n}} = \lim \frac{n^2}{3n^2+5} = \frac{1}{3},$$

which is less than 1. Consequently, by Cauchy's Root Test, the series $\sum_{n=1}^{\infty} \frac{n^{2n}}{(3n^2+5)^n}$ is convergent. By the General Comparison Test, the same happens to the series $\sum_{n=1}^{\infty} \left| \frac{n^{2n} \sin(n^3)}{(3n^2+5)^n} \right|$. The original series is absolutely convergent since the series of modules is convergent, considering Definition 2.4.1.

g) Let $a_n = \arcsin\left(\frac{n-1}{n^2+1}\right)$. Since $0 < \frac{n-1}{n^2+1} < 1$, $\forall n > 2$, we have that $a_n > 0$, and therefore, the series $\sum_{n=3}^{\infty} (-1)^n a_n$ is alternating. Let us start by studying the series of modules $\sum_{n=3}^{+\infty} a_n$. The limit

$$\lim \frac{\arcsin\left(\dfrac{n-1}{n^2+1}\right)}{\dfrac{n-1}{n^2+1}} = 1$$

belongs to $\mathbb{R}^+$, and the series $\sum_{n=3}^{\infty} \frac{n-1}{n^2+1}$ is divergent (compare it with the harmonic series). We can then conclude that the series of modules is divergent. Consequently, if convergent, the series under study is conditionally convergent. Since the series is alternating, let us check if we are in the conditions of Leibniz's test:

(i) $\arcsin\left(\dfrac{n-1}{n^2+1}\right) > 0$, $\forall n > 2$.

(ii) $\lim \arcsin\left(\dfrac{n-1}{n^2+1}\right) = 0$.

(iii) Let f be the real function of a real variable defined by $f(x) = \arcsin\left(\dfrac{x-1}{x^2+1}\right)$. This function is decreasing on $[3, +\infty[$, because

$$f'(x) = \frac{-x^2+2x+1}{(x^2+1)\sqrt{x^4+x^2+2x}} < 0, \quad \forall x \in [3, +\infty[.$$

Therefore, the sequence

$$\left(\arcsin\left(\frac{n-1}{n^2+1}\right) \right)_{n \in \mathbb{N} \setminus \{1,2\}}$$

is decreasing.

As we are in the conditions of Leibniz's test, we can conclude that the alternating series is convergent. The study done on the series of modules shows that the series is conditionally convergent.

h) The series $\sum_{n=1}^{\infty} \frac{4 \times 7 \times \cdots \times (3n+1)}{8 \times 11 \times \cdots \times (3n+5)}$ is of positive terms. Let us use D'Alembert's test to study it. If a_n is the general term of the series, then:

$$\lim \frac{a_{n+1}}{a_n} = \lim \frac{\dfrac{4 \times 7 \times \cdots \times (3n+1)(3n+4)}{8 \times 11 \times \cdots \times (3n+5)(3n+8)}}{\dfrac{4 \times 7 \times \cdots \times (3n+1)}{8 \times 11 \times \cdots \times (3n+5)}} = \lim \frac{3n+4}{3n+8} = 1.$$

Since this limit is 1, we cannot draw any conclusions using D'Alembert's test. Therefore, let us use Raabe's test, which is a common procedure when D'Alembert's test is inconclusive. In this case,

$$\lim n\left(\frac{a_n}{a_{n+1}} - 1\right) = \lim n\left(\frac{3n+8}{3n+4} - 1\right) = \lim \frac{4n}{3n+4} = \frac{4}{3}.$$

As this value is greater than 1, we can conclude that the series is convergent. Moreover, since it is a series of positive terms, it converges absolutely.

i) Consider the two series $\sum_{n=1}^{\infty} \frac{\arctan(n^3)}{\sqrt{n}+n^2}$ and $\sum_{n=0}^{\infty} \frac{2^n(2n)!}{3^n(2n+1)!}$. The first one is of positive terms, and we compare it with the p-series $\sum_{n=1}^{\infty} \frac{1}{n^2}$, which we know to be convergent. Indeed, the limit

$$\lim \frac{\dfrac{\arctan(n^3)}{\sqrt{n}+n^2}}{\dfrac{1}{n^2}} = \lim \frac{n^2}{\sqrt{n}+n^2} \arctan(n^3) = \frac{\pi}{2}$$

is finite and different from zero, so, by Corollary 2 of the General Comparison Test, the series $\sum_{n=1}^{\infty} \frac{\arctan(n^3)}{\sqrt{n}+n^2}$ is convergent.

The second series is also of positive terms. We can study it using D'Alembert's test. Let a_n be the general term of the series:

$$\lim \frac{a_{n+1}}{a_n} = \lim \frac{\dfrac{2^{n+1}(2n+2)!}{3^{n+1}(2n+3)!}}{\dfrac{2^n(2n)!}{3^n(2n+1)!}} = \lim \frac{2^{n+1}(2n+2)!\,3^n(2n+1)!}{2^n(2n)!\,3^{n+1}(2n+3)!} = \lim \frac{2(2n+1)}{3(2n+3)} = \frac{2}{3}.$$

This limit is less than 1; therefore, by D'Alembert's test, the second series is convergent.

The original series is the sum of the two series studied, so it is convergent. Since it is of positive terms, the series is absolutely convergent.

j) The series $\sum_{n=1}^{\infty} \frac{\sin\left(\frac{1}{\sqrt{n}}\right)}{n+\sqrt{n}}$ consists of positive terms, since $0 < \frac{1}{\sqrt{n}} < \frac{\pi}{2}$, $\forall n \in \mathbb{N}$. To check its convergence, let us compare it with the series $\sum_{n=1}^{\infty} \frac{1}{n^{3/2}}$, which we know to be convergent because it is a p-series with

2.7. Solved Exercises

$p = \frac{3}{2}$. Taking the limit, we have

$$\lim \frac{\frac{\sin\left(\frac{1}{\sqrt{n}}\right)}{n+\sqrt{n}}}{\frac{1}{n^{3/2}}} = \lim \frac{\frac{\sin\left(\frac{1}{\sqrt{n}}\right)}{n+\sqrt{n}}}{\frac{1}{n\sqrt{n}}} = \lim \frac{\sin\left(\frac{1}{\sqrt{n}}\right)}{\frac{1}{\sqrt{n}}} \cdot \frac{n}{n+\sqrt{n}} = 1.$$

Since the limit is finite and different from zero, by Corollary 2 of the General Comparison Test, the series $\sum_{n=1}^{\infty} \frac{\sin\left(\frac{1}{\sqrt{n}}\right)}{n+\sqrt{n}}$ is convergent. Moreover, as a series of positive terms, it converges absolutely.

k) The series $\sum_{n=1}^{\infty} \left(n \sin\left(\frac{1}{n}\right) - (n+2) \sin\left(\frac{1}{n+2}\right) \right)$ is telescopic. We can write the sequence of partial sums:

$$S_1 = \sin(1) - 3 \sin\left(\frac{1}{3}\right)$$

$$S_2 = \sin(1) - 3 \sin\left(\frac{1}{3}\right) + 2 \sin\left(\frac{1}{2}\right) - 4 \sin\left(\frac{1}{4}\right)$$

$$S_3 = \sin(1) - 3 \sin\left(\frac{1}{3}\right) + 2 \sin\left(\frac{1}{2}\right) - 4 \sin\left(\frac{1}{4}\right) + 3 \sin\left(\frac{1}{3}\right) - 5 \sin\left(\frac{1}{5}\right)$$

$$= \sin(1) + 2 \sin\left(\frac{1}{2}\right) - 4 \sin\left(\frac{1}{4}\right) - 5 \sin\left(\frac{1}{5}\right)$$

$$S_4 = \sin(1) + 2 \sin\left(\frac{1}{2}\right) - 4 \sin\left(\frac{1}{4}\right) - 5 \sin\left(\frac{1}{5}\right) + 4 \sin\left(\frac{1}{4}\right) - 6 \sin\left(\frac{1}{6}\right)$$

$$= \sin(1) + 2 \sin\left(\frac{1}{2}\right) - 5 \sin\left(\frac{1}{5}\right) - 6 \sin\left(\frac{1}{6}\right)$$

$$\vdots$$

$$S_n = \sin(1) + 2 \sin\left(\frac{1}{2}\right) - (n+1) \sin\left(\frac{1}{n+1}\right) - (n+2) \sin\left(\frac{1}{n+2}\right)$$

$$\vdots$$

As

$$\lim n \sin\left(\frac{1}{n}\right) = \lim \frac{\sin\left(\frac{1}{n}\right)}{\frac{1}{n}} = 1$$

we have

$$\lim S_n = \sin(1) + 2 \sin\left(\frac{1}{2}\right) - 2.$$

The series is convergent and

$$\sum_{n=1}^{\infty} \left(n \sin\left(\frac{1}{n}\right) - (n+2) \sin\left(\frac{1}{n+2}\right) \right) = \sin(1) + 2 \sin\left(\frac{1}{2}\right) - 2.$$

To determine whether it is absolutely convergent, we will study the sign of the general term. For this, we will prove that the sequence of general term $n \sin\left(\frac{1}{n}\right)$ is increasing. Let $f(x) = x \sin\left(\frac{1}{x}\right)$, $x \geq 1$. We have
$$f'(x) = \sin\left(\frac{1}{x}\right) - \frac{1}{x}\cos\left(\frac{1}{x}\right)$$
and, therefore, $f'(x) = 0 \Leftrightarrow \tan\left(\frac{1}{x}\right) = \frac{1}{x}$. However, this equation has no zeros. In fact, $\tan\left(\frac{1}{x}\right) > \frac{1}{x}$, $\forall x \geq 1$. Then $f'(x) > 0$, $\forall x \geq 1$, which implies that f is increasing. As $f(n) = n \sin\left(\frac{1}{n}\right)$, we can conclude that the sequence is increasing and, therefore, the series is of negative terms. Since $\sum_{n=1}^{\infty} |a_n| = -\sum_{n=1}^{\infty} a_n$, the given series is absolutely convergent.

l) The series $\sum_{n=1}^{\infty} \frac{2^n + 3}{(n+1)!}$ is of positive terms. We can establish the following inequality:

$$\frac{2^n + 3}{(n+1)!} \leq 2\frac{2^n}{(n+1)!}, \quad \forall n \geq 2.$$

To determine the convergence of the series $\sum_{n=1}^{\infty} \frac{2^n}{(n+1)!}$, we can apply D'Alembert's test. Let a_n be the general term of the series:

$$\lim \frac{a_{n+1}}{a_n} = \lim \frac{\frac{2^{n+1}}{(n+2)!}}{\frac{2^n}{(n+1)!}} = \lim \frac{2}{n+2} = 0.$$

This limit is less than 1, indicating that the series is convergent according to D'Alembert's test. Using the General Comparison Test, we can affirm that the series under study is convergent and, hence, absolutely convergent.

m) The series $\sum_{n=1}^{\infty} \frac{\arctan(n+1) - \arctan(n)}{n^2}$ is of positive terms because $\arctan(x)$ is an increasing function on $\mathbb{R}$. Let us compare this series with the series $\sum_{n=1}^{\infty} \frac{1}{n^2}$, which we know is convergent as it is a p-series with $p = 2$. By Corollary 3 of the General Comparison Test, given the value of the limit

$$\lim \frac{\frac{\arctan(n+1) - \arctan(n)}{n^2}}{\frac{1}{n^2}} = \lim \left(\arctan(n+1) - \arctan(n)\right) = 0,$$

we can conclude, as we are comparing with a convergent series, that the series under study is convergent. As it is a series of positive terms, it converges absolutely.

n) The general term of the series $\sum_{n=1}^{\infty} \frac{\log(n)}{n \sin(\frac{1}{n})}$ is not a null sequence. In fact,

2.7. Solved Exercises 153

$$\lim \frac{\log(n)}{n\sin\left(\frac{1}{n}\right)} = \lim \log(n) \cdot \frac{\frac{1}{n}}{\sin\left(\frac{1}{n}\right)} = +\infty.$$

Therefore, by Theorem 2.2.1, the series diverges.

o) Consider the series $\sum_{n=1}^{\infty} \left(\log(2)\right)^n \log\left(\frac{n+1}{n}\right)$. It consists of positive terms because $\frac{n+1}{n} > 1$, $\forall n \in \mathbb{N}$.

Let us analyze the series $\sum_{n=1}^{\infty} \left(\log(2)\right)^n$. It converges because it is a geometric series with $0 < r = \log(2) < 1$.
The limit

$$\lim \frac{\left(\log(2)\right)^n \log\left(\frac{n+1}{n}\right)}{\left(\log(2)\right)^n} = \lim \log\left(\frac{n+1}{n}\right) = 0$$

allows us to use Corollary 3 of the General Comparison Test to conclude that the series under study is convergent because we are comparing it with a convergent series. As the series is of positive terms, it is absolutely convergent.

p) Given that $n \geq 3$ implies $\log(n) > 1$, then $\log(\log(n)) > 0$ so $n\log(n)\left(\log(\log(n))\right)^p > 0$, $\forall p \in \mathbb{R}$. Therefore, the series is of positive terms.

Let us assume that $p \leq 0$. If $n > e^e$ (for example, if $n \geq 3^3 = 27$), then $\log(\log(n)) > 1$. Thus, $\left(\log(\log(n))\right)^{-p} \geq 1$, with equality holding for $p = 0$. Consequently, we can write:

$$\frac{1}{n\log(n)\left(\log(\log(n))\right)^p} = \frac{\left(\log(\log(n))\right)^{-p}}{n\log(n)} \geq \frac{1}{n\log(n)}.$$

Since the series $\sum_{n=27}^{+\infty} \frac{1}{n\log(n)}$ is divergent as seen in Example 2.5.2, we can apply the General Comparison Test to deduce that the series $\sum_{n=27}^{+\infty} \frac{1}{n\log(n)\left(\log(\log(n))\right)^p}$ is divergent. Therefore, the series $\sum_{n=3}^{+\infty} \frac{1}{n\log(n)\left(\log(\log(n))\right)^p}$ is also divergent.

Now, suppose that $p > 0$. With the application of the Integral Test in mind, let us consider the function

$$f(x) = \frac{1}{x\log(x)\left(\log(\log(x))\right)^p}$$

and let $g(x) = x\log(x)\left(\log(\log(x))\right)^p$. We know that $g(x) > 0$, $\forall x \in [3, +\infty[$. Additionally, since the logarithm function is continuous and increasing, g is continuous and increasing on $[3, +\infty[$, so f is positive, continuous, and decreasing on $[3, +\infty[$. We can now study the improper integral $\int_{3}^{+\infty} f(x)\,dx$.
If $p \neq 1$,

$$\int_{3}^{y} \frac{1}{x\log(x)\left(\log(\log(x))\right)^p}\,dx = \left[\frac{1}{-p+1}\left(\log(\log(x))\right)^{-p+1}\right]_{3}^{y}$$

$$= \frac{1}{-p+1}\left(\left(\log(\log(y))\right)^{-p+1} - \left(\log(\log(3))\right)^{-p+1}\right).$$

If $p = 1$,
$$\int_3^y \frac{1}{x\log(x)\log(\log(x))}\, dx = \Big[\log\big(\log(\log(x))\big)\Big]_3^y = \log\big(\log(\log(y))\big) - \log\big(\log(\log(3))\big).$$

We can then determine that
$$\lim_{y \to +\infty} \int_3^y \frac{1}{x\log(x)\left(\log(\log(x))\right)^p} = \begin{cases} +\infty, & \text{if } 0 < p \leq 1 \\ \dfrac{1}{p-1}\left(\log(\log(3))\right)^{-p+1} & \text{if } p > 1. \end{cases}$$

This means that the improper integral $\int_3^{+\infty} f(x)\, dx$ is convergent if $p > 1$ and divergent if $0 < p \leq 1$. By the Integral Test, the series under consideration is convergent if $p > 1$ and divergent if $p \leq 1$.

Conclusion: The series $\sum_{n=3}^{+\infty} \dfrac{1}{n\log(n)\left(\log(\log(n))\right)^p}$ is convergent if and only if $p > 1$.

12. a) The series $\sum_{n=1}^{\infty} \dfrac{(n+1)^n}{3^n n!}$ is of positive terms. To analyze this series, we will use D'Alembert's test. Let a_n be the general term of the series:

$$\lim \frac{a_{n+1}}{a_n} = \lim \frac{\dfrac{(n+2)^{n+1}}{3^{n+1}(n+1)!}}{\dfrac{(n+1)^n}{3^n n!}} = \frac{1}{3}\lim \frac{n+2}{n+1}\left(\frac{n+2}{n+1}\right)^n = \frac{e}{3}.$$

Since this limit is less than 1, according to D'Alembert's test, the series converges.

b) In the previous item, we proved that the series $\sum_{n=1}^{\infty} \dfrac{(n+1)^n}{3^n n!}$ is convergent, which implies, by Theorem 2.2.1, that its general term is a null sequence. Therefore, $\lim \dfrac{(n+1)^n}{3^n n!} = 0$.

13. We will study the convergence of the series of modules, $\sum_{n=1}^{\infty} |a_n|$. We will use the Root Test or one of its corollaries, to do this. We have

$$\sqrt[n]{|a_n|} = \begin{cases} \sqrt[n]{\left(\dfrac{n}{n+1}\right)^{n^2}} = \left(\dfrac{n}{n+1}\right)^n, & \text{if } n = 3k,\ k \in \mathbb{N} \\ \sqrt[n]{\dfrac{1}{n!}} = \dfrac{1}{\sqrt[n]{n!}}, & \text{if } n = 3k+1,\ k \in \mathbb{N}_0 \\ \sqrt[n]{\left|\dfrac{(-1)^n}{(n+1)^n}\right|} = \dfrac{1}{n+1}, & \text{if } n = 3k+2,\ k \in \mathbb{N}_0. \end{cases}$$

The sublimits of $\sqrt[n]{|a_n|}$ are $\dfrac{1}{e}$ and 0; therefore, $\overline{\lim} \sqrt[n]{|a_n|} = \dfrac{1}{e}$. Since this value is less than 1, we can affirm, by Corollary 1 of the Root Test, that the series $\sum_{n=1}^{\infty} |a_n|$ is convergent. Consequently, the series $\sum_{n=1}^{\infty} a_n$ is absolutely convergent.

2.7. Solved Exercises 155

14. One initial approach is to analyze the general term of the series $\sum_{n=1}^{\infty} a_n$. Since $\lim a_{2n} = \lim a_{2n+1} = +\infty$, then $\lim a_n = +\infty$, and by Theorem 2.2.1, the series diverges.

    Another approach:
    $$\frac{a_{n+1}}{a_n} = \begin{cases} \dfrac{n+2}{n}, & \text{if } n \text{ is even} \\ \dfrac{n+1}{n-1}, & \text{if } n \text{ is odd, } n > 1. \end{cases}$$

    We find that $\dfrac{a_{n+1}}{a_n} > 1$, $\forall n \in \mathbb{N} \setminus \{1\}$, and therefore by the Ratio Test, the series diverges.

15. If (a_n) is a sequence of positive terms such that the series $\sum_{n=1}^{\infty} n a_n$ is convergent, then the sequence (a_n) is bounded. We can prove this by reductio ad absurdum: If (a_n) was not bounded, then $\lim (n a_n)$ would not be zero, and in this case, the series $\sum_{n=1}^{\infty} n a_n$ would not be convergent. It follows from the fact that (a_n) is bounded that
    $$\lim \frac{a_n^2}{n a_n} = \lim \frac{a_n}{n} = 0.$$

    By Corollary 3 of the General Comparison Test, the series $\sum_{n=1}^{\infty} a_n^2$ is convergent.

16. As the series $\sum_{n=1}^{\infty} a_n$ converges, $\lim a_n = 0$. By hypothesis, $a_n > 0$, $\forall n \in \mathbb{N}$, so $\log(1 + a_n) > 0$. Thus, we have two series of positive terms. The limit
    $$\lim \frac{\log(1 + a_n)}{a_n} = 1$$

    is finite and different from zero; therefore, by Corollary 2 of the General Comparison Test, the series $\sum_{n=1}^{\infty} \log(1 + a_n)$ also converges.

    Note: The limit $\lim_{x \to 0} \dfrac{\log(1 + x)}{x}$ is an indeterminate form of type $\dfrac{0}{0}$. Applying L'Hôpital's Rule, we get
    $$\lim_{x \to 0} \frac{(\log(1 + x))'}{(x)'} = \lim_{x \to 0} \frac{1}{1 + x} = 1$$

    therefore, $\lim_{x \to 0} \dfrac{\log(1 + x)}{x} = 1$, which implies that $\lim \dfrac{\log(1 + a_n)}{a_n} = 1$.

17. a) The series $\sum_{n=1}^{\infty} (-1)^{n-1} a_n$ is alternating. Leibniz's test cannot be applied because the sequence (a_n) does not decrease. In fact, if n is even,
    $$a_{n+1} - a_n = \frac{1}{n+1} - \frac{1}{n^2} = \frac{n^2 - n - 1}{n^2(n+1)} > 0.$$

b) Let us prove that the series $\sum_{n=1}^{\infty}(-1)^{n-1}a_n$ is divergent. If it was convergent, then, by Theorem 2.2.4, any series obtained from it by association of its terms would be convergent. Consider the series $\sum_{n=1}^{\infty}b_n$, where (b_n) is the sequence defined by

$$b_n = a_{2n-1} - a_{2n} = \frac{1}{2n-1} - \frac{1}{(2n)^2} = \frac{(2n)^2 - 2n + 1}{(2n)^2(2n-1)} = \frac{4n^2 - 2n + 1}{4n^2(2n-1)}.$$

The limit
$$\lim \frac{\frac{4n^2 - 2n + 1}{4n^2(2n-1)}}{\frac{1}{n}} = \lim \frac{4n^2 - 2n + 1}{4n(2n-1)} = \frac{1}{2}$$

is finite and different from zero; therefore, based on Corollary 2 of the General Comparison Test, the series $\sum_{n=1}^{\infty}b_n$ diverges because the harmonic series is divergent. However, we obtained the series $\sum_{n=1}^{\infty}b_n$ by association of the terms of the original series. Hence, the original series must be divergent.

18. a) Suppose that the series $\sum_{n=1}^{\infty}b_n$ is convergent. Then, its general term is a null sequence, which implies that all its subsequences have limit zero. Therefore, we have $\lim b_{2n-1} = 0$, or equivalently, $\lim a_n = 0$. Conversely, let $\lim a_n = 0$ and (S_n) be the sequence of partial sums of the series $\sum_{n=1}^{\infty}b_n$. We can see that $S_{2n} = 0$ and $S_{2n-1} = a_n$ for all $n \in \mathbb{N}$. Since $\lim a_n = 0$, it follows that $\lim S_n = 0$. By definition, if the sequence of partial sums of a series has a finite limit, the series converges. As a result, it can be inferred that the series $\sum_{n=1}^{\infty}b_n$ converges.

b) Let S_n and S_n^* denote the general terms of the sequences of partial sums of the series $\sum_{n=1}^{\infty}|b_n|$ and $\sum_{n=1}^{\infty}|a_n|$, respectively. We can observe that

$$S_1 = |b_1| = |a_1| = S_1^*$$
$$S_2 = |b_1| + |b_2| = |a_1| + |a_1| = 2S_1^*$$
$$S_3 = |b_1| + |b_2| + |b_3| = |a_1| + |a_1| + |a_2| = S_1^* + S_2^*$$
$$S_4 = |b_1| + |b_2| + |b_3| + |b_4| = |a_1| + |a_1| + |a_2| + |a_2| = 2S_2^*$$
$$S_5 = |b_1| + |b_2| + |b_3| + |b_4| + |b_5| = |a_1| + |a_1| + |a_2| + |a_2| + |a_3| = S_2^* + S_3^*$$
$$\vdots$$

So, we can infer that
$$S_{2n} = 2S_n^* \quad \text{and} \quad S_{2n+1} = S_n^* + S_{n+1}^*.$$

If the series $\sum_{n=1}^{\infty}b_n$ converges absolutely, then the limits $\lim S_{2n}$ and $\lim S_{2n+1}$ exist, are finite, and equal. Since $\lim S_n^* = \frac{1}{2}\lim S_{2n}$, we can conclude that the series $\sum_{n=1}^{\infty}a_n$ converges absolutely.

2.7. Solved Exercises

Reciprocally, if the series $\sum_{n=1}^{\infty} a_n$ converges absolutely, then $\lim S_n^*$ exists and is finite. This implies that $\lim S_{2n+1} = \lim(S_n^* + S_{n+1}^*) = \lim 2S_n^* = \lim S_{2n}$, and we can state that the series $\sum_{n=1}^{\infty} b_n$ converges absolutely.

19. If the series $\sum_{n=1}^{\infty} a_n$ is convergent, then the sequence of its partial sums, (S_n), is also convergent. This implies that all its subsequences are convergent and have the same limit. Since the sequence (a_n) is decreasing, we have

$$S_{2n} - S_n = a_{2n} + \cdots + a_{n+1} \geq a_{2n} + \cdots + a_{2n} = n\, a_{2n},$$

and as $\lim(S_{2n} - S_n) = 0$ and $a_n \geq 0$, we can conclude that $\lim n\, a_{2n} = 0$, which implies that $\lim 2n\, a_{2n} = 0$. Similarly, we have

$$S_{2n+1} - S_{n+1} = a_{2n+1} + \cdots + a_{n+2} \geq a_{2n+1} + \cdots + a_{2n+1} = n\, a_{2n+1}.$$

As $\lim(S_{2n+1} - S_{n+1}) = 0$ and (a_n) is a sequence of nonnegative terms, we conclude that $\lim n\, a_{2n+1} = 0$. This also implies that $\lim(2n+1)\, a_{2n+1} = 0$ because $\lim a_n = 0$ since, by hypothesis, the series is convergent.

We have shown that the subsequence of even order terms of $(n\, a_n)$ has the same limit as its subsequence of odd order terms, which is zero. Hence $\lim n\, a_n = 0$.

20. We can use the inequality $ab \leq \dfrac{a^2 + b^2}{2}$, $\forall\, a, b \in \mathbb{R}$, along with the fact that the sequence (a_n) is of positive terms, to establish the following expression:

$$0 \leq \frac{1}{n^p}\sqrt{a_n} \leq \frac{\frac{1}{n^{2p}} + a_n}{2} = \frac{1}{2}\left(\frac{1}{n^{2p}} + a_n\right).$$

By hypothesis, the series $\sum_{n=1}^{\infty} a_n$ is convergent, and if $p > \dfrac{1}{2}$, then $2p > 1$, which implies that the p-series $\sum_{n=1}^{\infty} \dfrac{1}{n^{2p}}$ is convergent. The sum of these two series is convergent, and given the previous inequality, we can state by the General Comparison Test the convergence of the series $\sum_{n=1}^{\infty} \dfrac{\sqrt{a_n}}{n^p}$.

21. Consider the series $\sum_{n=2}^{\infty} \dfrac{1}{n-1}$, which is divergent as it is the harmonic series. Let $b_n = \dfrac{1}{n-1}$. Then

$$\frac{b_{n+1}}{b_n} = \frac{\frac{1}{n}}{\frac{1}{n-1}} = \frac{n-1}{n} = 1 - \frac{1}{n} \leq \frac{a_{n+1}}{a_n}, \quad \forall n \geq 2.$$

By Corollary 5 of the General Comparison Test, we can determine that the series $\sum_{n=1}^{\infty} a_n$ is divergent.

22. The functions g and g' are, by hypothesis, continuous and positive on $[p, +\infty[$, so $\dfrac{g'}{g}$ is continuous and positive. In addition, $\left(\dfrac{g'(x)}{g(x)}\right)' = \dfrac{g''(x)g(x) - (g'(x))^2}{(g(x))^2} < 0$, so $\dfrac{g'}{g}$ is decreasing on $[p, +\infty[$. We are in the conditions of the Integral's Test. But

$$\int_p^y \frac{g'(x)}{g(x)}\, dx = \log(g(y)) - \log(g(p));$$

therefore, the improper integral $\displaystyle\int_p^{+\infty} \dfrac{g'(x)}{g(x)}\, dx$ is convergent if and only if there exists and is finite $\displaystyle\lim_{x\to +\infty} \log(g(x))$. By the continuity of the logarithm function, $\displaystyle\lim_{x\to +\infty} \log(g(x))$ exists finite if and only if there exists finite and different from zero the $\displaystyle\lim_{x\to +\infty} g(x)$. But, by hypothesis, g is positive and increasing (g' is positive), so it cannot have a limit 0 when x tends to $+\infty$.

Conclusion: The series $\displaystyle\sum_{n=p}^{+\infty} \dfrac{g'(n)}{g(n)}$ is convergent if and only if $\displaystyle\lim_{x\to +\infty} g(x)$ exists and is finite.

23. We can prove by induction that (v_n) is a sequence of positive terms. Applying D'Alembert's test, we find that the series $\displaystyle\sum_{n=2}^{\infty} v_n$ is convergent because

$$\lim \frac{v_{n+1}}{v_n} = \lim \sin\left(\frac{\pi}{n}\right) = 0 < 1.$$

Furthermore,

$$\lim \frac{\arctan(v_n)}{v_n} = 1,$$

which implies by Corollary 2 of the General Comparison Test that the series $\displaystyle\sum_{n=2}^{\infty} \arctan(v_n)$ is convergent.

24. Let us look at the numbers of the construction of the Sierpinski triangle in a table.

Number of squares	Number of squares removed	Length of the edges	Area of each square	Area of the squares removed
1	0	1	1	0
3	1	$\dfrac{1}{2}$	$\left(\dfrac{1}{2}\right)^2$	$1 \times \left(\dfrac{1}{2}\right)^2$
3^2	3	$\dfrac{1}{2^2}$	$\left(\dfrac{1}{2^2}\right)^2$	$3 \times \left(\dfrac{1}{2^2}\right)^2$
3^3	3^2	$\dfrac{1}{2^3}$	$\left(\dfrac{1}{2^3}\right)^2$	$3^2 \times \left(\dfrac{1}{2^3}\right)^2$
$\vdots$	$\vdots$	$\vdots$	$\vdots$	$\vdots$

2.7. Solved Exercises

So the total area of the squares removed is the sum of the geometric series

$$A = \sum_{n=0}^{\infty} 3^n \left(\frac{1}{2^{n+1}}\right)^2 = \frac{1}{4} \sum_{n=0}^{\infty} \left(\frac{3}{4}\right)^n = 1.$$

Since the initial square has area 1, we conclude that the area of the Sierpinski triangle is zero.

25. The sum of the areas of the circles is given by the series $\pi r_1^2 + \sum_{n=2}^{\infty} 3\pi r_n^2$, where r_1 is the radius of the largest circle, r_2 is the radius of the circles tangent to it, etc.

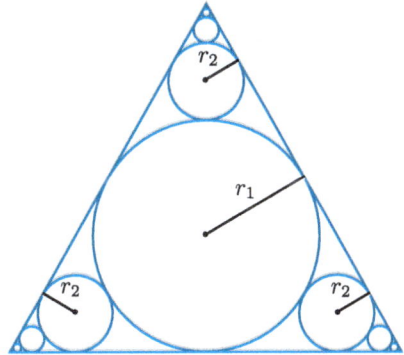
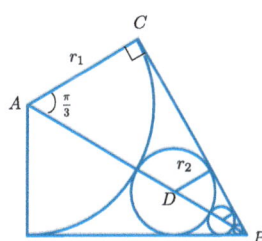

We need to find the expression of r_n as a function of r_1. In the figure, the angle $\angle BAC$ measures $\frac{\pi}{3}$ radians and $\cos\left(\frac{\pi}{3}\right) = \frac{r_1}{|AB|}$, and therefore, $\frac{1}{2} = \frac{r_1}{|AB|}$. Similarly, $\frac{1}{2} = \frac{r_2}{|DB|}$. Then $|AB| = 2r_1$ and $|DB| = 2r_2$. As a result, we have the equality $2r_1 = r_1 + r_2 + 2r_2$, from which we conclude that $r_2 = \frac{1}{3}r_1$. Similarly, we can show that $r_3 = \frac{1}{3}r_2 = \frac{1}{9}r_1$ and, more generally, $r_n = \left(\frac{1}{3}\right)^{n-1} r_1$, $n \geq 2$.

The series now takes the form

$$\pi r_1^2 + \sum_{n=2}^{\infty} 3\pi r_n^2 = \pi r_1^2 + \sum_{n=2}^{\infty} 3\pi \left(\frac{1}{3^{n-1}}\right)^2 r_1^2 = \pi r_1^2 \left(1 + 3 \sum_{n=2}^{\infty} \frac{1}{9^{n-1}}\right) = \frac{11}{8} \pi r_1^2.$$

What is the value of r_1? Considering that $|BC| = \frac{1}{2}$, $|AB| = 2r_1$, and $\sin\left(\frac{\pi}{3}\right) = \frac{|BC|}{|AB|}$, we have $r_1 = \frac{1}{2\sqrt{3}}$. Finally, the sum of the areas of the circles is

$$\pi r_1^2 + \sum_{n=2}^{\infty} 3\pi r_n^2 = \frac{11\pi}{96}.$$

26. The area of the colored region can be written as $A = \sum_{n=1}^{\infty} (A_n - B_n)$, where A_n is the area of the circle and B_n is the area of the square inscribed in that circle. Let r_n be the radius of the nth circle and l_n be the length of the side of the inscribed square.

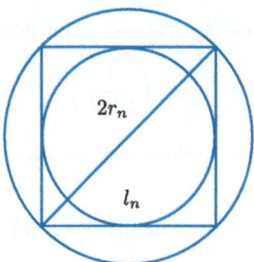

From Pythagoras's Theorem, it follows that $l_n^2 + l_n^2 = (2r_n)^2$, that is, $l_n^2 = 2r_n^2$. We can write the series in the form

$$A = \sum_{n=1}^{\infty} (\pi r_n^2 - l_n^2) = \sum_{n=1}^{\infty} (\pi - 2) \, r_n^2.$$

What is the relationship between r_1 and r_n? We know that $2r_2 = l_1$, so $4r_2^2 = l_1^2$. As $l_1^2 = 2r_1^2$, we have that $r_2^2 = \dfrac{1}{2}r_1^2$. By induction, it is easy to show that $r_n^2 = \dfrac{1}{2^{n-1}}r_1^2$. Thus,

$$A = \sum_{n=1}^{\infty} (\pi - 2) \, r_n^2 = \sum_{n=1}^{\infty} (\pi - 2) \, \frac{1}{2^{n-1}} r_1^2 = 2(\pi - 2) \, r_1^2.$$

As $r_1 = 1$, we conclude that $A = 2(\pi - 2)$.

2.8 Proposed Exercises

1. Find the general term and the sum of each of the following series:

   a) $\dfrac{1}{3} + \dfrac{1}{8} + \dfrac{1}{15} + \dfrac{1}{24} + \cdots$

   b) $\dfrac{1}{1\times 2\times 3} + \dfrac{1}{2\times 3\times 4} + \dfrac{1}{3\times 4\times 5} + \cdots$

   c) $\dfrac{1}{1\times 2\times 3\times 4} + \dfrac{1}{2\times 3\times 4\times 5} + \dfrac{1}{3\times 4\times 5\times 6} + \cdots$

2. Find the sum of the series:

   a) $\displaystyle\sum_{n=1}^{\infty} (-1)^n \dfrac{2n+1}{n(n+1)}$

   b) $\displaystyle\sum_{n=1}^{\infty} \dfrac{1}{2^n} \tan\left(\dfrac{a}{2^n}\right)$, $a \in \mathbb{R} \setminus \{k\pi, k \in \mathbb{Z}\}$.

   Hint: $\tan\left(\dfrac{x}{2}\right) = \cot\left(\dfrac{x}{2}\right) - 2\cot(x)$

   c) $\displaystyle\sum_{n=1}^{\infty} \dfrac{\sqrt{n+1} - \sqrt{n}}{\sqrt{n^2 + n}}$

   d) $\displaystyle\sum_{n=1}^{\infty} \dfrac{(-1)^n - 8}{3^n}$

   e) $\displaystyle\sum_{n=1}^{\infty} \left(\left(1 - \dfrac{1}{n}\right)^n - \left(1 - \dfrac{1}{n+1}\right)^{n+1}\right)$

   f) $\displaystyle\sum_{n=1}^{\infty} \left(2^{-n} - 2^{-3n}\right)$

   g) $\displaystyle\sum_{n=3}^{\infty} \dfrac{\arctan(n) + \arctan(-n+2)}{\arctan(n)\arctan(n-2)}$

3. Let $\displaystyle\sum_{n=1}^{\infty} a_n$ be a convergent series. Show that the series $\displaystyle\sum_{n=1}^{\infty} \dfrac{a_n^3 + 5n}{\sqrt{n^2 + 1}}$ is divergent.

4. Find the values of x for which the following series converge and, when possible, evaluate the sum of the series:

   a) $\displaystyle\sum_{n=0}^{\infty} \dfrac{8^n}{(x+1)^{3n}}$

   b) $\displaystyle\sum_{n=1}^{\infty} (|x| - 1)^n$

   c) $\displaystyle\sum_{n=0}^{\infty} (-1)^n x^{2n+1}$

5. Show that if $\displaystyle\sum_{n=0}^{\infty} a_n = A \in \mathbb{R}$, then
$$\sum_{n=1}^{\infty} (a_{n-1} + a_n + a_{n+1}) = 3A - a_1 - 2a_0.$$

6. Test for convergence or divergence the following series. If convergence occurs, indicate whether it is conditional or absolute:

   a) $\displaystyle\sum_{n=1}^{\infty} \dfrac{(-1)^{n+1}}{n}$

   b) $\displaystyle\sum_{n=1}^{\infty} \dfrac{(-1)^{n-1}}{n^2}$

   c) $\displaystyle\sum_{n=1}^{\infty} (-1)^n \dfrac{n^2}{1 + n^2}$

7. Consider the series $\displaystyle\sum_{n=1}^{\infty} \dfrac{(-1)^n}{(n+1)^2}$:

   a) Verify that it is convergent.

   b) Calculate its sum with an error less than $\dfrac{1}{10000}$.

8. Let $\sum a_n$ and $\sum b_n$ be two convergent series, $\sum c_n$ and $\sum d_n$ be two divergent series, and $\alpha \neq 0$ be a real number. What can be said about the convergence of the following series?

   a) $\sum (a_n + b_n)$ e) $\sum (a_n c_n)$

   b) $\sum (a_n b_n)$ f) $\sum (\alpha c_n)$

   c) $\sum (\alpha a_n)$ g) $\sum (c_n + d_n)$

   d) $\sum (a_n + c_n)$ h) $\sum (c_n d_n)$

9. Test the convergence of the following series using a comparison test:

   a) $\displaystyle\sum_{n=1}^{\infty} \dfrac{1}{n^3 + 3}$

b) $\sum_{n=1}^{\infty} \sqrt{\dfrac{n+1}{n^2+1}}$

c) $\sum_{n=1}^{\infty} \dfrac{1}{\sqrt{n(n^2+1)}}$

d) $\sum_{n=1}^{\infty} \dfrac{n\sqrt{n}}{(n+1)^3 \sqrt{n^3+1}}$

e) $\sum_{n=1}^{\infty} \left(n - \sqrt{n^2-1}\right)$

f) $\sum_{n=1}^{\infty} \left(\sin\left(\dfrac{\pi}{n}\right)\right)^2$

g) $\sum_{n=2}^{\infty} \dfrac{n+\sqrt{n}}{n^2-n}$

h) $\sum_{n=1}^{\infty} \dfrac{n^2+1}{n^3+1}$

10. Using the Cauchy Condensation Test or the Integral Test, determine the convergence of the following series:

a) $\sum_{n=2}^{\infty} \dfrac{1}{n(\log(n))^2}$

b) $\sum_{n=2}^{\infty} \dfrac{\log(3n)}{n^2}$

c) $\sum_{n=2}^{\infty} n\, e^{-n^2}$

d) $\sum_{n=2}^{\infty} \dfrac{\arctan(n)}{n^2+1}$

e) $\sum_{n=1}^{\infty} \dfrac{1}{\sqrt{n+1}-1}$

f) $\sum_{n=1}^{\infty} \left(\dfrac{1}{2}\right)^{\sqrt{n}}$

11. Use the Ratio Test or By the continuity of the logarithm function to determine the convergence or divergence of the following series:

a) $\sum_{n=1}^{\infty} \dfrac{n}{3^n}$

b) $\sum_{n=1}^{\infty} \dfrac{n!}{3^n}$

c) $\sum_{n=1}^{\infty} \dfrac{n^3}{n!}$

d) $\sum_{n=1}^{\infty} \dfrac{n^n}{(2n)!}$

e) $\sum_{n=1}^{\infty} \dfrac{n!}{n^2}$

f) $\sum_{n=1}^{\infty} 2^{-n+(-1)^n}\, 3^{-n+(-1)^{n+1}}$

g) $\sum_{n=1}^{\infty} \dfrac{n!}{3 \times 5 \times 7 \times \cdots \times (2n-1)}$

h) $\sum_{n=1}^{\infty} \dfrac{2 \times 4 \times 6 \times \cdots \times (2n)}{2 \times 5 \times 8 \times \cdots \times (3n-1)}$

12. Use the Root Test or Cauchy's Root Test to determine the convergence or divergence of the following series:

a) $\sum_{n=1}^{\infty} \dfrac{1}{(n+1)^n}$

b) $\sum_{n=2}^{\infty} \dfrac{n^{n/2}}{(\log(n))^n}$

c) $\sum_{n=1}^{\infty} \dfrac{k^n}{n!},\ k \in \mathbb{R}$

d) $\sum_{n=1}^{\infty} \left(\dfrac{n+1}{n}\right)^{n^2}$

e) $\sum_{n=1}^{\infty} 2^{-2n+(-1)^n n}$

f) $\sum_{n=1}^{\infty} \dfrac{1}{n^n}\left(\dfrac{1}{n}-1\right)^n$

g) $\sum_{n=1}^{\infty} \left(\sin\left(\dfrac{\pi}{n}\right)\right)^n$

13. Test the convergence of the following series using Raabe's test:

a) $\sum_{n=1}^{\infty} \dfrac{n}{(2n+1)!}$

2.8. Proposed Exercises

b) $\sum_{n=1}^{\infty} \dfrac{1}{n(n+1)(n+2)}$

c) $\sum_{n=1}^{\infty} \dfrac{\sqrt{n-1}}{n}$

d) $\sum_{n=1}^{\infty} \dfrac{(2n+1)!}{n \times n!}$

14. Investigate the convergence or divergence of the following series. If convergence occurs, indicate whether it is conditional or absolute:

a) $\sum_{n=0}^{\infty} \dfrac{1}{\sqrt{n^2+1}}$

b) $\sum_{n=2}^{\infty} \dfrac{1}{\sqrt{n(n^2-1)}}$

c) $\sum_{n=1}^{\infty} \dfrac{e^n n^n}{n!}$

d) $\sum_{n=1}^{\infty} \dfrac{2}{n^2+p^2}$, $p \in \mathbb{R}$

e) $\sum_{n=1}^{\infty} \dfrac{\cos(n\frac{\pi}{2})}{n^2}$

f) $\sum_{n=0}^{\infty} \dfrac{5^{2n}}{(n+1)!}$

g) $\sum_{n=3}^{\infty} \left(\tan\left(\dfrac{\pi}{n}\right)\right)^n$

h) $\sum_{n=1}^{\infty} \dfrac{1}{n}\cos\left(\dfrac{n\pi}{2}\right)$

i) $\sum_{n=1}^{\infty} \dfrac{1}{n^{(1+\frac{1}{n})}}$

j) $\sum_{n=1}^{\infty} \dfrac{n^2+2n+1}{3n^2+2}$

k) $\sum_{n=1}^{\infty} \dfrac{\left(\sin\left(\frac{3\pi}{2}\right)\right)^n}{n^2+1}$

l) $\sum_{n=2}^{\infty} \dfrac{1}{n^n \log(n)}$

m) $\sum_{n=2}^{\infty} \dfrac{1}{n^{\left(1+\frac{1}{\log(n)}\right)}}$

n) $\sum_{n=2}^{\infty} n^e \log(n)$

o) $\sum_{n=0}^{\infty} \dfrac{2^n}{(2n+1)!}$

p) $\sum_{n=1}^{\infty} \dfrac{2\cos(n\theta)}{n^{5/2}}$, $\theta \in \mathbb{R}$

q) $\sum_{n=0}^{\infty} \dfrac{3^n n^{2n}}{(2n)!}$

r) $\sum_{n=1}^{\infty} \dfrac{\log(n!) + n!}{n^n + 2^n}$

s) $\sum_{n=0}^{\infty} (-1)^n \dfrac{\sqrt[3]{n^2+1}}{n+3}$

15. Examine the convergence or divergence of the following series. If convergence occurs, indicate whether it is conditional or absolute:

a) $\sum_{n=1}^{\infty} \dfrac{\sin(n+1)}{n^2 \log(n+1)}$

b) $\sum_{n=3}^{\infty} \left(\dfrac{1}{3^n} + \dfrac{(n!)^2}{(2n)!}\right)$

c) $\sum_{n=1}^{\infty} \dfrac{1}{\sqrt{(n+1)(n+2)}}$

d) $\sum_{n=3}^{\infty} \cos(n\pi) \tan\left(\dfrac{e}{n}\right)$

e) $\sum_{n=1}^{\infty} \dfrac{1+(-1)^n\, n}{n^2+5}$

f) $\sum_{n=1}^{\infty} \left(\dfrac{2n}{4n+1}\right)^{3n-1}$

g) $\sum_{n=1}^{\infty} \left(\dfrac{n!}{n^n} 2^n + \dfrac{1}{n^2+n}\right)$

h) $\sum_{n=1}^{\infty} \dfrac{(n+p)!}{n!(n+q)!}$, $p, q \in \mathbb{N}$

i) $\sum_{n=1}^{\infty} \dfrac{n!}{(\pi+1)(\pi+2)\ldots(\pi+n)}$

j) $\sum_{n=1}^{\infty} \left(\dfrac{(-1)^n}{n(n+1)} + \dfrac{1}{\sqrt{3^n}}\right)$

k) $\sum_{n=0}^{\infty} (-1)^n \dfrac{2n+3}{(n+1)(n+2)}$

l) $\sum_{n=0}^{\infty} \left(\frac{a}{a+3}\right)^n$, $a \in \mathbb{R} \setminus \{-3\}$

m) $\sum_{n=0}^{\infty} \frac{1}{2^n + a}$, $a \in \mathbb{R}_0^+$

n) $\sum_{n=2}^{\infty} \log \left|\log\left(\frac{1}{n!}\right)\right|$

o) $\sum_{n=0}^{\infty} \frac{(3n)!}{27^n (n!)^2}$

p) $1 + \frac{1}{2} \cdot \frac{1}{3} + \frac{1 \times 3}{2 \times 4} \cdot \frac{1}{5} + \frac{1 \times 3 \times 5}{2 \times 4 \times 6} \cdot \frac{1}{7} + \cdots$

q) $\sum_{n=1}^{\infty} \frac{\sqrt{2n-1} \log(4n+1)}{n(n+1)}$

r) $\sum_{n=1}^{\infty} \frac{3 \times 5 \times 7 \times \cdots \times (2n+1)}{2^n (2n-1) n!}$

s) $\sum_{n=1}^{\infty} \frac{(-3)^n}{3 \times 5 \times 7 \times \cdots \times (2n+1)}$

t) $\sum_{n=1}^{\infty} \frac{2 + \sin^3(n+1)}{2^n + n^2}$

u) $\sum_{n=1}^{\infty} \frac{1}{2^n - 1 + \sin^2(n^3)}$

16. Indicate for which values of α the following series are conditionally or absolutely convergent:

a) $\sum_{n=1}^{\infty} (1 + \sin(\alpha))^n$

b) $\sum_{n=0}^{\infty} (-1)^n \frac{1}{(n+1)^\alpha}$

17. a) Test for convergence the series
$$\sum_{n=1}^{\infty} e^{2n} \left(\frac{n}{n+3}\right)^{n^2}.$$

b) Based on the previous item, indicate the limit of the sequence $\left(e^{2n}\left(\frac{n}{n+3}\right)^{n^2}\right)$.

18. Let $\sum a_n$ and $\sum b_n$ be two convergent series of positive terms. Show that the series $\sum \sqrt{a_n b_n}$ also converges.

Hint: Prove that $\frac{a_n + b_n}{2} \geq \sqrt{a_n b_n}$.

19. Knowing that $\sum a_n$ is convergent and of positive terms and $b_n > 0$, $\forall n \in \mathbb{N}$, what can be said about the convergence of the series $\sum \frac{a_n}{1+b_n}$?

20. Knowing that $\sum a_n$ and $\sum b_n$ are convergent series of positive terms, study the convergence of the following series:

a) $\sum \left(\frac{1}{a_n} + \frac{1}{b_n}\right)$

b) $\sum \frac{n+1}{n} a_n$

21. Let $\sum a_n$ be a divergent series, $a_n \geq 0$, and s_n be the sum of its first n terms. Show that the series
$$\sum \left(\sqrt{s_{n+1}} - \sqrt{s_n}\right)$$
is divergent.

22. Prove that the series $\sum \frac{a_0 n^p + \cdots + a_p}{b_0 n^q + \cdots + b_q}$ in which a_0, ..., a_p, b_0, ..., b_q are real numbers and $a_0 > 0$, $b_0 > 0$, is convergent if and only if $q - p > 1$.

23. Study the conditional and absolute convergence of the series:

a) $\sum_{n=1}^{\infty} \frac{a^n}{(1+a)(1+a^2)\ldots(1+a^n)}$, $a > 0$

b) $\sum_{n=1}^{\infty} \frac{(\alpha+1)(\alpha+2)\ldots(\alpha+n)}{(\beta+1)(\beta+2)\ldots(\beta+n)}$, $\beta \in \mathbb{R} \setminus \mathbb{Z}^-$:

  i. If $\alpha \in \mathbb{R} \setminus \mathbb{Z}^-$
  ii. If $\alpha \in \mathbb{Z}^-$

24. Let $u_n > 0$ and $\frac{u_{n+1}}{u_n} \leq 1 - \frac{2}{n} + \frac{1}{n^2}$ $\forall n \in \mathbb{N} \setminus \{1\}$. Show that $\sum u_n$ is convergent.

25. Consider the series
$$\sum_{n=0}^{\infty} \frac{(-1)^n}{n!} \quad \text{and} \quad \sum_{n=0}^{\infty} \frac{1}{(n+1)\sqrt{n+1}}:$$

a) Calculate the partial sum of order three of Cauchy's product of the two series.

b) Study the convergence of the product series.

2.8. Proposed Exercises

26. The figure shows a sequence of cubes constructed as follows: The edge of the larger cube has length 1, and the length of the edge of each of the following cubes is half the length of the edge of the previous cube. Find the sum of the volumes of the cubes.

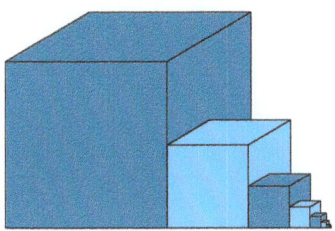

27. The figure shows a sequence of squares constructed as follows: Start with a square of side length 4. Now, connect the midpoints of the sides of this square to form another square inside it. Repeat this process with the new square, connecting its midpoints to form another square. Keep repeating this process infinitely, and we will obtain infinite squares within the first one.

Determine the sum of the areas of all the squares.

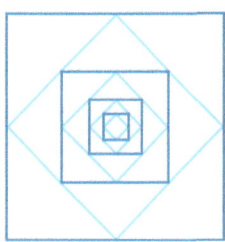

CHAPTER 3: Series of Functions

It is possible to express various functions as a combination of "simpler" functions, such as a series of monomials (known as Taylor series) or trigonometric functions (known as Fourier series). These function series are useful not only in approximating other functions but also in other areas of mathematics, such as differential equations.

3.1 Introduction: Sequences of Functions

In many Analysis issues, it is interesting to consider sequences of functions of the form $f_1(x), f_2(x), \ldots, f_n(x), \ldots$, and naturally the question of passing to the limit arises.

> **Definition 3.1.1** Let (f_n) be a sequence of functions, $f_n : D \subset \mathbb{R} \to \mathbb{R}$. The sequence (f_n) **converges at a point** $a \in D$ if the numerical sequence $(f_n(a))$ is convergent.
>
> If the sequence (f_n) converges at all points of D, we can define a function $f : D \to \mathbb{R}$ by
> $$f(x) = \lim_{n \to +\infty} f_n(x),$$
> which is called the **limit of** (f_n) on D. We also say that (f_n) **converges pointwise** to f on D.

Example 3.1.1 The sequence of functions $\left(1+\dfrac{x}{n}\right)^n$, defined on $\mathbb{R}$, converges whatever $x \in \mathbb{R}$ (see Fig. 3.1). The limit function is $f(x) = e^x$:

$$\lim \left(1+\frac{x}{n}\right)^n = e^x \quad \forall x \in \mathbb{R}.$$

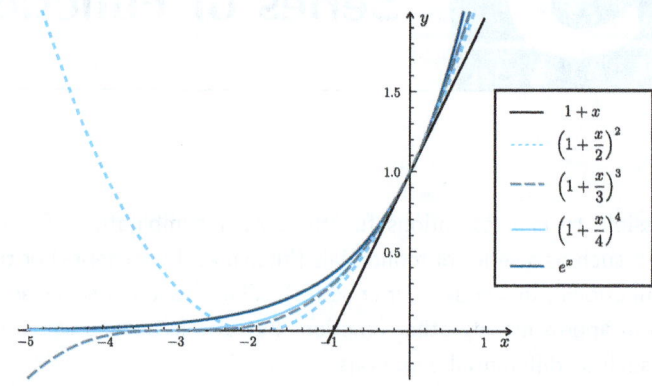

Figure 3.1: The function $f(x) = e^x$ and the first terms of the sequence

Example 3.1.2 Consider the functions $f_n(x) = x^n$, $n \in \mathbb{N}$, on the interval $[0, 1]$. They are continuous, and the limit function exists, but it is not continuous:

$$f(x) = \lim x^n = \begin{cases} 0, & \text{if } 0 \leq x < 1 \\ 1, & \text{if } x = 1 \end{cases}$$

(see Fig. 3.2).

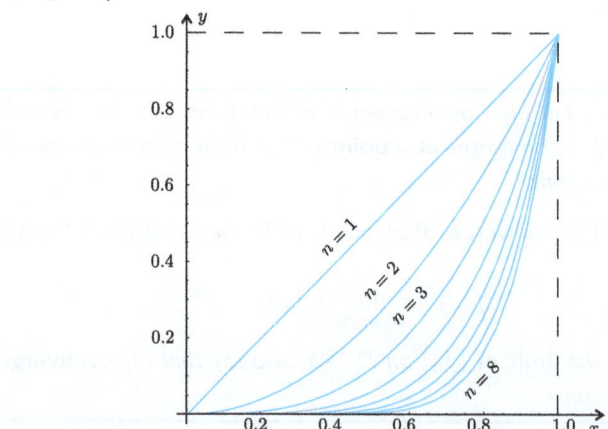

Figure 3.2: Sequence of continuous functions converging to a discontinuous function

3.1. Introduction: Sequences of Functions 169

Based on Example 3.1.2, we can observe that a sequence of continuous functions may converge pointwise to a discontinuous function. We will see later that even if all the integrals in question exist, the integral of the limit function may differ from the limit of the integrals of the sequence's functions. This means that the limit does not always commute with the integral. However, there is a stronger form of convergence known as uniform convergence, which guarantees both the continuity of the limit of a sequence of continuous functions and the ability to interchange limit and integration symbols.

> **Definition 3.1.2** *The sequence of functions (f_n)* **converges uniformly** *to f on $D \subset \mathbb{R}$ ($D \neq \varnothing$) if*
>
> $$\forall \delta > 0 \ \exists p \in \mathbb{N} : n > p \Rightarrow |f_n(x) - f(x)| < \delta, \ \forall x \in D.$$
>
> *This condition is equivalent to*
>
> $$\forall \delta > 0 \ \exists p \in \mathbb{N} : \ \forall n > p, \ \sup_{x \in D} |f_n(x) - f(x)| < \delta,$$
>
> *that is,*
>
> $$\lim_{n \to +\infty} \sup_{x \in D} |f_n(x) - f(x)| = 0.$$

It is evident that if a sequence of functions (f_n) converges uniformly to f on D, then it also converges pointwise to f on D. The converse is not true. If we take the sequence (f_n) from Example 3.1.2, which converges pointwise to f on $[0, 1]$, we have

$$\sup_{x \in [0,1]} |f_n(x) - f(x)| = 1, \ \forall n \in \mathbb{N},$$

so (f_n) does not converge uniformly to f on $[0, 1]$.

Note that if (f_n) converges uniformly to f on D, then (f_n) converges uniformly to f on any subset of D.

The definition of uniform convergence means that, for every fixed $\delta > 0$, there is an order from which the images of all functions are in the δ neighborhood of $f(x)$ for any $x \in D$, that is, there is an order from which the images of all functions are in the region of the plane defined by $f(x) - \delta$ and $f(x) + \delta$, in the set D. This is illustrated in Fig. 3.3.

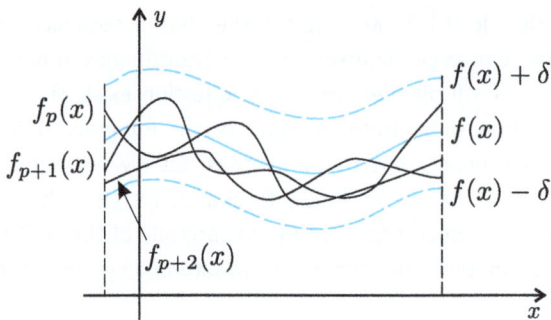

Figure 3.3: An illustration of uniform convergence

Obviously, there are sequences of continuous functions that converge pointwise to continuous functions, but not uniformly, as shown in the following example.

Example 3.1.3 Let $\alpha \in \mathbb{R}$. Consider the sequence of functions of general term
$$f_n(x) = x\, n^\alpha e^{-nx}, \quad x \in \mathbb{R}_0^+.$$
This sequence converges pointwise to the function (Fig. 3.4)
$$f(x) = 0, \quad \forall x \in \mathbb{R}_0^+.$$
However, this convergence is not uniform if $\alpha \geq 1$: Note that
$$\sup_{x \in \mathbb{R}_0^+} |f_n(x) - f(x)| = \max_{x \in \mathbb{R}_0^+} f_n(x) = f_n\left(\frac{1}{n}\right) = \frac{n^{\alpha-1}}{e},$$
which implies that
$$\lim \sup_{x \in \mathbb{R}_0^+} |f_n(x) - f(x)| = \lim \frac{n^{\alpha-1}}{e} = \begin{cases} +\infty, & \text{if } \alpha > 1 \\ \dfrac{1}{e}, & \text{if } \alpha = 1 \\ 0, & \text{if } \alpha < 1. \end{cases}$$

Figure 3.5 illustrates the uniform convergence of the sequence of functions
$$f_n(x) = x\, n^{1/2} e^{-nx}.$$
Remark the variation of $p \in \mathbb{N}$ in relation to $\delta > 0$.

3.1. Introduction: Sequences of Functions 171

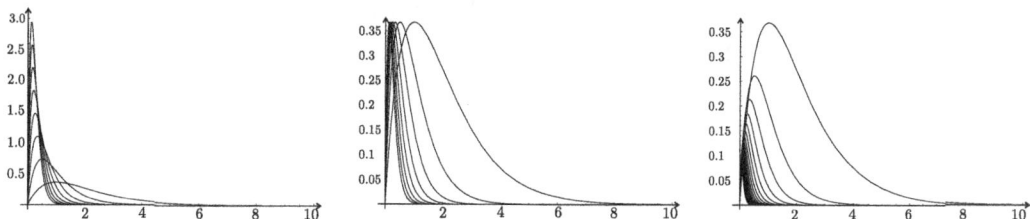

Figure 3.4: The sequence of functions $\left(x\, n^\alpha e^{-nx}\right)$, $\alpha > 1$, $\alpha = 1$, and $\alpha < 1$

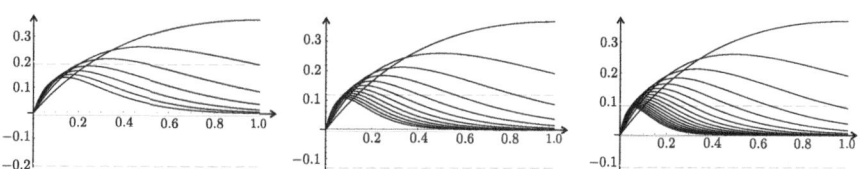

Figure 3.5: The sequence of functions $\left(x\, n^{1/2} e^{-nx}\right)$

3.2 Pointwise and Uniform Convergence of Series of Functions

The concepts of pointwise and uniform convergence extend to series of functions.

Definition 3.2.1 *Let (f_n) be a sequence of functions, $f_n : X \subset \mathbb{R} \to \mathbb{R}$. The **series of general term** f_n is the sequence of functions (S_n) defined by*
$$S_n(x) = f_1(x) + f_2(x) + \cdots + f_n(x), \ \forall x \in X;$$
the series is also represented by $\sum_{n=1}^{\infty} f_n$ or by $\sum f_n$.

Definition 3.2.2 *The series $\sum f_n$ **converges at a point** $a \in X$ if the numerical series $\sum f_n(a)$ is convergent.*

*If the series is convergent at all points of $D \subset X$, we can define a function $f : D \to \mathbb{R}$ that corresponds to the sum of the series $\sum f_n(x)$ at each point $x \in D$; the function f is called the **sum function of the series**. We also say that $\sum f_n(x)$ converges pointwise to f on D.*

Example 3.2.1 Consider the series $\displaystyle\sum_{n=0}^{\infty} \frac{x^2}{(1+x^2)^n}$.

If $x = 0$, all terms are zero; therefore, the series is convergent.
If $x \neq 0$, we can write
$$\sum_{n=0}^{\infty} \frac{x^2}{(1+x^2)^n} = x^2 \sum_{n=0}^{\infty} \frac{1}{(1+x^2)^n} = x^2 \sum_{n=0}^{\infty} \left(\frac{1}{1+x^2}\right)^n,$$
and this is a geometric series with ratio $r = \dfrac{1}{1+x^2}$; as $|r| < 1$, the series is convergent and
$$\sum_{n=0}^{\infty} \frac{x^2}{(1+x^2)^n} = x^2 \cdot \frac{1}{1 - \dfrac{1}{1+x^2}} = 1 + x^2.$$

The sum function is
$$f(x) = \begin{cases} 1 + x^2, & \text{if } x \neq 0 \\ 0, & \text{if } x = 0. \end{cases}$$

3.2. Pointwise and Uniform Convergence of Series of Functions 173

Example 3.2.2 Consider the series $\sum_{n=0}^{\infty} \dfrac{x^n}{n!}$. We can use a numerical series test to study the pointwise convergence of series of functions. In this case, we will apply D'Alembert's test to study the series

$$\sum_{n=0}^{\infty} \left| \dfrac{x^n}{n!} \right|.$$

$$\lim \dfrac{\left|\dfrac{x^{n+1}}{(n+1)!}\right|}{\left|\dfrac{x^n}{n!}\right|} = \lim \dfrac{|x|}{n+1} = 0, \quad \forall x \in \mathbb{R}.$$

Thus, we conclude that the original series is absolutely convergent $\forall x \in \mathbb{R}$, defining a function f on $\mathbb{R}$. We will see later that $f(x) = e^x$, that is,

$$\sum_{n=0}^{\infty} \dfrac{x^n}{n!} = e^x, \quad \forall x \in \mathbb{R}.$$

Example 3.2.3 Consider the series $\sum_{n=0}^{\infty} x\,(1-x)^n$, $x \in [0, 1]$.

If $x = 0$, all terms are zero. Therefore, the series is convergent.
If $x \neq 0$, as the series $\sum_{n=0}^{\infty} (1-x)^n$ is geometric with ratio $r = 1 - x$ and $|r| < 1$ since $x \in\,]0, 1]$, the series converges. In this case,

$$\sum_{n=0}^{\infty} x\,(1-x)^n = x \cdot \dfrac{1}{1-(1-x)} = 1.$$

Then, the series $\sum_{n=0}^{\infty} x\,(1-x)^n$, $x \in [0,1]$, converges pointwise to the function

$$f(x) = \begin{cases} 1, & \text{if } 0 < x \leq 1 \\ 0, & \text{if } x = 0. \end{cases}$$

> **Definition 3.2.3** *We say that a series $\sum f_n$* **converges uniformly** *to the function f on $D \subset \mathbb{R}$, $D \neq \emptyset$, if*
>
> $$\forall \delta > 0 \ \exists p \in \mathbb{N} : n > p \Rightarrow \left| f(x) - \sum_{i=1}^{n} f_i(x) \right| < \delta, \ \forall x \in D.$$
>
> *This condition is equivalent to*
>
> $$\forall \delta > 0 \ \exists p \in \mathbb{N} : n > p \Rightarrow \sup_{x \in D} |f(x) - S_n(x)| < \delta,$$
>
> *that is,*
>
> $$\lim_{n \to +\infty} \sup_{x \in D} |f(x) - S_n(x)| = 0.$$

Note: Uniform convergence implies pointwise convergence, but the opposite is not necessarily true. This means that if a series of functions converges uniformly, it will also converge pointwise. However, if a series of functions converges pointwise, it does not necessarily converge uniformly.

Example 3.2.4 We saw that the series $\sum_{n=0}^{\infty} \dfrac{x^2}{(1+x^2)^n}$ is pointwise convergent to the function f defined by

$$f(x) = \begin{cases} 1 + x^2, & \text{if } x \neq 0 \\ 0, & \text{if } x = 0. \end{cases}$$

However, this series is not uniformly convergent on $[-1, 1]$. In fact,

$$\lim_{n \to +\infty} \sup_{x \in [-1,1]} |f(x) - S_n(x)|$$
$$= \lim_{n \to +\infty} \sup_{\substack{x \in [-1,1] \\ x \neq 0}} \left| 1 + x^2 - (1+x^2)\left(1 - \frac{1}{(1+x^2)^{n+1}}\right) \right|$$
$$= \lim_{n \to +\infty} \sup_{\substack{x \in [-1,1] \\ x \neq 0}} \left| \frac{1}{(1+x^2)^n} \right| = \lim_{n \to +\infty} \sup_{\substack{x \in [-1,1] \\ x \neq 0}} \frac{1}{(1+x^2)^n} = 1.$$

The function f and some partial sums are illustrated in Fig. 3.6.

3.2. Pointwise and Uniform Convergence of Series of Functions

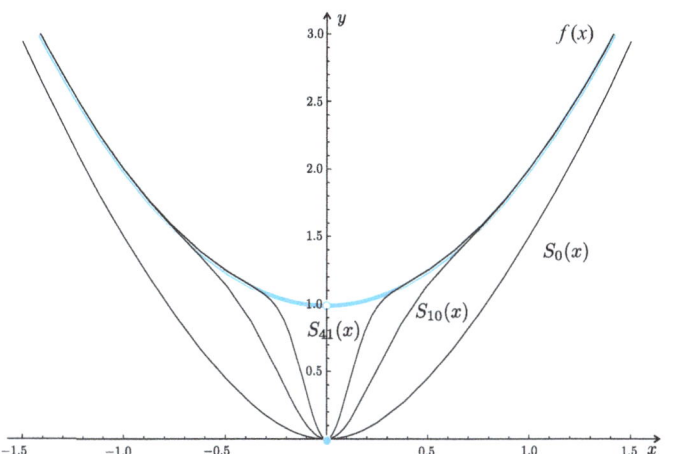

Figure 3.6: An illustration of the sum function and some partial sums of Example 3.2.4

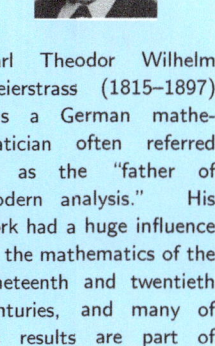

Karl Theodor Wilhelm Weierstrass (1815–1897) was a German mathematician often referred to as the "father of modern analysis." His work had a huge influence on the mathematics of the nineteenth and twentieth centuries, and many of his results are part of any Calculus course, such as, for example, the Weierstrass' test for series and his example of a continuous function that is not differentiable at any point. (Source of image: Dibner Library of the History of Science and Technology, Smithsonian Libraries)

The following theorem allows us to test the uniform convergence of a series of functions without knowing its sum function.

Theorem 3.2.1 *It is a necessary and sufficient condition for the series $\sum f_n$ to be uniformly convergent on $D \subset \mathbb{R}$ that*

$$\forall \delta > 0 \ \exists p \in \mathbb{N}: \ m > n > p \Rightarrow \left| \sum_{r=n+1}^{m} f_r(x) \right| < \delta, \quad \forall x \in D.$$

Theorem 3.2.2 (Weierstrass' Test) *If there exists a convergent numerical series with positive terms, $\sum a_n$, such that*

$$|f_n(x)| \leq a_n, \ \forall x \in D, \forall n \in \mathbb{N},$$

then the series $\sum f_n$ is uniformly convergent on D.

Proof: We know by Theorem 2.2.3 that $\sum a_n$ converges if and only if

$$\forall \delta > 0 \ \exists p \in \mathbb{N}: \ m > n > p \ \Rightarrow |a_{n+1} + \cdots + a_m| < \delta.$$

Let $\delta > 0$ be arbitrary. We have

$$|f_{n+1}(x)+\cdots+f_m(x)| \leq |f_{n+1}(x)|+\cdots+|f_m(x)|$$
$$\leq a_{n+1}+\cdots+a_m \quad \forall x \in D$$
$$= |a_{n+1}+\cdots+a_m|, \text{ since } a_n > 0, \forall n \in \mathbb{N}.$$

Then
$$\exists p \in \mathbb{N}: \ m > n > p \Rightarrow |f_{n+1}(x)+\cdots+f_m(x)| < \delta,$$
or still,
$$\exists p \in \mathbb{N}: \ m > n > p \Rightarrow \left|\sum_{r=n+1}^{m} f_r(x)\right| < \delta.$$

From Theorem 3.2.1 comes the desired result. ∎

Example 3.2.5 Let k be a constant such that $|k|<1$. The series $\sum_{n=1}^{\infty} f_n(x)$, where
$$f_n(x) = \frac{1 \times 3 \times \cdots \times (2n-1)}{2 \times 4 \times \cdots \times 2n} \cdot k^{2n} \big(\sin(x)\big)^{2n},$$
is uniformly convergent on $\mathbb{R}$. In fact,
$$|f_n(x)| = \left|\frac{1 \times 3 \times \cdots \times (2n-1)}{2 \times 4 \times \cdots \times 2n} \cdot k^{2n}\big(\sin(x)\big)^{2n}\right|$$
$$\leq \frac{1 \times 3 \times \cdots \times (2n-1)}{2 \times 4 \times \cdots \times 2n} \cdot k^{2n}, \quad \forall x \in \mathbb{R},$$
and the numerical series
$$\sum_{n=1}^{\infty} \frac{1 \times 3 \times \cdots \times (2n-1)}{2 \times 4 \times \cdots \times 2n} \cdot k^{2n}$$
is convergent. To verify this, we apply D'Alembert's test:
$$\lim \frac{\frac{1 \times 3 \times \cdots \times (2n-1)(2n+1)}{2 \times 4 \times \cdots \times 2n(2n+2)} \cdot k^{2n+2}}{\frac{1 \times 3 \times \cdots \times (2n-1)}{2 \times 4 \times \cdots \times 2n} \cdot k^{2n}} = \lim \frac{2n+1}{2n+2} \cdot k^2 = k^2 < 1.$$

3.2. Pointwise and Uniform Convergence of Series of Functions

Example 3.2.6 Since $\left|\dfrac{\sin(nx)}{n^2}\right| \leq \dfrac{1}{n^2}$, $\forall x \in \mathbb{R}$, and the series $\sum_{n=1}^{\infty} \dfrac{1}{n^2}$ is convergent, the series $\sum_{n=1}^{\infty} \dfrac{\sin(nx)}{n^2}$ is uniformly convergent on $\mathbb{R}$.

Note: The Weierstrass' test is a sufficient but not necessary condition for the uniform convergence of a series of functions: there exist uniformly convergent series whose general term cannot be bounded as required by the Weierstrass' test. It is important to note that this test implies absolute convergence of the series of functions.

Example 3.2.7 For each $x \in \mathbb{R}$, the series $\sum_{n=1}^{\infty} (-1)^n \dfrac{x^2 + n}{n^2}$ is alternating, and using Leibniz's test, we can prove that it is convergent. However, it is not absolutely convergent because

$$\left|(-1)^n \dfrac{x^2+n}{n^2}\right| = \dfrac{x^2+n}{n^2} \geq \dfrac{1}{n}, \quad \forall x \in \mathbb{R},$$

and the series $\sum_{n=1}^{\infty} \dfrac{1}{n}$ is divergent. This makes it impossible to use the Weierstrass' test to determine the uniform convergence of the series. Instead, we have to study it directly.

For each $x \in \mathbb{R}$, let $S_n(x)$ be the partial sum of order n of the series $S(x) = \sum_{n=1}^{\infty} (-1)^n \dfrac{x^2+n}{n^2}$. By Corollary 1 of Theorem 2.3.2,

$$|S(x) - S_n(x)| \leq \dfrac{x^2+n+1}{(n+1)^2}.$$

If $X \subset \mathbb{R}$ is a bounded set, that is, if there exists $M > 0$ such that $|x| \leq M$, $\forall x \in X$, then

$$|S(x) - S_n(x)| \leq \dfrac{M^2+n+1}{(n+1)^2}, \quad \forall x \in X;$$

therefore (S_n) converges uniformly to S on X, that is, the given series converges uniformly on any bounded subset of $\mathbb{R}$.

The importance of uniform convergence lies in the fact that a series that converges uniformly behaves similarly to the sum of a finite number of functions. For instance, when we add a finite number of continuous functions, the result is also a continuous function. Similarly, with a series of functions that converge uniformly, we obtain the following outcome:

> **Theorem 3.2.3** *If the functions $f_1, f_2, \ldots, f_n, \ldots$ are continuous on D and the series $\sum f_n$ converges uniformly to f on D, then f is continuous on D.*

Proof: Let x_0 be an arbitrary point of D. We want to prove that

$$\forall \delta > 0 \ \exists \varepsilon > 0: \ |x - x_0| < \varepsilon \Rightarrow |f(x) - f(x_0)| < \delta.$$

We can write

$$f(x) - f(x_0) = f(x) - S_n(x) + S_n(x) - S_n(x_0) + S_n(x_0) - f(x_0),$$

which implies that

$$|f(x) - f(x_0)| \leq |f(x) - S_n(x)| + |S_n(x) - S_n(x_0)| + |S_n(x_0) - f(x_0)|.$$

Let $\delta > 0$. As the series converges uniformly to f, we know that

$$\exists p \in \mathbb{N}: \ \forall n > p, \ |f(x) - S_n(x)| < \frac{\delta}{3}, \ \forall x \in D.$$

As $f_1, f_2, \ldots, f_n, \ldots$ are continuous, S_n is a continuous function, so

$$\exists \varepsilon > 0: \ |x - x_0| < \varepsilon \Rightarrow |S_n(x) - S_n(x_0)| < \frac{\delta}{3}.$$

Then

$$\exists \varepsilon > 0: \ |x - x_0| < \varepsilon \Rightarrow |f(x) - f(x_0)| < \delta.$$

The result follows from the arbitrariness of x_0. ∎

Example 3.2.8 Consider the function $f(x) = \sum_{n=1}^{\infty} \dfrac{\cos(nx)}{\sqrt{n^3 + 1}}$. The terms of the series are functions with domain $\mathbb{R}$. The function f is the sum of the series; therefore, its domain is the subset of $\mathbb{R}$ where the series converges. We have the inequality

$$\left| \frac{\cos(nx)}{\sqrt{n^3 + 1}} \right| \leq \frac{1}{\sqrt{n^3 + 1}}, \quad \forall x \in \mathbb{R}, \tag{3.1}$$

3.2. Pointwise and Uniform Convergence of Series of Functions

since $|\cos(x)| \leq 1$, $\forall x \in \mathbb{R}$.

Let us study the series of positive terms $\sum_{n=1}^{\infty} \dfrac{1}{\sqrt{n^3+1}}$, comparing it with the series $\sum_{n=1}^{\infty} \dfrac{1}{n^{3/2}}$, which we know is convergent. As the limit

$$\lim \frac{\dfrac{1}{\sqrt{n^3+1}}}{\dfrac{1}{n^{3/2}}} = \lim \sqrt{\dfrac{n^3}{n^3+1}} = 1$$

is finite and different from zero, we conclude that the series $\sum_{n=1}^{\infty} \dfrac{1}{\sqrt{n^3+1}}$ is convergent. By inequality (3.1), the series $\sum_{n=1}^{\infty} \left|\dfrac{\cos(nx)}{\sqrt{n^3+1}}\right|$ converges, $\forall x \in \mathbb{R}$, so the series $\sum_{n=1}^{\infty} \dfrac{\cos(nx)}{\sqrt{n^3+1}}$ is convergent, $\forall x \in \mathbb{R}$, that is, the domain of f is $\mathbb{R}$. To conclude that f is continuous on $\mathbb{R}$, it is sufficient that the following two conditions are satisfied:

(i) The functions $f_n(x) = \dfrac{\cos(nx)}{\sqrt{n^3+1}}$ are continuous on $\mathbb{R}$.

(ii) The series $\sum_{n=1}^{\infty} \dfrac{\cos(nx)}{\sqrt{n^3+1}}$ is uniformly convergent to f on $\mathbb{R}$.

The functions $\cos(nx)$ are continuous on $\mathbb{R}$, $\forall n \in \mathbb{N}$, so the first condition is true. The second condition is a consequence of the inequality (3.1) and the Weierstrass' test. As the series converges to f, this function is continuous.

Note: If the sum of a series of functions is not continuous, this indicates that either the individual functions $f_1, f_2, \ldots, f_n, \ldots$ are not continuous or the series is not uniformly convergent. Therefore, if $f_1, f_2, \ldots, f_n, \ldots$ are continuous functions and f is not continuous, we can be sure that the convergence is not uniform.

Example 3.2.9 Let us consider the series $\sum_{n=0}^{\infty} \dfrac{x^2}{(1+x^2)^n}$, on the interval $[-a, a]$, $a > 0$. We proved in Example 3.2.1 that this series converges pointwise to the function

$$f(x) = \begin{cases} 1 + x^2, & \text{if } x \neq 0 \\ 0, & \text{if } x = 0. \end{cases}$$

Since f is discontinuous at $x = 0$ and $f_n(x) = \dfrac{x^2}{(1+x^2)^n}$ is continuous, $\forall n \in \mathbb{N}$, the series is not uniformly convergent.

Theorem 3.2.4 *Let $a, b \in \mathbb{R}$, $a < b$. If the functions $f_1, f_2, \ldots, f_n, \ldots$ are continuous on $[a, b]$ and the series $\sum\limits_{n=1}^{\infty} f_n$ converges uniformly to f on $[a, b]$, then*

$$\int_a^b f(x)\, dx = \sum_{n=1}^{\infty} \int_a^b f_n(x)\, dx.$$

(The series is integrable term by term.)

Proof: By Theorem 3.2.3, f is continuous on $[a, b]$; therefore, it is integrable on $[a, b]$. By hypothesis, the functions $f_1, f_2, \ldots, f_n, \ldots$ are continuous on $[a, b]$, which implies that they are integrable on this interval. We intend to prove that the sequence (S_n^*) of the partial sums of the series $\sum\limits_{n=1}^{\infty} \int_a^b f_n(x)\, dx$ has limit $\int_a^b f(x)\, dx$, that is,

$$\forall \delta > 0 \;\; \exists p \in \mathbb{N}: \; n > p \Rightarrow \left| S_n^* - \int_a^b f(x)\, dx \right| < \delta.$$

Let $\delta > 0$ be arbitrary.

$$\left| S_n^* - \int_a^b f(x)\, dx \right| = \left| \sum_{i=1}^n \int_a^b f_i(x)\, dx - \int_a^b f(x)\, dx \right|$$

$$= \left| \int_a^b \left(\sum_{i=1}^n f_i(x)\, dx \right) - \int_a^b f(x)\, dx \right|$$

3.2. Pointwise and Uniform Convergence of Series of Functions

$$= \left| \int_a^b \left(\sum_{i=1}^n f_i(x) - f(x) \right) dx \right| \leq \int_a^b \left| \sum_{i=1}^n f_i(x) - f(x) \right| dx$$

$$= \int_a^b |S_n(x) - f(x)| \, dx.$$

But the series $\sum_{n=1}^\infty f_n(x)$ converges uniformly to f on $[a,b]$; therefore,

$$\exists p \in \mathbb{N}: \ n > p \Rightarrow |f(x) - S_n(x)| < \frac{\delta}{b-a}, \ \forall x \in [a,b],$$

which implies that

$$\exists p \in \mathbb{N}: \ n > p \Rightarrow \int_a^b |S_n(x) - f(x)| \, dx < \int_a^b \frac{\delta}{b-a} \, dx = \delta. \ \blacksquare$$

Example 3.2.10 Let us consider the series $\sum_{n=1}^\infty \frac{e^{-nx}}{2^n}$, on $[0,1]$.

$$\left| \frac{e^{-nx}}{2^n} \right| = \frac{1}{e^{nx} 2^n} \leq \frac{1}{2^n}, \ \forall x \in [0,1].$$

The series $\sum_{n=1}^\infty \frac{1}{2^n}$ is geometric with ratio $\frac{1}{2}$; therefore, it is convergent. According to Weierstrass' test, the original series is uniformly convergent on $[0,1]$. By Theorem 3.2.4,

$$\int_0^1 \sum_{n=1}^\infty \frac{e^{-nx}}{2^n} \, dx = \sum_{n=1}^\infty \int_0^1 \frac{e^{-nx}}{2^n} \, dx$$

$$= \sum_{n=1}^\infty \frac{1}{2^n} \left[-\frac{e^{-nx}}{n} \right]_0^1 = \sum_{n=1}^\infty \frac{1-e^{-n}}{n 2^n}.$$

Example 3.2.11 The series $\sum_{n=0}^\infty \frac{x^n}{n!}$ is uniformly convergent on every interval $[a,b]$, $a,b \in \mathbb{R}$, because on this interval

$$\left|\frac{x^n}{n!}\right| \leq \frac{M^n}{n!}, \quad \text{where } M = \max(|a|,|b|),$$

and the series $\sum_{n=0}^{\infty} \frac{M^n}{n!}$ is convergent. As $f_n(x) = \frac{x^n}{n!}$ is continuous $\forall n \in \mathbb{N}$, the series $\sum_{n=0}^{\infty} \frac{x^n}{n!}$ is integrable term by term and

$$\int_a^b \sum_{n=0}^{\infty} \frac{x^n}{n!} \, dx = \sum_{n=0}^{\infty} \int_a^b \frac{x^n}{n!} \, dx$$

$$= \sum_{n=0}^{\infty} \frac{1}{n!} \left[\frac{x^{n+1}}{n+1}\right]_a^b = \sum_{n=0}^{\infty} \frac{b^{n+1} - a^{n+1}}{(n+1)!}.$$

Note: A series can be integrable term by term on an interval $[a,b]$ even if it is not uniformly convergent on $[a,b]$.

Example 3.2.12 The series $x + \sum_{n=2}^{\infty}(x^n - x^{n-1})$ is convergent on $[0,1]$ to the function f defined by

$$f(x) = \begin{cases} 0, & \text{if } 0 \leq x < 1 \\ 1, & \text{if } x = 1. \end{cases}$$

Since f_n is continuous on $[0,1]$, $\forall n \in \mathbb{N}$, and f is discontinuous on this interval, the series is not uniformly convergent. However, $\int_0^1 f(x)\, dx = 0$ and

$$\sum_{n=1}^{\infty} \int_0^1 f_n(x)\, dx = \int_0^1 x\, dx + \sum_{n=2}^{\infty} \int_0^1 (x^n - x^{n-1})\, dx$$

$$= \left[\frac{x^2}{2}\right]_0^1 + \sum_{n=2}^{\infty} \left[\frac{x^{n+1}}{n+1} - \frac{x^n}{n}\right]_0^1 = \frac{1}{2} + \sum_{n=2}^{\infty}\left(\frac{1}{n+1} - \frac{1}{n}\right)$$

$$= \frac{1}{2} - \sum_{n=2}^{\infty}\left(\frac{1}{n} - \frac{1}{n+1}\right) = 0.$$

3.2. Pointwise and Uniform Convergence of Series of Functions

> **Corollary 1** Let $a, b \in \mathbb{R}$, $a < b$. If the functions $f_1, f_2, \ldots, f_n, \ldots$ are continuous on $[a, b]$ and the series $\sum_{n=1}^{\infty} f_n$ converges uniformly to f on $[a, b]$, then
> $$\int_a^x f(t)\,dt = \sum_{n=1}^{\infty} \int_a^x f_n(t)\,dt, \quad \forall x \in [a, b].$$
> (*The series is integrable term by term.*)

> **Corollary 2** Let $a, b \in \mathbb{R}$, $a < b$. If the series $\sum_{n=1}^{\infty} f_n$ converges pointwise on the interval $[a, b]$ to a function f, if on this interval the derivatives f_n' exist and are continuous, and if the series $\sum_{n=1}^{\infty} f_n'$ converges uniformly on $[a, b]$, then f is differentiable on $[a, b]$ and
> $$f'(x) = \sum_{n=1}^{\infty} f_n'(x), \quad \forall x \in [a, b].$$
> (*The series is differentiable term by term.*)

<u>Proof</u>: Let $g(x) = \sum_{n=1}^{\infty} f_n'(x)$, $x \in [a, b]$. By Corollary 1,

$$\int_a^x g(t)\,dt = \sum_{n=1}^{\infty} \int_a^x f_n'(t)\,dt = \sum_{n=1}^{\infty} \left[f_n(t)\right]_a^x = \sum_{n=1}^{\infty} \left(f_n(x) - f_n(a)\right),$$

that is,

$$\int_a^x g(t)\,dt = f(x) - f(a);$$

therefore,

$$f(x) = \int_a^x g(t)\,dt + f(a),$$

which implies that $f'(x) = g(x)$, that is, $f'(x) = \sum_{n=1}^{\infty} f_n'(x)$. ∎

Example 3.2.13 Consider the series $\sum_{n=0}^{\infty} x^n$. It is a geometric series of ratio x. The series converges if and only if $|x| < 1$ and, in this case,

$$\sum_{n=0}^{\infty} x^n = \frac{1}{1-x}.$$

Let $0 < r < 1$. Then $|x^n| \leq r^n$, $\forall x \in [-r, r]$. As $\sum_{n=0}^{\infty} r^n$ is a convergent numerical series, then $\sum_{n=0}^{\infty} x^n$ is uniformly convergent on $[-r, r]$. Furthermore

$$\int_{-r}^{r} \frac{1}{1-x} \, dx = \int_{-r}^{r} \sum_{n=0}^{\infty} x^n \, dx = \sum_{n=0}^{\infty} \int_{-r}^{r} x^n \, dx$$

$$\Leftrightarrow \left[-\log|1-x|\right]_{-r}^{r} = \sum_{n=0}^{\infty} \left[\frac{x^{n+1}}{n+1}\right]_{-r}^{r}$$

$$\Leftrightarrow -\log|1-r| + \log|1+r| = \sum_{n=0}^{\infty} \left(\frac{r^{n+1} - (-r)^{n+1}}{n+1}\right)$$

$$\Leftrightarrow \log\left(\frac{1+r}{1-r}\right) = \sum_{n=0}^{\infty} \frac{r^{n+1}}{n+1}(1 - (-1)^{n+1}) = 2 \sum_{n=0}^{\infty} \frac{r^{2n+1}}{2n+1}.$$

Differentiating term by term the series $\sum_{n=0}^{\infty} x^n$, we obtain $\sum_{n=0}^{\infty} (n+1) x^n$. Since

$$\lim \frac{|(n+2)x^{n+1}|}{|(n+1)x^n|} = \lim \frac{(n+2)|x|}{n+1} = |x|,$$

by D'Alembert's test, the series converges if $|x| < 1$ and diverges if $|x| > 1$; if $|x| = 1$, we have the divergent series $\sum_{n=0}^{\infty} (n+1)$ and $\sum_{n=0}^{\infty} (-1)^n (n+1)$.

The series $\sum_{n=0}^{\infty} (n+1) x^n$ is uniformly convergent on every interval $[-r, r]$ if $0 < r < 1$ because $|(n+1)x^n| \leq (n+1)r^n$ and the series $\sum_{n=0}^{\infty} (n+1) r^n$ is

3.2. Pointwise and Uniform Convergence of Series of Functions

convergent. We can then write

$$\left(\frac{1}{1-x}\right)' = \left(\sum_{n=0}^{\infty} x^n\right)' = \sum_{n=0}^{\infty} (n+1)\, x^n, \quad |x| < 1$$

$$\Leftrightarrow \quad \frac{1}{(1-x)^2} = \sum_{n=0}^{\infty} (n+1)\, x^n, \quad |x| < 1.$$

3.3 Power Series

Definition 3.3.1 *Let $x_0 \in \mathbb{R}$ be arbitrary. A series of the form*

$$\sum_{n=0}^{\infty} a_n (x - x_0)^n$$

*with $a_n \in \mathbb{R}$, $\forall n \in \mathbb{N}$, is called a **power series** of $x - x_0$.*

Notes:

1. By making $y = x - x_0$, a power series can always be reduced to the form $\sum_{n=0}^{\infty} a_n y^n$.

2. To ensure accuracy, the power series should be written as

$$a_0 + \sum_{n=1}^{+\infty} a_n (x - x_0)^n.$$

At $x = x_0$, the sum of the series is a_0 (all terms are 0 for $n \geq 1$). This is the meaning of the series that appears in the definition, which we write in this way to simplify. When evaluating the indeterminate form $a_0 \, 0^0$ at $x = x_0$, it should be interpreted as equal to a_0.

Theorem 3.3.1 *Let $\dfrac{1}{\overline{\lim} \sqrt[n]{|a_n|}} = r$. If $r \in \mathbb{R}^+$, the power series $\sum_{n=0}^{\infty} a_n x^n$ is absolutely convergent at each point $x \in \,]-r, r[\,$ and divergent at each point $x \in \,]-\infty, -r[\, \cup \,]r, +\infty[$. If $r = +\infty$, then the power series is absolutely convergent whatever $x \in \mathbb{R}$. If $r = 0$, the series converges if $x = 0$ and diverges if $x \neq 0$.*

<u>Proof</u>: Consider the series $\sum_{n=0}^{\infty} |a_n x^n|$. Let $r \in \mathbb{R}^+$. Using the fact that

$$\overline{\lim} \sqrt[n]{|a_n x^n|} = |x| \, \overline{\lim} \sqrt[n]{|a_n|},$$

3.3. Power Series

we can apply Corollary 1 of the Root Test to state that if $|x| \, \overline{\lim} \, \sqrt[n]{|a_n|} < 1$ (i.e., if $|x| < r$), the series converges; therefore, the series $\sum_{n=0}^{\infty} a_n \, x^n$ converges absolutely.

If $|x| \, \overline{\lim} \, \sqrt[n]{|a_n|} > 1$ (i.e., if $|x| > r$), then, by the reasoning used in Corollary 1 of the Root Test, there exists a subsequence of $\left(|a_n \, x^n|\right)$ that takes values greater than or equal to 1, which implies that the sequence $\left(|a_n \, x^n|\right)$ does not tend to zero; thus the sequence $\left(a_n \, x^n\right)$ does not tend to zero, and the series $\sum_{n=0}^{\infty} a_n \, x^n$ diverges.

If $r = 0$, then $\overline{\lim} \, \sqrt[n]{|a_n|} = +\infty$. It follows that, whatever $x \in \mathbb{R}$, $x \neq 0$, $|x| \, \overline{\lim} \, \sqrt[n]{|a_n|} = +\infty > 1$, and as before, the series $\sum_{n=0}^{\infty} a_n \, x^n$ diverges. Obviously, if $x = 0$, all terms are zero; thus, the series is convergent.

If $r = +\infty$, then $\overline{\lim} \, \sqrt[n]{|a_n|} = 0$. Therefore, whatever $x \in \mathbb{R}$, $|x| \, \overline{\lim} \, \sqrt[n]{|a_n|} = 0 < 1$, and we can conclude that the series $\sum_{n=0}^{\infty} a_n \, x^n$ converges absolutely. ∎

Definition 3.3.2 *Under the conditions of Theorem 3.3.1, r is called the* **radius of convergence** *of the series, and the interval $\,]-r, r[\,$ is the* **interval of convergence**.

Corollary 1 *If $\lim \left|\dfrac{a_n}{a_{n+1}}\right| = r$ is a positive real number, zero, or $+\infty$, then the radius of convergence of the power series is r.*

Proof: It is sufficient to note that by Theorem 1.1.11,

$$\lim \left|\frac{a_n}{a_{n+1}}\right| = \lim \frac{|a_n|}{|a_{n+1}|} = \frac{1}{\lim \sqrt[n]{|a_n|}} = \frac{1}{\overline{\lim} \, \sqrt[n]{|a_n|}} = r. \blacksquare$$

Note that there always exists $\overline{\lim} \, \sqrt[n]{|a_n|}$; that is, every power series has a radius of convergence. However, $\lim \left|\dfrac{a_n}{a_{n+1}}\right|$ may not exist, as illustrated in the following example.

Example 3.3.1 Consider the series $\sum_{n=0}^{\infty} a_n x^n = \sum_{n=0}^{\infty} \left(3+(-1)^n\right)^n x^n$. As

$$\frac{a_n}{a_{n+1}} = \begin{cases} 2^{n-1}, & \text{if } n \text{ is even} \\ \frac{1}{2^{n+2}}, & \text{if } n \text{ is odd,} \end{cases}$$

$\lim \left|\frac{a_n}{a_{n+1}}\right|$ does not exist, but $r = \dfrac{1}{\overline{\lim} \sqrt[n]{|a_n|}} = \dfrac{1}{4}$.

The interval of convergence of the series is $\left]-\frac{1}{4}, \frac{1}{4}\right[$. The series converges absolutely on the interval $\left]-\frac{1}{4}, \frac{1}{4}\right[$ and diverges on $\left]-\infty, -\frac{1}{4}\right[\cup \left]\frac{1}{4}, +\infty\right[$.

Example 3.3.2 Let us calculate the radius of convergence of the series $\sum_{n=1}^{\infty} \frac{x^n}{n^n}$:

$$r = \frac{1}{\overline{\lim} \sqrt[n]{|a_n|}} = \frac{1}{\overline{\lim} \sqrt[n]{\frac{1}{n^n}}} = \frac{1}{\overline{\lim} \frac{1}{n}} = +\infty.$$

The radius of convergence of the series is infinite; therefore, the series is absolutely convergent $\forall x \in \mathbb{R}$.

Example 3.3.3 The series $\sum_{n=0}^{\infty} n!\, x^n$ has a radius of convergence $r = 0$:

$$r = \lim \left|\frac{a_n}{a_{n+1}}\right| = \lim \frac{n!}{(n+1)!} = \lim \frac{1}{n+1} = 0,$$

that is, the series only converges at $x = 0$.

Example 3.3.4 Consider the series $\sum_{n=1}^{\infty} \frac{x^n}{n}$. Taking into account that

$$r = \lim \left|\frac{a_n}{a_{n+1}}\right| = \lim \frac{\frac{1}{n}}{\frac{1}{n+1}} = \lim \frac{n+1}{n} = 1,$$

3.3. Power Series

the interval of convergence is $]-1,1[$. The series converges absolutely on the interval $]-1,1[$ and diverges on $]-\infty,-1[\cup]1,+\infty[$.

Example 3.3.5 Consider the series $\sum_{n=0}^{\infty} \frac{(-1)^n}{2^n}(x+1)^n$. Let $y = x+1$. The power series of y

$$\sum_{n=0}^{\infty} \frac{(-1)^n}{2^n} y^n$$

has radius of convergence

$$r = \lim \left| \frac{a_n}{a_{n+1}} \right| = \lim \left| \frac{\frac{(-1)^n}{2^n}}{\frac{(-1)^{n+1}}{2^{n+1}}} \right| = 2.$$

Then the power series of y converges absolutely if $y \in]-2,2[$ and diverges if $y \in]-\infty,-2[\cup]2,+\infty[$. The power series of $x+1$ converges absolutely if $x \in]-3,1[$ and diverges if $x \in]-\infty,-3[\cup]1,+\infty[$.

Example 3.3.6 Consider the series $\sum_{n=0}^{\infty}(-1)^n \frac{x^{2n+1}}{(2n+1)!}$. Let $y = x^2$. The series $\sum_{n=0}^{\infty}(-1)^n \frac{y^n}{(2n+1)!}$ has radius of convergence

$$r = \lim \left| \frac{\frac{(-1)^n}{(2n+1)!}}{\frac{(-1)^{n+1}}{(2n+3)!}} \right| = \lim \frac{(2n+3)!}{(2n+1)!} = \lim (2n+3)(2n+2) = +\infty;$$

therefore, the series is absolutely convergent for all $y \in \mathbb{R}_0^+$, and the original series is absolutely convergent for all $x \in \mathbb{R}$ (note that it is enough to multiply the power series of x by x).

Note: The preceding theorem omits any reference to the convergence or divergence of the power series at the endpoints of the convergence interval $]-r,r[$, $r \in \mathbb{R}^+$. It is possible for the series to converge at both endpoints, converge at one and diverge at the other, or diverge at both. Therefore, we must always analyze the series at $x = r$ and $x = -r$.

In the case of Example 3.3.4, the convergence interval is $]-1,1[$:

- If $x = -1$, we obtain the series $\sum_{n=1}^{\infty} \frac{(-1)^n}{n}$, which is conditionally convergent.
- If $x = 1$, we obtain the series $\sum_{n=1}^{\infty} \frac{1}{n}$, which is divergent.

We conclude that the series diverges on $]-\infty, -1[\cup [1, +\infty[$ and converges on $[-1, 1[$.

Theorem 3.3.2 *If the radius of convergence of the series $\sum_{n=0}^{\infty} a_n x^n$ is $r > 0$ and if $0 < \rho < r$, then the series is uniformly convergent on $[-\rho, \rho]$.*

<u>Proof</u>: By hypothesis, $|a_n x^n| \leq |a_n| \rho^n = |a_n \rho^n|$, $\forall x \in [-\rho, \rho]$. The series $\sum_{n=0}^{\infty} |a_n \rho^n|$ is convergent by Theorem 3.3.1, since $\rho \in]-r, r[$.

According to Weierstrass' test, the series $\sum_{n=0}^{\infty} a_n x^n$ is uniformly convergent on $[-\rho, \rho]$. ∎

Corollary 1 *A power series is uniformly convergent on every closed interval $[a, b]$ contained within its interval of convergence and*

$$\int_a^b \sum_{n=0}^{\infty} a_n x^n \, dx = \sum_{n=0}^{\infty} a_n \frac{b^{n+1} - a^{n+1}}{n+1}.$$

<u>Proof</u>: If $[a, b] \subset]-r, r[$, then there exists $\rho > 0$ such that

$$[a, b] \subset [-\rho, \rho] \subset]-r, r[.$$

By the theorem, the series is uniformly convergent on $[-\rho, \rho]$ and also on $[a, b]$. Then we can integrate the series term by term on $[a, b]$:

$$\int_a^b \sum_{n=0}^{\infty} a_n x^n \, dx = \sum_{n=0}^{\infty} \int_a^b a_n x^n \, dx = \sum_{n=0}^{\infty} a_n \int_a^b x^n \, dx$$

$$= \sum_{n=0}^{\infty} a_n \frac{b^{n+1} - a^{n+1}}{n+1}. \quad \blacksquare$$

3.3. Power Series

Example 3.3.7 Let us evaluate $\int_0^1 f(x)\,dx$, where $f(x) = \sum_{n=0}^{\infty}(-1)^n \dfrac{x^{2n}}{(2n)!}$.

Let $y = x^2$. The series $\sum_{n=0}^{\infty}(-1)^n \dfrac{y^n}{(2n)!}$ has an infinite radius of convergence:

$$r = \lim \left| \dfrac{\dfrac{(-1)^n}{(2n)!}}{\dfrac{(-1)^{n+1}}{(2n+2)!}} \right| = \lim \dfrac{(2n+2)!}{(2n)!} = \lim (2n+2)(2n+1) = +\infty,$$

which implies that the original series converges for all x in $\mathbb{R}$; therefore, it is uniformly convergent on $[0, 1]$ and integrable term by term on that interval:

$$\int_0^1 \sum_{n=0}^{\infty}(-1)^n \dfrac{x^{2n}}{(2n)!}\,dx = \sum_{n=0}^{\infty} \int_0^1 (-1)^n \dfrac{x^{2n}}{(2n)!}\,dx$$

$$= \sum_{n=0}^{\infty}(-1)^n \dfrac{1}{(2n)!} \left[\dfrac{x^{2n+1}}{2n+1} \right]_0^1 = \sum_{n=0}^{\infty} \dfrac{(-1)^n}{(2n+1)!}.$$

Theorem 3.3.3 *Every power series with radius of convergence $r > 0$ is term by term differentiable on the interval of convergence, that is,*

$$\left(\sum_{n=0}^{\infty} a_n x^n \right)' = \sum_{n=1}^{\infty} n\, a_n\, x^{n-1}, \quad \forall x \in\,]-r, r[.$$

<u>Proof</u>: Consider the series $\sum_{n=0}^{\infty} a_n x^n$. It is pointwise convergent on $]-r, r[$; the functions $(a_n x^n)' = n a_n x^{n-1}$ are continuous on $]-r, r[$, $\forall n \in \mathbb{N}$. Moreover, $\sum_{n=0}^{\infty} n a_n x^{n-1}$ is a power series with radius of convergence r:

$$\dfrac{1}{\overline{\lim}\sqrt[n]{|na_n|}} = \dfrac{1}{\overline{\lim}\sqrt[n]{n}\sqrt[n]{|a_n|}} = \dfrac{1}{\overline{\lim}\sqrt[n]{|a_n|}} = r;$$

therefore, it is uniformly convergent on $[a, b] \subset\,]-r, r[$.
Thus, by Corollary 2 of Theorem 3.2.4,

$$\left(\sum_{n=0}^{\infty} a_n x^n \right)' = \sum_{n=1}^{\infty} n\, a_n\, x^{n-1}, \quad \forall x \in\,]-r, r[. \quad \blacksquare$$

Note: If the power series $\sum_{n=0}^{\infty} a_n x^n$ has radius of convergence r, then both the series of derivatives and the series of indefinite integrals have the same radius of convergence, r.

Example 3.3.8 Let us consider the power series $\sum_{n=1}^{\infty}(-1)^{n+1}\dfrac{(x-5)^n}{n\,5^n}$.

Let $y = x - 5$. The series $\sum_{n=1}^{\infty}(-1)^{n+1}\dfrac{y^n}{n\,5^n}$ has radius of convergence

$$r = \lim \left|\dfrac{\dfrac{(-1)^{n+1}}{n\,5^n}}{\dfrac{(-1)^{n+2}}{(n+1)\,5^{n+1}}}\right| = \lim \dfrac{(n+1)\,5^{n+1}}{n\,5^n} = 5,$$

which implies the absolute convergence of the original series on the interval $]0, 10[$.

- If $x = 0$, we obtain the series $\sum_{n=1}^{\infty}\dfrac{-1}{n}$, which is divergent.

- If $x = 10$, we obtain the series $\sum_{n=1}^{\infty}\dfrac{(-1)^{n+1}}{n}$, which is conditionally convergent.

Then, the original series converges absolutely on the interval $]0, 10[$, converges conditionally at $x = 10$, and diverges on $]-\infty, 0] \cup]10, +\infty[$.

The series of derivatives is

$$\left(\sum_{n=1}^{\infty}(-1)^{n+1}\dfrac{(x-5)^n}{n\,5^n}\right)' = \sum_{n=1}^{\infty}(-1)^{n+1}\,n\,\dfrac{(x-5)^{n-1}}{n\,5^n}$$

$$= \sum_{n=1}^{\infty}(-1)^{n+1}\dfrac{(x-5)^{n-1}}{5^n}.$$

The interval of convergence of this series is $]0, 10[$.

- If $x = 0$, we obtain the series $\sum_{n=1}^{\infty}\dfrac{1}{5}$ which is divergent.

- If $x = 10$, we obtain the series $\sum_{n=1}^{\infty}\dfrac{(-1)^{n+1}}{5}$ which is divergent.

3.4 Taylor Series and Maclaurin Series

Let I be an interval and $f : I \subset \mathbb{R} \to \mathbb{R}$ a function of class C^n on I. Let $x_0 \in I$. We know that there exists $0 < \theta < 1$ such that

$$f(x) = f(x_0) + f'(x_0)(x - x_0)$$
$$+ f''(x_0)\frac{(x - x_0)^2}{2!} + \cdots + f^{(n-1)}(x_0)\frac{(x - x_0)^{n-1}}{(n - 1)!} + R_n(x),$$

where $R_n(x) = f^{(n)}\bigl(x_0 + \theta(x - x_0)\bigr)\dfrac{(x - x_0)^n}{n!}$. This is the Taylor's formula for f, of order n, with Lagrange remainder around the point x_0.

Suppose that $f \in C^\infty(I)$. The power series

$$\sum_{n=0}^{\infty} \frac{f^{(n)}(x_0)}{n!}(x - x_0)^n$$

is called the **Taylor series of** f **at** x_0.

If $x_0 = 0 \in I$, the Taylor series is called **Maclaurin series** and is written as follows:

$$\sum_{n=0}^{\infty} \frac{f^{(n)}(0)}{n!} x^n.$$

Brook Taylor (1685–1731) was an English mathematician. His work titled *Methodus Incrementorum Directa et Inversa* (London, 1715) added a new branch to Mathematics known as "calculus of finite differences." Among other applications, he used it to determine the shape of the movement of a vibrating string, reducing it to mechanical principles. This work also contains the famous formula known as Taylor's Formula. (Source of image: Line engraving after Richard Earlom (1743–1822))

Example 3.4.1 Let us determine the Maclaurin series of $f(x) = \sin(x)$. We know that $f \in C^\infty(\mathbb{R})$ and $f^{(n)}(x) = \sin\left(x + \dfrac{n\pi}{2}\right)$, $\forall n \in \mathbb{N}$. Then $f^{(n)}(0) = \sin\left(\dfrac{n\pi}{2}\right)$, and therefore, the Maclaurin series of f is

$$x - \frac{x^3}{3!} + \frac{x^5}{5!} - \frac{x^7}{7!} + \cdots = \sum_{n=0}^{\infty}(-1)^n \frac{x^{2n+1}}{(2n+1)!}.$$

We saw in Example 3.3.6 that this series is convergent on $\mathbb{R}$.

Example 3.4.2 Consider the function $f(x) = (1 + x)^\alpha$, $\alpha \in \mathbb{R}$, $x > -1$. By the rules of differentiation, $f \in C^\infty(]-1, +\infty[)$ and

$$f^{(n)}(x) = \alpha(\alpha - 1)\ldots(\alpha - n + 1)(1 + x)^{\alpha - n}, \forall n \geq 1.$$

Therefore, $f^{(n)}(0) = \alpha(\alpha - 1)\ldots(\alpha - n + 1)$, and its Maclaurin series is

$$1 + \alpha x + \frac{\alpha(\alpha - 1)}{2!}x^2 + \cdots + \frac{\alpha(\alpha - 1)\ldots(\alpha - n + 1)}{n!}x^n + \cdots =$$

Collin Maclaurin (1698–1764) was a Scottish mathematician. In 1717, at the age of 19 years, he was appointed professor in Aberdeen. In 1719, he published the work *Organic Geometry*, a text that can be considered the most important of his works and contains, among others, an original method of generating conics. He developed Newton's work in calculus, geometry, and gravitation. (Source of image: Pencil and chalk on paper by David Steuart Erskine (1742–1829), Scottish National Gallery)

$$= 1 + \sum_{n=1}^{\infty} \frac{\alpha(\alpha-1)\ldots(\alpha-n+1)}{n!} x^n.$$

If $\alpha \in \mathbb{N}$, the series is reduced to the development of Newton's binomial. Suppose that $\alpha \notin \mathbb{N}_0$. Then let us study the convergence of the series. The radius of convergence is

$$\lim \left| \frac{\frac{\alpha(\alpha-1)\ldots(\alpha-n+1)}{n!}}{\frac{\alpha(\alpha-1)\ldots(\alpha-n+1)(\alpha-n)}{(n+1)!}} \right| = \lim \frac{(n+1)!}{n!} \frac{1}{|\alpha-n|}$$

$$= \lim \frac{n+1}{|\alpha-n|} = 1.$$

Therefore, the series is absolutely convergent on $]-1, 1[$ and divergent on $]-\infty, -1[\cup]1, +\infty[$.

This series is usually referred to as the **binomial series**.

A fundamental question in the Taylor series development of an indefinitely differentiable function is as follows:

Is there a neighborhood V of x_0 such that

$$f(x) = f(x_0) + f'(x_0)(x-x_0)$$
$$+ f''(x_0)\frac{(x-x_0)^2}{2!} + \cdots + f^{(n-1)}(x_0)\frac{(x-x_0)^{n-1}}{(n-1)!} + \cdots, \forall x \in V?$$

In other words, does the Taylor series expansion of f at x_0 converge for all $x \in V$ and sum up to $f(x)$?

In reality, the mere existence of the derivatives $f^{(n)}(x_0)$ for all natural values of n, although it allows writing the Taylor series of f at x_0, does not guarantee that, in some neighborhood of x_0, the equality is satisfied:

$$f(x) = \sum_{n=0}^{\infty} \frac{f^{(n)}(x_0)}{n!} (x-x_0)^n, \qquad (3.2)$$

as can be seen in the following example.

Example 3.4.3 Consider the function

$$f(x) = \begin{cases} e^{-1/x^2}, & \text{if } x \neq 0 \\ 0, & \text{if } x = 0. \end{cases}$$

3.4. Taylor Series and Maclaurin Series

As $f^{(n)}(0) = 0$, $\forall n \in \mathbb{N}$, the Maclaurin series of f is written as follows:

$$0 + 0x + 0x^2 + \cdots,$$

which converges to the zero function on $\mathbb{R}$. Therefore, the only point where f equals the sum of the series is 0, since $f(x) \neq 0$ if $x \neq 0$.

To ensure equality (3.2), what additional conditions must be imposed on a function f that is assumed to be indefinitely differentiable on a neighborhood of x_0? We can easily answer this question, by considering Taylor's formula for f. Specifically, if $S_n(x)$ denotes the partial sum of order n of the Taylor series of f at $x_0 \in I$, then we have $R_n(x) = f(x) - S_n(x)$, which satisfies the following result:

> **Theorem 3.4.1** *It is a necessary and sufficient condition for an indefinitely differentiable function, $f : I \to \mathbb{R}$, to be the sum of its Taylor series in a neighborhood, V, of $x_0 \in I$, that*
>
> $$\lim R_n(x) = 0, \quad \forall x \in V.$$

In practice, sufficient conditions are used:

> **Theorem 3.4.2** *Let $f : I \to \mathbb{R}$ be an indefinitely differentiable function, and suppose that there are constants $M, k \geq 0$ such that, in a neighborhood, V, of x_0,*
>
> $$\left| f^{(n)}(x) \right| \leq Mk^n, \quad \forall x \in V, \forall n \in \mathbb{N}.$$
>
> *Then, f is the sum of its Taylor series on V.*

Proof: We know that the expression of the Lagrange remainder, $R_n(x)$, is

$$R_n(x) = f^{(n)}\left(x_0 + \theta\left(x - x_0\right)\right) \frac{(x - x_0)^n}{n!}, \quad 0 < \theta < 1.$$

Then

$$|R_n(x)| \leq M \frac{(k|x - x_0|)^n}{n!}, \quad x \in V;$$

since the series of general term $\dfrac{(k|x - x_0|)^n}{n!}$ is convergent, this sequence has limit 0. The desired result is an immediate consequence of Theorem 3.4.1. ∎

> **Corollary 1** *If there exists $M \geq 0$ such that on the neighborhood V of x_0,*
> $$\left|f^{(n)}(x)\right| \leq M \quad \forall x \in V, \ \forall n \in \mathbb{N},$$
> *then f is the sum of its Taylor series on V.*

Example 3.4.4 Consider the function $f(x) = \sin(x)$. In Example 3.4.1, we found that the Maclaurin series of f converges absolutely on $\mathbb{R}$. We know that
$$R_n(x) = \frac{\sin\left(\theta x + \frac{n\pi}{2}\right)}{n!} x^n, \quad 0 < \theta < 1,$$
from which
$$0 \leq |R_n(x)| = \frac{|\sin\left(\theta x + \frac{n\pi}{2}\right)|}{n!} |x|^n \leq \frac{|x|^n}{n!}.$$

But $\lim \dfrac{|x|^n}{n!} = 0$, $\forall x \in \mathbb{R}$, as it is the general term of a convergent series, which implies that
$$\sin(x) = \sum_{n=0}^{\infty} (-1)^n \frac{x^{2n+1}}{(2n+1)!}, \quad \forall x \in \mathbb{R}.$$

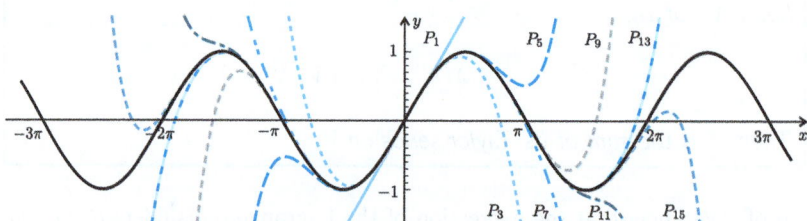

Figure 3.7: Some Taylor polynomials of the function $f(x) = \sin(x)$

In Fig. 3.7, we can see how the partial sums are increasingly better approximations of the function $\sin(x)$, thus illustrating the convergence of the series. The polynomial $\sum_{k=0}^{n} (-1)^k \dfrac{x^{2k+1}}{(2k+1)!}$ is denoted by P_{2n+1} in the figure. The black line represents the graph of the function $\sin(x)$.

3.4. Taylor Series and Maclaurin Series

Example 3.4.5 If $f(x) = e^x$, we obtain $f^{(n)}(x) = e^x$, $\forall n \in \mathbb{N}$. Then its Maclaurin series development is

$$\sum_{n=0}^{\infty} \frac{f^{(n)}(0)}{n!} x^n = \sum_{n=0}^{\infty} \frac{x^n}{n!}.$$

We know that this series is absolutely convergent on $\mathbb{R}$ defining a function g on $\mathbb{R}$. We intend to prove that $f(x) = g(x)$, $\forall x \in \mathbb{R}$. For this purpose, we will show that the Lagrange remainder of the Maclaurin formula of the function f tends to 0 on $\mathbb{R}$.

$$R_n(x) = \frac{f^{(n)}(\theta x)}{n!} x^n = \frac{e^{\theta x}}{n!} x^n, \quad 0 < \theta < 1,$$

which implies that, taking into account that $e^{\theta x} \leq e^{|x|}$ since f is an increasing function,

$$0 \leq |R_n(x)| \leq \frac{e^{|x|}}{n!} |x|^n.$$

The series of general term $\dfrac{e^{|x|}}{n!} |x|^n$ is convergent, $\forall x \in \mathbb{R}$; therefore,

$$\lim \frac{e^{|x|}}{n!} |x|^n = 0, \quad \forall x \in \mathbb{R},$$

which allows us to conclude that

$$e^x = \sum_{n=0}^{\infty} \frac{x^n}{n!}, \quad \forall x \in \mathbb{R}.$$

The Maclaurin series developments in the first two examples were obtained directly from the formula:

$$f(0) + f'(0)\, x + \frac{f''(0)}{2!} x^2 + \cdots,$$

by evaluating all derivatives of the function at zero. Since this process is quite laborious, it is not commonly employed. Instead, well-known developments are often utilized in practice. Furthermore, the next theorem guarantees that these developments are indeed the Taylor series.

Theorem 3.4.3 *Every power series of $x - x_0$ is the Taylor series (around x_0) of the function defined by it. In particular, the power series development of $x - x_0$ is unique.*

Proof: By hypothesis, $f(x) = \sum_{n=0}^{\infty} a_n (x-x_0)^n$ in a neighborhood V of x_0, which implies that $f(x_0) = a_0$. Differentiating,

$$f'(x) = \sum_{n=1}^{\infty} n\, a_n (x-x_0)^{n-1},$$

and therefore, $f'(x_0) = a_1$. The nth order derivative is

$$f^{(n)}(x) = n!\, a_n + (n+1) \cdots 2\, a_{n+1}(x-x_0)$$

$$+(n+2)(n+1) \cdots 3\, a_{n+2}(x-x_0)^2 + \cdots,$$

and so $f^{(n)}(x_0) = n!\, a_n$, $n \in \mathbb{N}$. Therefore, we have

$$a_n = \frac{f^{(n)}(x_0)}{n!}, \quad \forall n \in \mathbb{N}. \quad \blacksquare$$

Example 3.4.6 Consider the function $f(x) = \dfrac{1}{2+3x}$. Taking into account that

$$\frac{1}{2+3x} = \frac{1}{2} \cdot \frac{1}{1-\left(-\frac{3}{2}x\right)}$$

and that

$$\frac{1}{1-x} = \sum_{n=0}^{\infty} x^n, \quad \forall x \in\,]-1,1[,$$

we can deduce that

$$\frac{1}{2} \cdot \frac{1}{1-\left(-\frac{3}{2}x\right)} = \frac{1}{2} \sum_{n=0}^{\infty} \left(-\frac{3}{2}x\right)^n,$$

and this equality is valid as long as $\left|-\dfrac{3}{2}x\right| < 1$, that is, $|x| < \dfrac{2}{3}$. Then, the Maclaurin series of f is

$$\sum_{n=0}^{\infty} (-1)^n \frac{3^n}{2^{n+1}} x^n, \quad |x| < \frac{2}{3}.$$

Example 3.4.7 Let $f(x) = \dfrac{1}{x^2-x-6} = \dfrac{1}{(x-3)(x+2)}$. We have

$$f(x) = \frac{1}{5}\left(\frac{1}{x-3} - \frac{1}{x+2}\right) = \frac{1}{5}\left(-\frac{1}{3} \cdot \frac{1}{1-\frac{x}{3}} - \frac{1}{2} \cdot \frac{1}{1-\left(-\frac{x}{2}\right)}\right).$$

3.4. Taylor Series and Maclaurin Series

After taking into account that

$$\frac{1}{1-\frac{x}{3}} = \sum_{n=0}^{\infty} \left(\frac{x}{3}\right)^n, \quad |x| < 3$$

and

$$\frac{1}{1-\left(-\frac{x}{2}\right)} = \sum_{n=0}^{\infty} \left(-\frac{x}{2}\right)^n = \sum_{n=0}^{\infty} (-1)^n \left(\frac{x}{2}\right)^n, \quad |x| < 2,$$

we can write the Maclaurin series of f as follows:

$$f(x) = \frac{1}{5}\left(-\frac{1}{3}\sum_{n=0}^{\infty}\left(\frac{x}{3}\right)^n - \frac{1}{2}\sum_{n=0}^{\infty}(-1)^n\left(\frac{x}{2}\right)^n\right)$$

$$= \sum_{n=0}^{\infty} \frac{1}{5}\left(\frac{(-1)^{n+1}}{2^{n+1}} - \frac{1}{3^{n+1}}\right) x^n, \quad |x| < 2.$$

Example 3.4.8 In Example 3.4.2 we developed the function

$$f(x) = (1+x)^\alpha, \quad \alpha \in \mathbb{R}, \quad x > -1,$$

into a Maclaurin series, thereby obtaining

$$1 + \sum_{n=1}^{\infty} \frac{\alpha(\alpha-1)\ldots(\alpha-n+1)}{n!} x^n,$$

convergent on the interval $]-1,1[$. Let

$$g(x) = 1 + \sum_{n=1}^{\infty} \frac{\alpha(\alpha-1)\ldots(\alpha-n+1)}{n!} x^n, \quad |x| < 1.$$

Next, we prove that $f(x) = g(x)$, $\forall x \in \,]-1,1[$, that is, f is the sum of its Maclaurin series in that interval.

Since a power series is differentiable term by term on the interval of convergence, we obtain

$$g'(x) = \sum_{n=1}^{\infty} \frac{\alpha(\alpha-1)\ldots(\alpha-n+1)}{(n-1)!} x^{n-1},$$

and multiplying by x

$$x\,g'(x) = \sum_{n=1}^{\infty} \frac{\alpha(\alpha-1)\ldots(\alpha-n+1)}{(n-1)!} x^n.$$

Then

$$g'(x) + x\,g'(x) =$$
$$= \sum_{n=1}^{\infty} \frac{\alpha(\alpha-1)\ldots(\alpha-n+1)}{(n-1)!} x^{n-1} + \sum_{n=1}^{\infty} \frac{\alpha(\alpha-1)\ldots(\alpha-n+1)}{(n-1)!} x^n$$
$$= \sum_{n=0}^{\infty} \frac{\alpha(\alpha-1)\ldots(\alpha-n)}{n!} x^n + \sum_{n=1}^{\infty} \frac{\alpha(\alpha-1)\ldots(\alpha-n+1)}{(n-1)!} x^n$$
$$= \alpha + \sum_{n=1}^{\infty} \left(\frac{\alpha(\alpha-1)\ldots(\alpha-n)}{n!} + \frac{\alpha(\alpha-1)\ldots(\alpha-n+1)}{(n-1)!} \right) x^n$$
$$= \alpha + \sum_{n=1}^{\infty} \frac{\alpha(\alpha-1)\ldots(\alpha-n+1)}{(n-1)!} \left(\frac{\alpha-n}{n} + 1 \right) x^n$$
$$= \alpha + \sum_{n=1}^{\infty} \frac{\alpha(\alpha-1)\ldots(\alpha-n+1)}{(n-1)!} \cdot \frac{\alpha}{n} x^n$$
$$= \alpha \left(1 + \sum_{n=1}^{\infty} \frac{\alpha(\alpha-1)\ldots(\alpha-n+1)}{n!} x^n \right)$$
$$= \alpha\,g(x),$$

that is,
$$(1+x)\,g'(x) = \alpha\,g(x). \tag{3.3}$$

Let us consider the function $\dfrac{g(x)}{(1+x)^\alpha}$ and calculate its derivative:

$$\left(\frac{g(x)}{(1+x)^\alpha} \right)' = \frac{g'(x)(1+x)^\alpha - \alpha(1+x)^{\alpha-1}g(x)}{(1+x)^{2\alpha}}$$
$$= \frac{(1+x)^{\alpha-1}\big((1+x)\,g'(x) - \alpha\,g(x)\big)}{(1+x)^{2\alpha}}.$$

By equality (3.3) the numerator of this fraction is zero; therefore,
$$\left(\frac{g(x)}{(1+x)^\alpha} \right)' = 0,$$

which implies that the function defined by
$$\frac{g(x)}{(1+x)^\alpha}$$

is constant on $]-1,1[$, that is, $g(x) = c\,(1+x)^\alpha$, for some constant c. Since $g(0) = 1$, we obtain $c = 1$, and we have
$$g(x) = (1+x)^\alpha, \quad \forall x \in\,]-1,1[.$$

3.5 Introduction to Fourier Series

Taylor series are not the only series used to approximate and/or represent functions. The Fourier series, which we introduce in this section, are another example of how series are applied to the study of functions.

First, we establish some definitions and results concerning periodic functions that will be useful in studying Fourier series (Fig. 3.8).

Definition 3.5.1 *We say that a function $f : \mathbb{R} \to \mathbb{R}$ is **periodic** with period $T > 0$ if $f(x+T) = f(x)$, $\forall x \in \mathbb{R}$.*

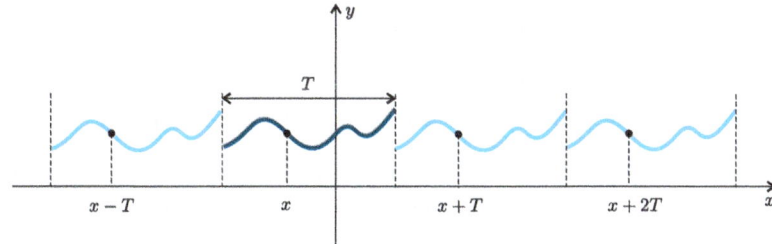

Figure 3.8: A periodic function with period T

The following two propositions are direct consequences of the definition.

Proposition 3.5.1 *Let $f : \mathbb{R} \to \mathbb{R}$ be a periodic function with period T. Then f is periodic with period kT, $\forall k \in \mathbb{N}$.*

Proposition 3.5.2 *Let f and g be periodic functions with period T, and $\alpha, \beta \in \mathbb{R}$. Then, the function $\alpha f + \beta g$ is periodic with period T.*

Consider the functions
$$a_k \cos(kx) + b_k \sin(kx), \ k = 1, 2, \ldots,$$
where $a_k, b_k \in \mathbb{R}, \forall k \in \mathbb{N}$. These are periodic functions with period $T_k = \dfrac{2\pi}{k}$, $k = 1, 2, \ldots$. Therefore, $T = 2\pi = kT_k$ is a common period to all. As a result, sums of the form
$$s_n(x) = A + \sum_{k=1}^{n} \big(a_k \cos(kx) + b_k \sin(kx) \big),$$

where A is a constant, are periodic functions with period $T = 2\pi$. These sums are usually referred to as **trigonometric polynomials**.

> **Definition 3.5.2** *A* **trigonometric series** *is a series of the form*
> $$\frac{a_0}{2} + \sum_{n=1}^{\infty} \left(a_n \cos(nx) + b_n \sin(nx)\right).$$
> *The constants a_0, a_n, and b_n, $n \in \mathbb{N}$, are the coefficients of the trigonometric series.*

Two questions arise at this point:

1. What conditions ensure the convergence of a series of this kind?

2. What conditions must a function meet to be the sum of a trigonometric series, and how can the coefficients be determined?

Let us begin by answering the final part of the second question, remarking that if the series converges, its sum is a periodic function with period 2π. Suppose that the trigonometric series converges and its sum is a function f, that is,

$$f(x) = \frac{a_0}{2} + \sum_{n=1}^{\infty} \left(a_n \cos(nx) + b_n \sin(nx)\right). \tag{3.4}$$

The objective is to find the coefficients a_0, a_n, and b_n, $n \in \mathbb{N}$. Suppose the function f is integrable and the series involved in this process is integrable term by term. By integrating the previous equality from $-\pi$ to π, we obtain

$$\int_{-\pi}^{\pi} f(x)\,dx = \int_{-\pi}^{\pi} \frac{a_0}{2}\,dx + \int_{-\pi}^{\pi} \sum_{n=1}^{\infty} \left(a_n \cos(nx) + b_n \sin(nx)\right)\,dx$$

$$= \pi a_0 + \sum_{n=1}^{\infty} \int_{-\pi}^{\pi} \left(a_n \cos(nx) + b_n \sin(nx)\right)\,dx$$

$$= \pi a_0 + \sum_{n=1}^{\infty} \left(a_n \int_{-\pi}^{\pi} \cos(nx)\,dx + b_n \int_{-\pi}^{\pi} \sin(nx)\,dx\right).$$

But

$$\int_{-\pi}^{\pi} \cos(nx)\,dx = \left[\frac{1}{n}\sin(nx)\right]_{-\pi}^{\pi} = 0 \tag{3.5}$$

and

$$\int_{-\pi}^{\pi} \sin(nx)\,dx = \left[-\frac{1}{n}\cos(nx)\right]_{-\pi}^{\pi} = 0; \tag{3.6}$$

3.5. Introduction to Fourier Series

therefore,
$$\int_{-\pi}^{\pi} f(x)\,dx = \pi a_0.$$

Solving for a_0, we get
$$a_0 = \frac{1}{\pi}\int_{-\pi}^{\pi} f(x)\,dx.$$

To calculate a_n, for $n \geq 1$, we multiply both sides of equality (3.4) by $\cos(mx)$ ($m \in \mathbb{N}$) and integrate term by term between $-\pi$ and π:

$$\int_{-\pi}^{\pi} f(x)\cos(mx)\,dx =$$

$$= \int_{-\pi}^{\pi} \left(\frac{a_0}{2} + \sum_{n=1}^{\infty}\left(a_n\cos(nx) + b_n\sin(nx)\right)\right)\cos(mx)\,dx$$

$$= \frac{a_0}{2}\int_{-\pi}^{\pi}\cos(mx)\,dx + \sum_{n=1}^{\infty}\int_{-\pi}^{\pi}\left(a_n\cos(nx) + b_n\sin(nx)\right)\cos(mx)\,dx$$

$$= \frac{a_0}{2}\int_{-\pi}^{\pi}\cos(mx)\,dx +$$

$$+ \sum_{n=1}^{\infty}\left(a_n\int_{-\pi}^{\pi}\cos(nx)\cos(mx)\,dx + b_n\int_{-\pi}^{\pi}\sin(nx)\cos(mx)\,dx\right).$$

From the trigonometric equalities

$$\cos(nx)\cos(mx) = \frac{1}{2}\Big(\cos\big((n+m)x\big) + \cos\big((n-m)x\big)\Big) \qquad (3.7)$$

$$\sin(nx)\cos(mx) = \frac{1}{2}\Big(\sin\big((n+m)x\big) + \sin\big((n-m)x\big)\Big), \qquad (3.8)$$

it follows
$$\int_{-\pi}^{\pi}\cos(nx)\cos(mx)\,dx = \begin{cases} 0, & \text{if } n \neq m \\ \pi, & \text{if } n = m \end{cases} \qquad (3.9)$$

and
$$\int_{-\pi}^{\pi}\sin(nx)\cos(mx)\,dx = 0. \qquad (3.10)$$

From the previous equalities, we obtain
$$\int_{-\pi}^{\pi} f(x)\cos(mx)\,dx = a_m\pi.$$

Solving for a_m and considering that $m = n$,

$$a_n = \frac{1}{\pi} \int_{-\pi}^{\pi} f(x) \cos(nx)\, dx.$$

To evaluate b_n, for $n \geq 1$, we multiply both sides of equality (3.4) by $\sin(mx)$ ($m \in \mathbb{N}$) and integrate term by term between $-\pi$ and π:

$$\int_{-\pi}^{\pi} f(x) \sin(mx)\, dx =$$

$$= \int_{-\pi}^{\pi} \left(\frac{a_0}{2} + \sum_{n=1}^{\infty} \big(a_n \cos(nx) + b_n \sin(nx)\big) \right) \sin(mx)\, dx$$

$$= \frac{a_0}{2} \int_{-\pi}^{\pi} \sin(mx)\, dx + \sum_{n=1}^{\infty} \int_{-\pi}^{\pi} \big(a_n \cos(nx) + b_n \sin(nx)\big) \sin(mx)\, dx$$

$$= \frac{a_0}{2} \int_{-\pi}^{\pi} \sin(mx)\, dx +$$

$$+ \sum_{n=1}^{\infty} \left(a_n \int_{-\pi}^{\pi} \cos(nx) \sin(mx)\, dx + b_n \int_{-\pi}^{\pi} \sin(nx) \sin(mx)\, dx \right).$$

From the trigonometric equality

$$\sin(nx) \sin(mx) = \frac{1}{2} \Big(\cos\big((n-m)x\big) - \cos\big((n+m)x\big) \Big), \qquad (3.11)$$

we conclude that

$$\int_{-\pi}^{\pi} \sin(nx) \sin(mx)\, dx = \begin{cases} 0, & \text{if } n \neq m \\ \pi, & \text{if } n = m, \end{cases} \qquad (3.12)$$

and we obtain

$$\int_{-\pi}^{\pi} f(x) \sin(mx)\, dx = b_m \pi.$$

Solving for b_m and taking into account that $m = n$,

$$b_n = \frac{1}{\pi} \int_{-\pi}^{\pi} f(x) \sin(nx)\, dx.$$

3.5. Introduction to Fourier Series

Definition 3.5.3 *Let f be a real-valued function of a real variable, periodic with period 2π, for which the coefficients a_0, a_n, and b_n, $n \in \mathbb{N}$, can be calculated as follows:*

$$a_0 = \frac{1}{\pi} \int_{-\pi}^{\pi} f(x)\,dx$$
$$a_n = \frac{1}{\pi} \int_{-\pi}^{\pi} f(x) \cos(nx)\,dx$$
$$b_n = \frac{1}{\pi} \int_{-\pi}^{\pi} f(x) \sin(nx)\,dx,$$

*and define the trigonometric series. This series is called the **Fourier series** of f. The coefficients are referred to as **Fourier coefficients** of f.*

Notes:

1. We will use the notation

$$f(x) \sim \frac{a_0}{2} + \sum_{n=1}^{\infty} \big(a_n \cos(nx) + b_n \sin(nx)\big)$$

   to mean that the series is the Fourier series of f, but we still do not know if it converges to f.

2. The Fourier series is well determined by the values of the coefficients a_0, a_n, b_n, $n \in \mathbb{N}$, calculated earlier. If we change the value of the function f at a finite number of points, the integrals that define the coefficients do not change. In particular, the value of the function at isolated points, including the endpoints of the interval, does not matter. As a result, we can define the function on the closed interval, the open interval, or the interval that is closed at one endpoint and open at the other.

We selected the interval $[-\pi, \pi]$ for convenience. In fact, as the trigonometric functions involved are periodic with a period of 2π, any interval of amplitude 2π can be used. It is essential to keep in mind Proposition 3.5.3, which can be easily proved by a change of variables.

Proposition 3.5.3 *Let $f : \mathbb{R} \to \mathbb{R}$ be a periodic function with period 2π. Then, for every $a, b \in \mathbb{R}$,*

$$\int_a^{a+2\pi} f(x)\,dx = \int_b^{b+2\pi} f(x)\,dx.$$

Joseph Fourier (1768–1830) was a mathematician and physicist born in Auxerre, France. He studied at the École Royale Militaire of the Benedictine Order, where he became a professor. During 1795, Fourier was a student of Lagrange, Laplace, and Monge at the École Normal in Paris, and in that same year, he became a professor at the École Polytechnique. In 1797, he succeeded Lagrange in the chair of Analysis and Mechanics. His most well-known work is *Théorie analytique de la chaleur*, published in 1822. Due to his research in partial differential equations, Fourier stated that "any" function could be represented by a series of elementary trigonometric functions—sines and cosines. This statement was so astonishing that scientists of his time did not believe it. (Source of image: Engraving by Amédée Félix Barthélemy Geille (1803–1843) after Julien-Léopold Boilly (1796–1874))

In the case where f is explicitly defined on the interval $[0, 2\pi]$ instead of $[-\pi, \pi]$, it may be more convenient to compute the coefficients using the following formulas (see Example 3.5.6):

$$a_0 = \frac{1}{\pi} \int_0^{2\pi} f(x)\, dx$$

$$a_n = \frac{1}{\pi} \int_0^{2\pi} f(x) \cos(nx)\, dx$$

$$b_n = \frac{1}{\pi} \int_0^{2\pi} f(x) \sin(nx)\, dx.$$

A problem that often appears in applications is to develop a function f in a Fourier series with the function only defined on the interval $[-\pi, \pi]$ and without reference to the periodicity of f. As the formulas to compute the coefficients involve only the interval $[-\pi, \pi]$, we can calculate the Fourier series of the function. The trigonometric functions intervening in the Fourier series are periodic with period 2π. If the series is convergent, the sum function, g, will also be periodic with period 2π:

$$g(x + 2\pi) = g(x), \quad \forall x \in \mathbb{R}.$$

When calculating the Fourier series of a function defined on $]-\pi, \pi]$, we are, in fact, determining the Fourier series of its periodic extension, whose definition we introduce next.

> **Proposition 3.5.4** Let $f :]-\pi, \pi] \to \mathbb{R}$. There exists a unique function $\tilde{f}$ periodic with period 2π, called the **periodic extension** of f, that satisfies $\tilde{f}(x) = f(x)$, for all $x \in]-\pi, \pi]$.

Given the expression of $f(x)$, the function $\tilde{f}$ is well defined by

$$\tilde{f}(x) = f(x), \quad \forall x \in]-\pi, \pi]$$
$$\tilde{f}(x + 2\pi) = \tilde{f}(x), \quad \forall x \in \mathbb{R}.$$

In some situations, specifying the analytical expression of $\tilde{f}$ may be useful.

3.5. Introduction to Fourier Series

Example 3.5.1 The periodic extension of $f(x) = x$, defined on $]-\pi, \pi]$, is the function $\tilde{f}$ represented in Fig. 3.9.

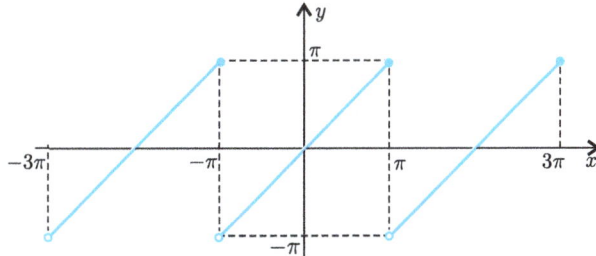

Figure 3.9: Periodic extension of $f(x) = x$, $x \in]-\pi, \pi]$

The analytical expression of $\tilde{f}$ is

$$\tilde{f}(x) = x - 2k\pi, \quad x \in](2k-1)\pi, (2k+1)\pi], \quad k \in \mathbb{Z}.$$

Example 3.5.2 The periodic extension of $f(x) = x^2$, defined on the interval $]-\pi, \pi]$, is the function $\tilde{f}$ represented in Fig. 3.10.

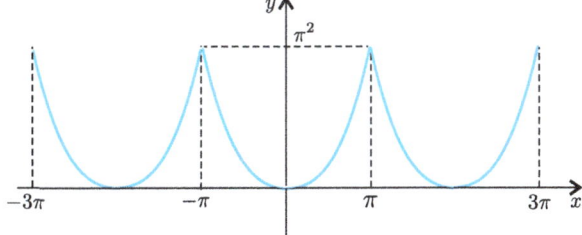

Figure 3.10: Periodic extension of $f(x) = x^2$, $x \in]-\pi, \pi]$

Its analytical expression is

$$\tilde{f}(x) = (x - 2k\pi)^2, \quad x \in](2k-1)\pi, (2k+1)\pi], \quad k \in \mathbb{Z}.$$

Note: If $f(-\pi^+) = f(\pi)$ and assuming that f is continuous on $]-\pi, \pi]$, its extension by periodicity is a continuous function on $\mathbb{R}$. If $f(-\pi^+) \neq f(\pi)$, the periodic extension is not continuous.

Before we study the properties of the Fourier series, let us examine some examples of how these series are calculated.

Example 3.5.3 Consider the real-valued function of a real variable

$$f(x) = \begin{cases} 1, & \text{if } -\pi \leq x < 0 \\ x, & \text{if } 0 \leq x \leq \pi. \end{cases}$$

The graph of f is presented in Fig. 3.11. The Fourier coefficients of f can be calculated using integration by parts.

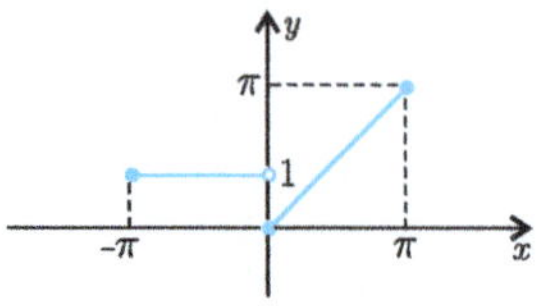

Figure 3.11: The graph of the function of Example 3.5.3

$$a_0 = \frac{1}{\pi} \int_{-\pi}^{\pi} f(x)\,dx = \frac{1}{\pi}\left(\int_{-\pi}^{0} 1\,dx + \int_{0}^{\pi} x\,dx\right)$$

$$= \frac{1}{\pi}\left([x]_{-\pi}^{0} + \left[\frac{x^2}{2}\right]_{0}^{\pi}\right) = 1 + \frac{\pi}{2};$$

$$a_n = \frac{1}{\pi}\left(\int_{-\pi}^{0} \cos(nx)\,dx + \int_{0}^{\pi} x\cos(nx)\,dx\right)$$

$$= \frac{1}{\pi}\left(\left[\frac{\sin(nx)}{n}\right]_{-\pi}^{0} + \left[x\frac{\sin(nx)}{n}\right]_{0}^{\pi} - \int_{0}^{\pi} \frac{\sin(nx)}{n}\,dx\right)$$

$$= \frac{1}{\pi}\left[\frac{\cos(nx)}{n^2}\right]_{0}^{\pi} = \frac{1}{\pi}\left(\frac{\cos(n\pi)}{n^2} - \frac{\cos(0)}{n^2}\right) = \frac{(-1)^n - 1}{\pi n^2};$$

$$b_n = \frac{1}{\pi}\left(\int_{-\pi}^{0} \sin(nx)\,dx + \int_{0}^{\pi} x\sin(nx)\,dx\right)$$

$$= \frac{1}{\pi}\left(\left[-\frac{\cos(nx)}{n}\right]_{-\pi}^{0} + \left[-x\frac{\cos(nx)}{n}\right]_{0}^{\pi} + \int_{0}^{\pi} \frac{\cos(nx)}{n}\,dx\right)$$

$$= \frac{1}{\pi}\left(-\frac{1}{n} + \frac{\cos(-n\pi)}{n} - \pi\frac{\cos(n\pi)}{n} + \left[\frac{\sin(nx)}{n^2}\right]_{0}^{\pi}\right)$$

$$= \frac{1}{\pi}\left(-\frac{1}{n} + \frac{(-1)^n}{n} - \frac{\pi(-1)^n}{n}\right) = \frac{(1-\pi)(-1)^n - 1}{\pi n}.$$

The Fourier series of f is

$$\frac{1}{2} + \frac{\pi}{4} + \sum_{n=1}^{\infty}\left(\frac{(-1)^n - 1}{\pi n^2}\cos(nx) + \frac{(1-\pi)(-1)^n - 1}{\pi n}\sin(nx)\right).$$

3.5. Introduction to Fourier Series

With this example, it is easy to verify that the sum of the series is not equal to the function at all points of the interval $[-\pi, \pi]$. In fact, at $x = -\pi$ and at $x = \pi$, we obtain the numerical series

$$\frac{1}{2} + \frac{\pi}{4} + \sum_{n=1}^{\infty} \frac{2}{\pi(2n-1)^2},$$

and $f(-\pi) = 1 \neq f(\pi) = \pi$.

Example 3.5.4 Let us consider the function $f(x) = x$ defined on $[-\pi, \pi]$ (Fig. 3.12). The Fourier coefficients of f can be computed using integration by parts and equalities (3.5) and (3.6).

$$a_0 = \frac{1}{\pi} \int_{-\pi}^{\pi} x \, dx = \frac{1}{\pi} \left[\frac{x^2}{2}\right]_{-\pi}^{\pi} = 0;$$

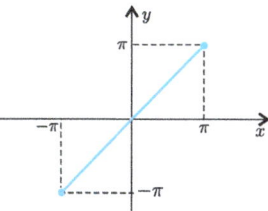

Figure 3.12: The graph of the function of Example 3.5.4

$$a_n = \frac{1}{\pi} \int_{-\pi}^{\pi} x \cos(nx) \, dx = \frac{1}{\pi} \left(\left[x \frac{\sin(nx)}{n} \right]_{-\pi}^{\pi} - \int_{-\pi}^{\pi} \frac{\sin(nx)}{n} \, dx \right) = 0;$$

$$b_n = \frac{1}{\pi} \int_{-\pi}^{\pi} x \sin(nx) \, dx = \frac{1}{\pi} \left(\left[-x \frac{\cos(nx)}{n} \right]_{-\pi}^{\pi} + \int_{-\pi}^{\pi} \frac{\cos(nx)}{n} \, dx \right)$$

$$= \frac{1}{\pi} \left(-\pi \frac{\cos(n\pi)}{n} - \pi \frac{\cos(-n\pi)}{n} \right) = (-1)^{n+1} \frac{2}{n}.$$

The Fourier series of f is $\sum_{n=1}^{\infty} (-1)^{n+1} \frac{2}{n} \sin(nx)$.

Example 3.5.5 Let us consider the function $f(x) = x^2$ defined on $[-\pi, \pi]$ (Fig. 3.13). Using the calculations from the previous example, we have

$$a_0 = \frac{1}{\pi} \int_{-\pi}^{\pi} x^2 \, dx = \frac{2\pi^2}{3};$$

$$a_n = \frac{1}{\pi} \int_{-\pi}^{\pi} x^2 \cos(nx) \, dx = \frac{1}{\pi} \left(\left[x^2 \frac{\sin(nx)}{n} \right]_{-\pi}^{\pi} - \int_{-\pi}^{\pi} 2x \frac{\sin(nx)}{n} \, dx \right)$$

$$= (-1)^n \frac{4}{n^2};$$

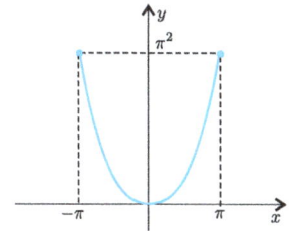

Figure 3.13: The graph of the function of Example 3.5.5

$$b_n = \frac{1}{\pi} \int_{-\pi}^{\pi} x^2 \sin(nx) \, dx$$

$$= \frac{1}{\pi} \left(\left[-x^2 \frac{\cos(nx)}{n} \right]_{-\pi}^{\pi} + \int_{-\pi}^{\pi} 2x \frac{\cos(nx)}{n} \, dx \right) = 0.$$

Thus, the Fourier series of f is

$$\frac{\pi^2}{3} + \sum_{n=1}^{\infty} (-1)^n \frac{4}{n^2} \cos(nx).$$

In the following example, we have a periodic function with period 2π, defined on $]0, 2\pi[$ (see Proposition 3.5.3).

Example 3.5.6 Consider the real-valued function of a real variable

$$f(x) = \begin{cases} \sin(x), & \text{if } 0 < x < \pi \\ 0, & \text{if } \pi \leq x < 2\pi. \end{cases}$$

The graph of f is shown in Fig. 3.14 Using equalities (3.8) and (3.11), we can calculate the Fourier coefficients as follows:

$$a_0 = \frac{1}{\pi} \int_0^{2\pi} f(x)\,dx = \frac{1}{\pi} \int_0^{\pi} \sin(x)\,dx = \frac{2}{\pi};$$

$$a_n = \frac{1}{\pi} \int_0^{2\pi} f(x) \cos(nx)\,dx = \frac{1}{\pi} \int_0^{\pi} \sin(x) \cos(nx)\,dx$$

$$= \begin{cases} 0, & \text{if } n \text{ is odd} \\ \dfrac{2}{\pi(1-n^2)}, & \text{if } n \text{ is even;} \end{cases}$$

$$b_n = \frac{1}{\pi} \int_0^{2\pi} f(x) \sin(nx)\,dx = \frac{1}{\pi} \int_0^{\pi} \sin(x) \sin(nx)\,dx$$

$$= \begin{cases} 0, & \text{if } n \neq 1 \\ \dfrac{1}{2}, & \text{if } n = 1. \end{cases}$$

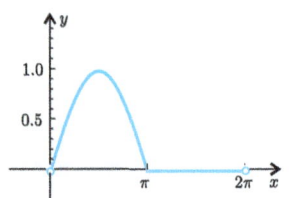

Figure 3.14: The graph of the function of Example 3.5.6

Thus, the Fourier series of f is

$$\frac{1}{\pi} + \frac{1}{2} \sin(x) + \sum_{n=1}^{\infty} \frac{2}{\pi(1-4n^2)} \cos(2nx).$$

3.5. Introduction to Fourier Series

Let us examine the issue of convergence of Fourier series. Generally, this topic can be quite complex. Even if we know that the series does converge, the issue of determining the sum function still remains. For instance, in Example 3.5.4, the series cannot converge to $f(x) = x$ for all values of x. When $x = \pi$, each term in the series is zero, causing the series to converge to zero, which is different from $f(\pi) = \pi$.

We now introduce some concepts that will allow us to approach the study of the convergence of the Fourier series, which is an essential problem of this theory.

Definition 3.5.4 *A function f is* **piecewise continuous** *on an interval $[a, b]$ if it is continuous on that interval or continuous except at a finite number of points $x_1 \leq x_2 \leq \cdots \leq x_n$, $x_i \in [a, b]$, $\forall i \in \{1, \ldots, n\}$, and exist, at each point of discontinuity, the left- and right-hand limits*

$$f(x_k^-) = \lim_{x \to x_k^-} f(x) \quad \text{and} \quad f(x_k^+) = \lim_{x \to x_k^+} f(x).$$

A function f is piecewise continuous on $\mathbb{R}$ if it is piecewise continuous on every bounded interval of $\mathbb{R}$.

In Fig. 3.15 we can see an illustration of the definition above.

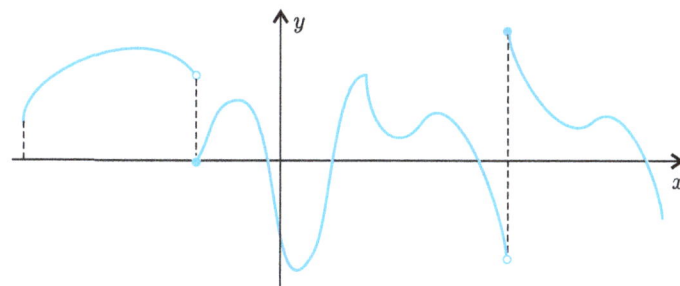

Figure 3.15: A piecewise continuous function

Definition 3.5.5 *A function f is **piecewise of class C^1**, or simply **piecewise C^1**, on an interval $[a,b]$ if it is continuously differentiable on that interval, or continuously differentiable except at a finite number of points $x_1 \leq x_2 \leq \cdots \leq x_n$, $x_i \in [a,b]$, $\forall i \in \{1,\ldots,n\}$, and exist, at each point x_k, the left- and right-hand limits*

$$f(x_k^-) = \lim_{x \to x_k^-} f(x) \quad \text{and} \quad f(x_k^+) = \lim_{x \to x_k^+} f(x)$$

and

$$f'(x_k^-) = \lim_{x \to x_k^-} f'(x) \quad \text{and} \quad f'(x_k^+) = \lim_{x \to x_k^+} f'(x).$$

A function f is piecewise C^1 on $\mathbb{R}$ if it is piecewise C^1 on every bounded interval of $\mathbb{R}$.

Theorem 3.5.1 *If the function f, periodic with period 2π, is piecewise C^1 on $[-\pi,\pi]$, its Fourier series converges at all points. At points of continuity of f, the sum of the series is equal to the value of the function. At each point of discontinuity of f, the sum of the series is equal to the arithmetic mean of the left- and right-hand limits of the function at that point.*

Example 3.5.7 In Example 3.5.4, we obtained the Fourier series of the function $f(x) = x$, $x \in \,]-\pi,\pi]$

$$\sum_{n=1}^{\infty} (-1)^{n+1} \frac{2}{n} \sin(nx).$$

In Example 3.5.1, we extended f by periodicity, defining a function $\tilde{f}$. From the previous theorem, we obtain

$$\sum_{n=1}^{\infty} \frac{2(-1)^{n+1}}{n} \sin(nx) = \begin{cases} \tilde{f}(x), & \text{if } x \neq (2k+1)\pi,\ k \in \mathbb{Z} \\ \dfrac{\tilde{f}(x^+) + \tilde{f}(x^-)}{2} = 0, & \text{if } x = (2k+1)\pi,\ k \in \mathbb{Z}. \end{cases}$$

Figure 3.16 shows some partial sums of the Fourier series of the function $\tilde{f}$.

3.5. Introduction to Fourier Series

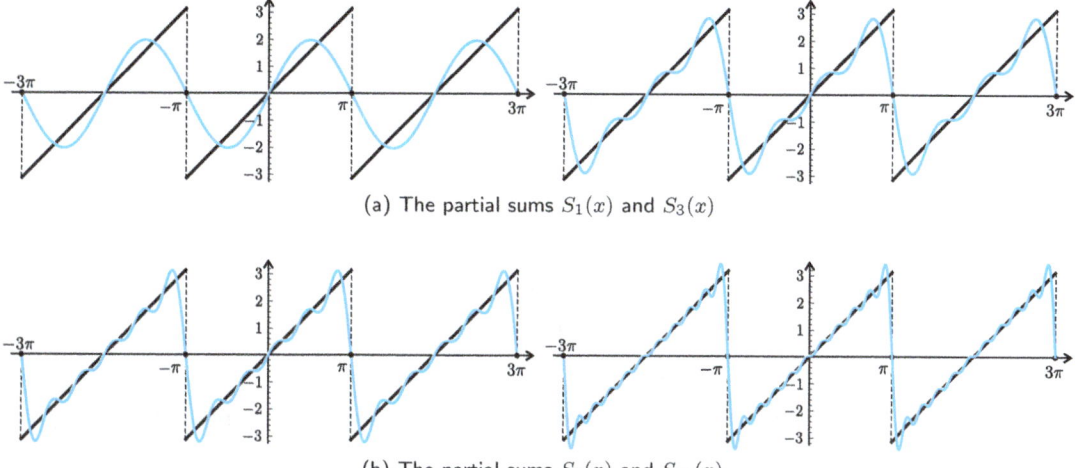

(a) The partial sums $S_1(x)$ and $S_3(x)$

(b) The partial sums $S_5(x)$ and $S_{10}(x)$

Figure 3.16: Partial sums of the Fourier series of the function from Example 3.5.7

Example 3.5.8 Consider the real-valued function of a real variable f defined by (Fig. 3.17)

$$f(x) = \begin{cases} 0, & \text{if } -\pi < x < 0 \\ 3, & \text{if } 0 < x < \pi. \end{cases}$$

The Fourier coefficients are

$$a_0 = \frac{1}{\pi}\int_{-\pi}^{\pi} f(x)\,dx = \frac{1}{\pi}\int_0^{\pi} 3\,dx = 3;$$

$$a_n = \frac{1}{\pi}\int_{-\pi}^{\pi} f(x)\cos(nx)\,dx = \frac{1}{\pi}\int_0^{\pi} 3\cos(nx)\,dx = 0;$$

$$b_n = \frac{1}{\pi}\int_{-\pi}^{\pi} f(x)\sin(nx)\,dx = \frac{1}{\pi}\int_0^{\pi} 3\sin(nx)\,dx = \begin{cases} \dfrac{6}{n\pi}, & \text{if } n \text{ is odd} \\ 0, & \text{if } n \text{ is even.} \end{cases}$$

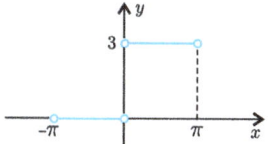

Figure 3.17: The graph of the function of Example 3.5.8

The Fourier series is

$$\frac{3}{2} + \sum_{n=1}^{\infty} \frac{6}{(2n-1)\pi} \sin\left((2n-1)x\right)$$

and converges pointwise to the function

$$\tilde{f}(x) = \begin{cases} 0, & \text{if } (2n-1)\pi < x < 2n\pi,\ n \in \mathbb{Z} \\ 3, & \text{if } 2n\pi < x < (2n+1)\pi,\ n \in \mathbb{Z} \\ \dfrac{3}{2}, & \text{if } x = n\pi,\ n \in \mathbb{Z}. \end{cases}$$

Figure 3.18 shows some partial sums of the Fourier series of the function f.

Example 3.5.9 In Example 3.5.3, we obtained the Fourier series for the real-valued function of a real variable

$$f(x) = \begin{cases} 1, & \text{if } -\pi \le x < 0 \\ x, & \text{if } 0 \le x \le \pi. \end{cases}$$

The Fourier series is given by

$$\frac{1}{2} + \frac{\pi}{4} + \sum_{n=1}^{\infty} \left(\frac{(-1)^n - 1}{\pi n^2} \cos(nx) + \frac{(1-\pi)(-1)^n - 1}{\pi n} \sin(nx) \right).$$

The series converges pointwise to the function

$$\tilde{f}(x) = \begin{cases} 1, & \text{if } (2k-1)\pi < x < 2k\pi,\ k \in \mathbb{Z} \\ \dfrac{1}{2}, & \text{if } x = 2k\pi,\ k \in \mathbb{Z} \\ x - 2k\pi, & \text{if } 2k\pi < x < (2k+1)\pi,\ k \in \mathbb{Z} \\ \dfrac{\pi + 1}{2}, & \text{if } x = (2k+1)\pi,\ k \in \mathbb{Z}. \end{cases}$$

The coefficients a_n of the function $f(x) = x$ are all zero. This is a consequence of the fact that f is odd.

Let us consider the following two results, which can be easily shown (see Figs. 3.19 and 3.20).

3.5. Introduction to Fourier Series

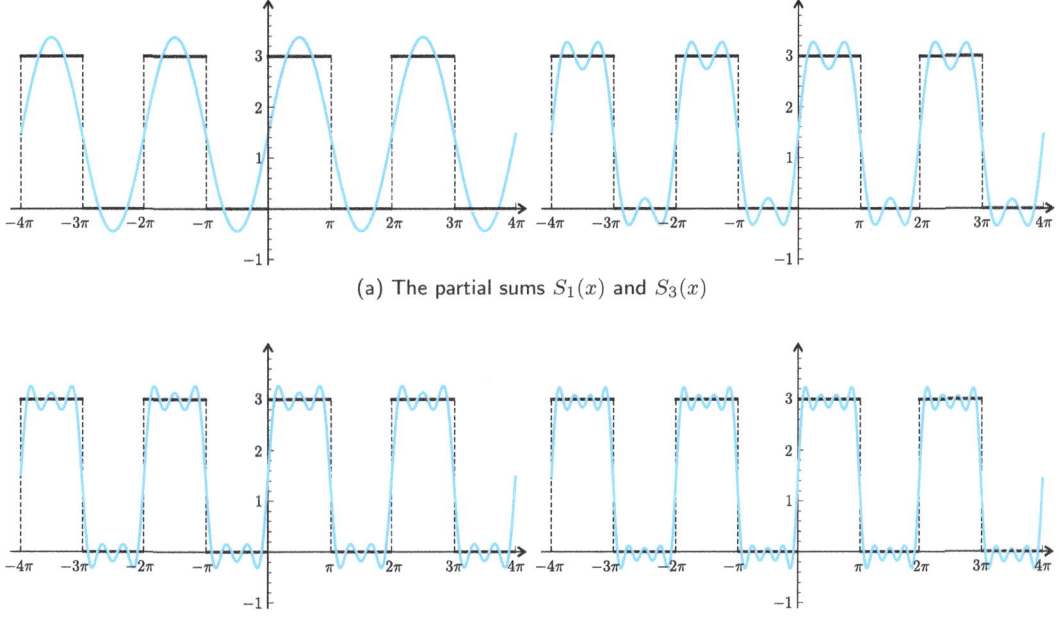

(a) The partial sums $S_1(x)$ and $S_3(x)$

(b) The partial sums $S_5(x)$ and $S_7(x)$

Figure 3.18: Partial sums of the Fourier series of the function from Example 3.5.8

Proposition 3.5.5 *The sum of two even functions is an even function, and the sum of two odd functions is an odd function. The product of two even functions or two odd functions is an even function. The product of an even function and an odd function is an odd function.*

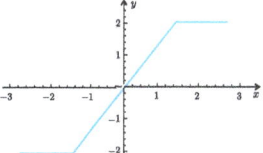

Figure 3.19: An odd function

Proposition 3.5.6 *Let $a \in \mathbb{R}^+$ and $f : \mathbb{R} \to \mathbb{R}$ be an integrable function on $[-a, a]$.*

a) *If f is even, then $\int_{-a}^{a} f(x)\,dx = 2\int_{0}^{a} f(x)\,dx$.*

b) *If f is odd, then $\int_{-a}^{a} f(x)\,dx = 0$.*

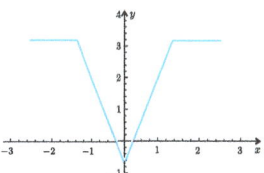

Figure 3.20: An even function

Let f be an even function. Taking into account that the functions $\cos(nx)$ are even and $\sin(nx)$ are odd, for all $n \in \mathbb{N}$, by Proposition 3.5.5, we find that $f(x)\cos(nx)$ are even and $f(x)\sin(nx)$ are odd, for all $n \in \mathbb{N}$.

By Proposition 3.5.6, the Fourier coefficients of f are

$$a_0 = \frac{1}{\pi}\int_{-\pi}^{\pi} f(x)\,dx = \frac{2}{\pi}\int_0^{\pi} f(x)\,dx;$$

$$a_n = \frac{1}{\pi}\int_{-\pi}^{\pi} f(x)\cos(nx)\,dx = \frac{2}{\pi}\int_0^{\pi} f(x)\cos(nx)\,dx;$$

$$b_n = \frac{1}{\pi}\int_{-\pi}^{\pi} f(x)\,\sin(nx)\,dx = 0.$$

We can state the following proposition:

> **Proposition 3.5.7** *Let f be an even function defined on $[-\pi, \pi]$. Then*
>
> $$f(x) \sim \frac{a_0}{2} + \sum_{n=1}^{\infty} a_n \cos(nx),$$
>
> *where* $a_n = \dfrac{2}{\pi}\displaystyle\int_0^{\pi} f(x)\cos(nx)\,dx$, $n \in \mathbb{N}_0$, *that is, f is represented by a Fourier cosine series.*

Example 3.5.10 Consider the even function $f(x) = |x|$ on the interval $[-\pi, \pi]$ (Fig. 3.21). The Fourier coefficients of f are

$$a_0 = \frac{2}{\pi}\int_0^{\pi} f(x)\,dx = \frac{2}{\pi}\int_0^{\pi} x\,dx = \pi;$$

$$a_n = \frac{2}{\pi}\int_0^{\pi} f(x)\cos(nx)\,dx = \frac{2}{\pi}\int_0^{\pi} x\cos(nx)\,dx$$

$$= \begin{cases} -\dfrac{4}{n^2\pi}, & \text{if } n \text{ is odd} \\ 0, & \text{if } n \text{ is even}, n \neq 0. \end{cases}$$

The analytical expression of the periodic extension of f to $\mathbb{R}$ is

$$\tilde{f}(x) = |x - 2k\pi|,\ x \in [(2k-1)\pi, (2k+1)\pi]\ \ k \in \mathbb{Z}.$$

The function is continuous on $\mathbb{R}$; therefore by Theorem 3.5.1,

$$\tilde{f}(x) = \frac{\pi}{2} - \sum_{n=1}^{\infty} \frac{4}{(2n-1)^2\pi}\cos((2n-1)x),\ \ \forall x \in \mathbb{R}.$$

3.5. Introduction to Fourier Series

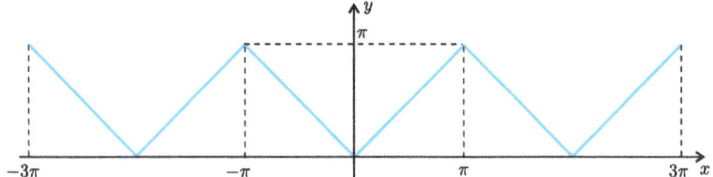

Figure 3.21: The graph of the extension of $f(x) = |x|$

Let f be an odd function. Considering that $\cos(nx)$ are even and $\sin(nx)$ are odd for all $n \in \mathbb{N}$, by Proposition 3.5.5, we obtain that $f(x)\cos(nx)$ are odd and $f(x)\sin(nx)$ are even for all $n \in \mathbb{N}$. By Proposition 3.5.6, the Fourier coefficients of f are

$$a_0 = \frac{1}{\pi} \int_{-\pi}^{\pi} f(x)\, dx = 0;$$

$$a_n = \frac{1}{\pi} \int_{-\pi}^{\pi} f(x)\cos(nx)\, dx = 0;$$

$$b_n = \frac{1}{\pi} \int_{-\pi}^{\pi} f(x)\sin(nx)\, dx = \frac{2}{\pi} \int_{0}^{\pi} f(x)\sin(nx)\, dx.$$

We can state the following proposition:

> **Proposition 3.5.8** Let f be an odd function defined on $[-\pi, \pi]$. Then
>
> $$f(x) \sim \sum_{n=1}^{\infty} b_n \sin(nx),$$
>
> where $b_n = \dfrac{2}{\pi} \int_0^{\pi} f(x)\sin(nx)\, dx$, $n \in \mathbb{N}$, that is, f is represented by a Fourier sine series.

Example 3.5.11 Consider the function f defined by

$$f(x) = \begin{cases} -\pi - x, & \text{if } -\pi \leq x < 0 \\ \pi - x, & \text{if } 0 \leq x < \pi. \end{cases}$$

It is an odd function (Fig. 3.22). The Fourier coefficients of f are

$$b_n = \frac{2}{\pi} \int_0^{\pi} f(x)\sin(nx)\, dx = \frac{2}{\pi} \int_0^{\pi} (\pi - x)\sin(nx)\, dx = \frac{2}{n}.$$

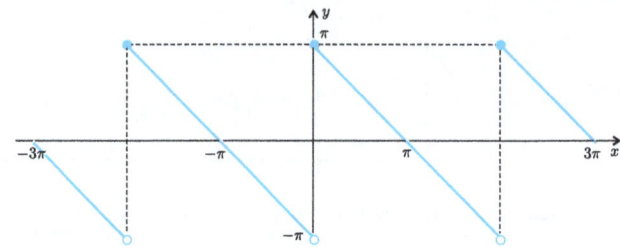

Figure 3.22: Graph of the extension of the function from Example 3.5.11

The analytical expression of the periodic extension of f to $\mathbb{R}$ is

$$\tilde{f}(x) = -x + (2k+1)\pi, \ x \in [2k\pi, (2k+2)\pi[\ \ k \in \mathbb{Z}.$$

It is a piecewise C^1 function on $\mathbb{R}$; therefore by Theorem 3.5.1,

$$\sum_{n=1}^{\infty} \frac{2}{n} \sin(nx) = \begin{cases} \tilde{f}(x), & \text{if } x \in \,]2k\pi, (2k+2)\pi[, \ k \in \mathbb{Z} \\ 0, & \text{if } x = 2k\pi, \ k \in \mathbb{Z}. \end{cases}$$

Sometimes, we must solve the problem of determining the development in a Fourier series of sines or cosines of a function defined on the interval $[0, \pi]$. To develop f in a Fourier cosine series, we consider an even extension of f from the interval $[0, \pi]$ to the interval $[-\pi, \pi]$ (see Fig. 3.23):

$$f_{even}(x) = \begin{cases} f(-x), & \text{if } -\pi < x < 0 \\ f(x), & \text{if } 0 \leq x < \pi. \end{cases}$$

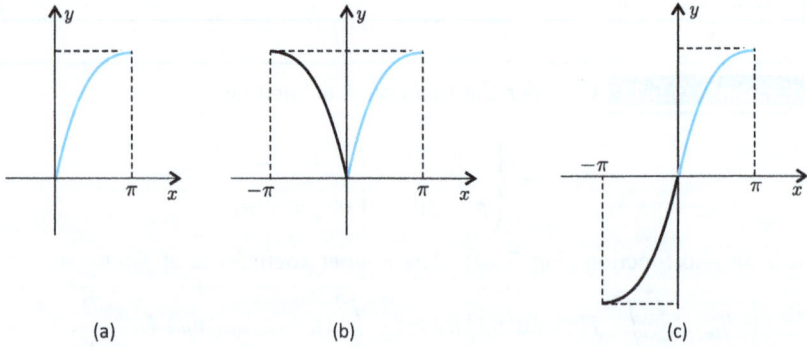

(a)   (b)   (c)

Figure 3.23: (a) The function f. (b) Even extension of f. (c) Odd extension of f

3.5. Introduction to Fourier Series

Then, we apply everything that has been stated for extensions as an even function. Therefore, the Fourier coefficients can be computed by the formulas of Proposition 3.5.7.

To develop f in a Fourier sine series, we take an odd extension of f from the interval $[0, \pi]$ to the interval $[-\pi, \pi]$ (see Fig. 3.23):

$$f_{odd}(x) = \begin{cases} -f(-x), & \text{if } -\pi < x < 0 \\ f(x), & \text{if } 0 \leq x < \pi. \end{cases}$$

Next, we apply the previous knowledge of extensions as an odd function. Consequently, the formulas of Proposition 3.5.8 can be used to determine the Fourier coefficients.

Example 3.5.12 Let us determine the Fourier sine series and the Fourier cosine series of the function f defined by $f(x) = x^3 - x$, on $[0, \pi[$ (Fig. 3.24). The analytical expression of the even extension to the interval $]-\pi, \pi[$ is

$$f_{even}(x) = \begin{cases} -x^3 + x, & \text{if } -\pi < x < 0 \\ x^3 - x, & \text{if } 0 \leq x < \pi. \end{cases}$$

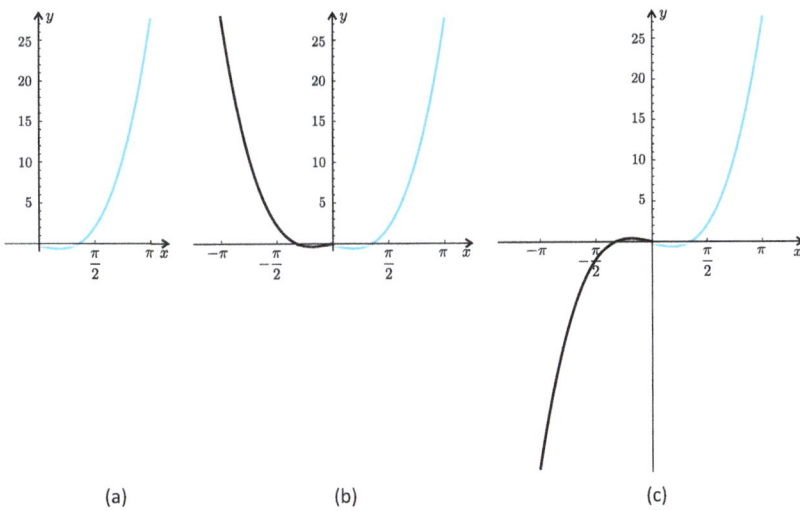

(a) (b) (c)

Figure 3.24: (a) The function f. (b) Even extension of f. (c) Odd extension of f

The coefficients of the Fourier cosine series are

$$a_0 = \frac{2}{\pi} \int_0^\pi (x^3 - x) \, dx = \frac{2}{\pi} \left[\frac{x^4}{4} - \frac{x^2}{2} \right]_0^\pi = \frac{\pi^3}{2} - \pi;$$

$$a_n = \frac{2}{\pi} \int_0^\pi f(x) \cos(nx) \, dx = \frac{2}{\pi} \int_0^\pi (x^3 - x) \cos(nx) \, dx$$

$$= \frac{2}{\pi} \left((-1)^n \frac{3\pi^2}{n^2} + \frac{6 + n^2}{n^4} (1 - (-1)^n) \right),$$

and the series is

$$\frac{\pi^3}{4} - \frac{\pi}{2} + \frac{2}{\pi} \sum_{n=1}^\infty \left((-1)^n \frac{3\pi^2}{n^2} + \frac{6 + n^2}{n^4} (1 - (-1)^n) \right) \cos(nx).$$

Note that the periodic extension of f_{even} is

$$\tilde{f}_{even}(x) = \begin{cases} -(x - 2k\pi)^3 + (x - 2k\pi), & \text{if } (2k-1)\pi < x < 2k\pi, \, k \in \mathbb{Z} \\ (x - 2k\pi)^3 - (x - 2k\pi), & \text{if } 2k\pi \leq x < (2k+1)\pi, \, k \in \mathbb{Z} \end{cases}$$

whose graph can be seen in Fig. 3.25.

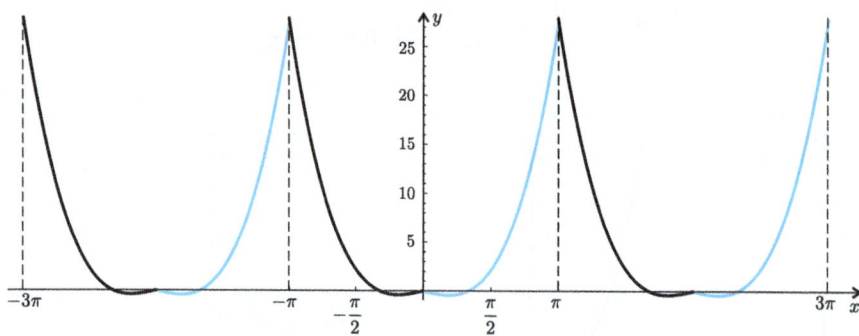

Figure 3.25: Periodic extension of the function f_{even}

Since $\tilde{f}_{even}$ is continuous on $\mathbb{R}$, Theorem 3.5.1 allows us to affirm that the following equality is satisfied, whatever $x \in \mathbb{R}$:

$$\tilde{f}_{even}(x) = \frac{\pi^3}{4} - \frac{\pi}{2} + \frac{2}{\pi} \sum_{n=1}^\infty \left((-1)^n \frac{3\pi^2}{n^2} + \frac{6 + n^2}{n^4} (1 - (-1)^n) \right) \cos(nx).$$

The analytical expression of the odd extension to the interval $]-\pi, \pi[$ is

$$\tilde{f}_{odd}(x) = x^3 - x, \; x \in \,]-\pi, \pi[.$$

3.5. Introduction to Fourier Series

The coefficients of the Fourier sine series are

$$b_n = \frac{2}{\pi} \int_0^\pi f(x) \sin(nx)\, dx = 2\,(-1)^n \left(\frac{1-\pi^2}{n} + \frac{6}{n^3} \right),$$

and the series is

$$2 \sum_{n=1}^\infty (-1)^n \left(\frac{1-\pi^2}{n} + \frac{6}{n^3} \right) \sin(nx).$$

The periodic extension of $\tilde{f}_{odd}$ has the analytical expression

$$\tilde{f}_{odd}(x) = (x - 2k\pi)^3 - x - 2k\pi, \ x \in\,](2k-1)\pi, (2k+1)\pi[, \ k \in \mathbb{Z},$$

and its graph is shown in Fig. 3.26.

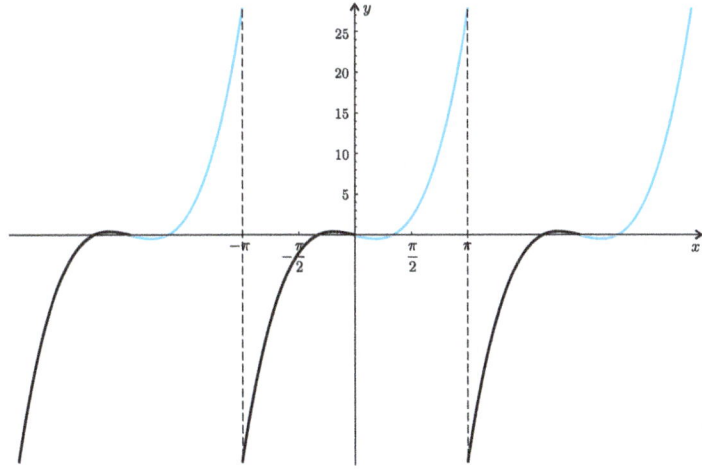

Figure 3.26: Periodic extension of the function f_{odd}

The function is continuous on $\mathbb{R} \setminus \{(2k+1)\pi, \ k \in \mathbb{Z}\}$, but it is piecewise C^1 on $\mathbb{R}$. Theorem 3.5.1 allows us to affirm that the equality is satisfied:

$$2 \sum_{n=1}^\infty (-1)^n \left(\frac{1-\pi^2}{n} + \frac{6}{n^3} \right) \sin(nx) = \begin{cases} \tilde{f}_{odd}(x), & \text{if } x \neq (2k+1)\pi \\ 0, & \text{if } x = (2k+1)\pi, \end{cases}$$

for all $k \in \mathbb{Z}$.

Thus far, we have dealt with periodic functions of period 2π. Let us now focus on functions f with period $T = 2l$, $l > 0$.

Let $x = \dfrac{lt}{\pi}$. The function $g(t) = f\left(\dfrac{lt}{\pi}\right)$ has period 2π. In fact,

$$g(t + 2\pi) = f\left(\frac{l(t + 2\pi)}{\pi}\right) = f\left(\frac{lt}{\pi} + 2l\right) = f\left(\frac{lt}{\pi}\right) = g(t).$$

The Fourier series of g is

$$\frac{a_0}{2} + \sum_{n=1}^{\infty} \left(a_n \cos(nt) + b_n \sin(nt)\right),$$

where

$$a_0 = \frac{1}{\pi} \int_{-\pi}^{\pi} g(t)\, dt;$$

$$a_n = \frac{1}{\pi} \int_{-\pi}^{\pi} g(t) \cos(nt)\, dt;$$

$$b_n = \frac{1}{\pi} \int_{-\pi}^{\pi} g(t) \sin(nt)\, dt.$$

Then the Fourier series of $f(x) = f\left(\dfrac{lt}{\pi}\right)$ is

$$\frac{a_0}{2} + \sum_{n=1}^{\infty} \left(a_n \cos\left(\frac{n\pi x}{l}\right) + b_n \sin\left(\frac{n\pi x}{l}\right)\right),$$

where

$$a_0 = \frac{1}{\pi} \int_{-\pi}^{\pi} g(t)\, dt = \frac{1}{\pi} \int_{-l}^{l} f(x) \cdot \frac{\pi}{l}\, dx. = \frac{1}{l} \int_{-l}^{l} f(x)\, dx\,;$$

$$a_n = \frac{1}{\pi} \int_{-\pi}^{\pi} g(t) \cos(nt)\, dt = \frac{1}{\pi} \int_{-l}^{l} f(x) \cos\left(\frac{n\pi x}{l}\right) \cdot \frac{\pi}{l}\, dx$$

$$= \frac{1}{l} \int_{-l}^{l} f(x) \cos\left(\frac{n\pi x}{l}\right) dx\,;$$

$$b_n = \frac{1}{\pi} \int_{-\pi}^{\pi} g(t) \sin(nt)\, dt = \frac{1}{\pi} \int_{-l}^{l} f(x) \sin\left(\frac{n\pi x}{l}\right) \cdot \frac{\pi}{l}\, dx$$

$$= \frac{1}{l} \int_{-l}^{l} f(x) \sin\left(\frac{n\pi x}{l}\right) dx.$$

3.5. Introduction to Fourier Series

Proposition 3.5.9 *If f is a periodic function with period $2l$, the coefficients of its Fourier series are given by*

$$a_0 = \frac{1}{l} \int_{-l}^{l} f(x)\, dx$$

$$a_n = \frac{1}{l} \int_{-l}^{l} f(x) \cos\left(\frac{n\pi x}{l}\right) dx$$

$$b_n = \frac{1}{l} \int_{-l}^{l} f(x) \sin\left(\frac{n\pi x}{l}\right) dx.$$

Example 3.5.13 Consider the function f, periodic with period 4, defined by

$$f(x) = \begin{cases} 0, & \text{if } -2 \leq x < 0 \\ 2, & \text{if } 0 \leq x < 2. \end{cases}$$

The graph of f is shown in Fig. 3.27. The coefficients of its Fourier series are given by

$$a_0 = \frac{1}{2} \int_{-2}^{2} f(x)\, dx = \frac{1}{2} \int_{0}^{2} 2\, dx = 2;$$

$$a_n = \frac{1}{2} \int_{-2}^{2} f(x) \cos\left(\frac{n\pi x}{2}\right) dx = \frac{1}{2} \int_{0}^{2} 2 \cos\left(\frac{n\pi x}{2}\right) dx = 0;$$

$$b_n = \frac{1}{2} \int_{-2}^{2} f(x) \sin\left(\frac{n\pi x}{2}\right) dx = \frac{1}{2} \int_{0}^{2} 2 \sin\left(\frac{n\pi x}{2}\right) dx$$

$$= \begin{cases} 0, & \text{if } n \text{ is even} \\ \dfrac{4}{n\pi}, & \text{if } n \text{ is odd.} \end{cases}$$

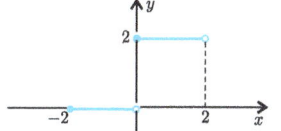

Figure 3.27: The graph of the function of Example 3.5.13

The Fourier series of f is

$$\frac{a_0}{2} + \sum_{n=1}^{\infty} \left(a_n \cos\left(\frac{n\pi x}{2}\right) + b_n \sin\left(\frac{n\pi x}{2}\right) \right)$$

$$= 1 + \sum_{n=1}^{\infty} \frac{4}{(2n-1)\pi} \sin\left(\frac{(2n-1)\pi x}{2}\right).$$

Example 3.5.14 Consider the function f defined by

$$f(x) = \begin{cases} x, & \text{if } 0 \leq x < 1 \\ 2-x, & \text{if } 1 \leq x < 2. \end{cases}$$

We want to develop f into a Fourier sine series. For this, we take an odd extension of f, which has period $2l = 4$ (Fig. 3.28).

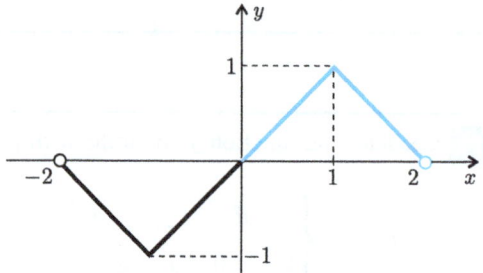

Figure 3.28: Odd extension of function f

The coefficients of its Fourier series are given by

$$b_n = \frac{1}{2}\int_{-2}^{2} f(x)\sin\left(\frac{n\pi x}{2}\right)dx = \int_{0}^{2} f(x)\sin\left(\frac{n\pi x}{2}\right)dx$$

$$= \frac{8}{(n\pi)^2}\sin\left(\frac{n\pi}{2}\right)$$

$$= \begin{cases} 0, & \text{if } n \text{ is even} \\ \frac{8}{(n\pi)^2}(-1)^{(n-1)/2}, & \text{if } n \text{ is odd.} \end{cases}$$

The Fourier sine series of f is

$$\sum_{n=1}^{\infty} b_n \sin\left(\frac{n\pi x}{2}\right) = \frac{8}{\pi^2}\sum_{n=1}^{\infty}\frac{(-1)^{n+1}}{(2n-1)^2}\sin\left(\frac{(2n-1)\pi x}{2}\right).$$

We saw in Theorem 3.5.1 sufficient conditions for the convergence of the Fourier series of a periodic function with period 2π. If f has period $2l$, the theorem can be easily generalized:

3.5. Introduction to Fourier Series

Theorem 3.5.2 *If the function f, periodic with period $2l$, is piecewise C^1 on $[-l, l]$, its Fourier series converges at all points. At points of continuity of f, the sum of the series is equal to the value of the function. At each point of discontinuity of f, the sum of the series is equal to the arithmetic mean of the left- and right-hand limits of the function at that point.*

We often want to differentiate and integrate term by term the Fourier series and, as we saw in Sect. 3.2, sufficient conditions to guarantee this procedure involve the concept of uniform convergence. We have the following results.

Theorem 3.5.3 *If the series $\sum_{n=1}^{\infty} (|a_n| + |b_n|)$ converges, then the trigonometric series*
$$\frac{a_0}{2} + \sum_{n=1}^{\infty} \left(a_n \cos\left(\frac{n\pi x}{l}\right) + b_n \sin\left(\frac{n\pi x}{l}\right) \right)$$
converges absolutely and uniformly on $\mathbb{R}$, and its sum is a continuous function. The trigonometric series is the Fourier series of the sum function.

Proof: It is a direct consequence of the Weierstrass' test because
$$\left| a_n \cos\left(\frac{n\pi x}{l}\right) + b_n \sin\left(\frac{n\pi x}{l}\right) \right| \leq \left| a_n \cos\left(\frac{n\pi x}{l}\right) \right| + \left| b_n \sin\left(\frac{n\pi x}{l}\right) \right|$$
$$\leq |a_n| + |b_n|, \quad \forall x \in \mathbb{R}. \quad \blacksquare$$

Theorem 3.5.4 *Let f be a periodic function with period $2l$, continuous, piecewise C^1. Then, its Fourier series converges absolutely and uniformly to f.*

Theorem 3.5.5 *If the Fourier series of a continuous function f is convergent, then the sum of the series is f.*

Theorem 3.5.6 *Let f be a periodic function with period $2l$, piecewise C^1 on $[-l, l]$. Let $x_1 \leq x_2 \leq \cdots \leq x_n$, $x_i \in \,]-l, l[$, $\forall i \in \{1, \ldots, n\}$, be the points of discontinuity of f. Let $[a, b] \subset \,]-l, l[$ such that $x_i \notin [a, b]$, $\forall i \in \{1, \ldots, n\}$. Then, its Fourier series converges uniformly to f on $[a, b]$.*

Theorem 3.5.7 *Let f be a piecewise continuous function on $[-l, l[$ and periodic with period $2l$. Then, the Fourier series of f is integrable term by term. More precisely, if*

$$f(x) \sim \frac{a_0}{2} + \sum_{n=1}^{\infty} \left(a_n \cos\left(\frac{n\pi x}{l}\right) + b_n \sin\left(\frac{n\pi x}{l}\right) \right)$$

and $a, b \in \mathbb{R}$, then

$$\int_a^b f(x)\, dx = \frac{a_0}{2}(b-a)$$
$$+ \sum_{n=1}^{\infty} \left(a_n \int_a^b \cos\left(\frac{n\pi x}{l}\right) dx + b_n \int_a^b \sin\left(\frac{n\pi x}{l}\right) dx \right).$$

It should be noted that the previous theorem remains valid even if the Fourier series is not pointwise convergent for the function f. By integrating term by term, a factor of $\frac{1}{n}$ is introduced in the series, which results in its convergence.

Theorem 3.5.8 *Let f be a continuous function on the interval $[-l, l]$, with $f(-l) = f(l)$, periodic with period $2l$ and such that f' is piecewise C^1 on $[-l, l]$. Then, the Fourier series of f' can be obtained by differentiating term by term the Fourier series of f. In addition, the Fourier series of f' converges to f' at the points of continuity of f', and at each point of discontinuity, the sum of the series is equal to the arithmetic mean of the left- and right-hand derivatives of the function f at that point.*

Note that the theorem regulating term by term differentiation is more demanding than that of term by term integration. Recall what we saw in Sect. 3.2: Term by term differentiation of a series of functions can only be guaranteed if the series of derivatives is uniformly convergent.

The previous theorems allow us to operate on the Fourier series similarly to what was done with the Taylor series.

Example 3.5.15 We saw in Example 3.5.4 that

$$x = \sum_{n=1}^{\infty} (-1)^{n+1} \frac{2}{n} \sin(nx), \quad -\pi < x < \pi,$$

3.5. Introduction to Fourier Series

that is,
$$\frac{x}{2} = \sum_{n=1}^{\infty} (-1)^{n+1} \frac{\sin(nx)}{n}, \quad -\pi < x < \pi.$$

We can integrate this series term by term, obtaining
$$\int_0^x \frac{t}{2}\, dt = \sum_{n=1}^{\infty} (-1)^{n+1} \int_0^x \frac{\sin(nt)}{n}\, dt, \quad -\pi < x < \pi$$
$$\Leftrightarrow \frac{x^2}{4} = \sum_{n=1}^{\infty} \frac{(-1)^{n+1}}{n^2} - \sum_{n=1}^{\infty} (-1)^{n+1} \frac{\cos(nx)}{n^2}, \quad -\pi < x < \pi.$$

The value of the numerical series can be computed by integrating this result between $-\pi$ and π:
$$\int_{-\pi}^{\pi} \frac{x^2}{4}\, dx = 2\pi \sum_{n=1}^{\infty} \frac{(-1)^{n+1}}{n^2} - \sum_{n=1}^{\infty} \frac{(-1)^{n+1}}{n^2} \int_{-\pi}^{\pi} \cos(nx)\, dx$$
$$= 2\pi \sum_{n=1}^{\infty} \frac{(-1)^{n+1}}{n^2}.$$

Therefore,
$$\frac{\pi^3}{6} = 2\pi \sum_{n=1}^{\infty} \frac{(-1)^{n+1}}{n^2},$$

that is,
$$\frac{\pi^2}{12} = \sum_{n=1}^{\infty} \frac{(-1)^{n+1}}{n^2}.$$

Thus, the Fourier series of $f(x) = x^2$ is
$$x^2 = \frac{\pi^2}{3} + 4 \sum_{n=1}^{\infty} (-1)^n \frac{\cos(nx)}{n^2}, \quad -\pi < x < \pi,$$

as we saw in Example 3.5.5.

The hypothesis about the continuity of the function in Theorem 3.5.8 is essential, as can be seen in the following example.

Example 3.5.16 Consider the development in Fourier series of the function $f(x) = x$ defined on $[-\pi, \pi]$, obtained in Example 3.5.4. We saw, in Example 3.5.7, that
$$x = \sum_{n=1}^{\infty} (-1)^{n+1} \frac{2}{n} \sin(nx), \quad -\pi < x < \pi.$$

We have $f'(x) = 1$, but

$$\sum_{n=1}^{\infty} \left((-1)^{n+1}\frac{2}{n}\sin(nx)\right)' = \sum_{n=1}^{\infty} (-1)^{n+1}\frac{2}{n} n\cos(nx)$$
$$= 2\sum_{n=1}^{\infty} (-1)^{n+1}\cos(nx),$$

and this series diverges whatever $x \in \mathbb{R}$ (the general term is not a null sequence).

This result does not contradict Theorem 3.5.8 because $f(-\pi) \neq f(\pi)$, which shows the need for the condition of equality. The extension by periodicity of f is discontinuous (see Example 3.5.1). We can observe that, in Theorem 3.5.8, the continuity of f on $[-l, l]$ and the condition $f(-l) = f(l)$ ensure the continuity of the extension by periodicity of f.

3.6 Solved Exercises

1. Consider the real-valued function f of a real variable, which is defined by
$$f(x) = \sum_{n=1}^{\infty} \frac{\sin(n^2 x)}{\sqrt[5]{n^7 + 3}}.$$

   a) Find the domain of f.

   b) Check if f is continuous on its domain.

2. Consider the real-valued function f of a real variable, which is defined by
$$f(x) = \sum_{n=0}^{\infty} \frac{1}{1 + (3x)^n}.$$

   a) Determine the domain of f.

   b) Show that f is continuous on $[1, +\infty[$.

3. Consider the real-valued function f of a real variable, which is defined by
$$f(x) = \sum_{n=1}^{\infty} \frac{x}{n^x}.$$

   a) Find the domain of f.

   b) Show that f is continuous on $]1, +\infty[$.

4. Consider the real-valued function f of a real variable, which is defined by
$$f(x) = \sum_{n=1}^{\infty} \frac{1}{(nx+1)\sqrt{n^3+1}}.$$

   Calculate $\int_1^2 f(x)\,dx$.

5. Consider the real-valued function f of a real variable, which is defined by
$$f(x) = \sum_{n=1}^{\infty} \frac{\cos(nx)}{n^2}.$$

   Show that
$$\int_0^{\pi/2} f(x)\,dx = \sum_{n=0}^{\infty} \frac{(-1)^n}{(2n+1)^3}.$$

6. Consider the real-valued function f of a real variable, which is defined by
$$f(x) = \sum_{n=0}^{\infty} \frac{e^{-nx}}{n^4 + 5}.$$

   Find $\int_0^1 f(x)\,dx$.

7. Consider the series of functions defined, for $x \geq 0$, by $\sum_{n=1}^{\infty} \left(x + \frac{1}{n}\right)^{n + \frac{x}{n}}$.

   a) Determine the values of x for which the series is convergent.

   b) Let $0 < \alpha < 1$. Show that the series converges uniformly on $[0, \alpha]$.

8. Let (a_n) be a sequence of positive terms such that
$$\lim \frac{a_{n+1}}{a_n} = \frac{1}{2}.$$

   If, for each $n \in \mathbb{N}$, $f_n : \mathbb{R} \to \mathbb{R}$ is a function such that
$$|f_n(x)| \leq \frac{1}{n}, \ \forall\, x \in \mathbb{R},$$

   show that the series of functions $\sum_{n=1}^{\infty} a_n f_n(x)$ converges uniformly on $\mathbb{R}$.

9. Let $f_n : [0,1] \to \mathbb{R}$, $n \in \mathbb{N}$, and $f : [0,1] \to \mathbb{R}$ be continuous functions such that
$$f(x) = \sum_{n=1}^{\infty} f_n(x), \ \ \forall x \in [0,1].$$

   Assuming that

   (i) $\int_0^1 f_n(x)\,dx = \frac{1}{n} - \frac{1}{n+1}$

   (ii) $\int_0^1 f(x)\,dx = 0$

   show that the series of functions does not converge uniformly on $[0,1]$.

10. Consider the real-valued function f of a real variable, which is defined by

$$f(x) = \sum_{n=1}^{\infty} \frac{\text{arccot}(nx)}{\sqrt{n^5}}.$$

a) Determine the domain of f.

b) Demonstrate that f is continuous on its domain.

c) Show that f is differentiable on its domain.

11. Let (a_n) be a sequence of positive terms. Show that $\sum_{n=0}^{\infty} a_n \cos(nx)$ converges uniformly on $\mathbb{R}$ if, and only if, $\sum_{n=0}^{\infty} a_n$ converges.

12. Consider a sequence of functions $(f_n)_{n \in \mathbb{N}}$, continuous on $\mathbb{R}$, such that

$$|f_n(x)| \leq a_n, \quad \forall n \in \mathbb{N}, \; \forall x \in \mathbb{R},$$

where a_n is the general term of a convergent series of positive terms. Let $(b_n)_{n \in \mathbb{N}}$ be a sequence such that the series $\sum_{n=1}^{\infty} (b_n - b_{n+1})$ converges. Show that the series of functions $\sum_{n=1}^{\infty} b_n f_n(x)$ defines a continuous function on $\mathbb{R}$.

13. a) Show that the series $\sum_{n=0}^{\infty} (-1)^n (1-x) x^n$ converges uniformly on $[0, 1]$.

b) Show that the series $\sum_{n=0}^{\infty} (1-x) x^n$ converges pointwise, but not uniformly, on $[0, 1]$.

14. Find the real values of x for which the following series are absolutely convergent, conditionally convergent, and divergent. Give a justification for the answers provided.

a) $\sum_{n=2}^{\infty} \dfrac{\pi^n x^n}{n \log^2(n)}$

b) $\sum_{n=1}^{\infty} \dfrac{2^n n^n}{(n+1)^{n+1}} x^n$

c) $\sum_{n=1}^{\infty} \dfrac{3^n \log(n)}{7^n n^2} x^n$

d) $\sum_{n=1}^{\infty} \dfrac{2 \times 5 \times \cdots \times (3n-1)}{(n+2)!} x^n$

e) $\sum_{n=1}^{\infty} \dfrac{\log(n)}{n} x^n$

f) $\sum_{n=0}^{\infty} \dfrac{3 n^2}{4^n (n^3 + 2)} x^n$

g) $\sum_{n=1}^{\infty} \dfrac{1 \times 4 \times \cdots \times n^2}{(2n)!} x^n$

h) $\sum_{n=1}^{\infty} n^n \left(\sin\left(\dfrac{1}{n^2} \right) \right)^n x^n$

i) $\sum_{n=1}^{\infty} \dfrac{1}{2^n n \left(\arctan(n) - \frac{\pi}{2} \right)} x^n$

j) $\sum_{n=1}^{\infty} \dfrac{3 \times 6 \times \cdots \times 3n}{(n+3)!} x^n$

k) $\sum_{n=0}^{\infty} (-1)^n \dfrac{n+1}{\sqrt{e^{2n}+5}} x^n$

l) $\sum_{n=1}^{\infty} \dfrac{1 \times 3 \times 5 \times \cdots \times (2n-1)}{2 \times 4 \times 6 \times \cdots \times 2n} \cdot \dfrac{1}{n} x^n$

15. Find the real values of x for which the following series are absolutely convergent, conditionally convergent, and divergent. Explain the reasoning behind the response.

a) $\sum_{n=1}^{\infty} \left(\dfrac{n}{2n+1} \right)^{2n} (x-1)^n$

b) $\sum_{n=2}^{\infty} (-1)^n \dfrac{(x-2)^n}{n \log(n)}$

c) $\sum_{n=1}^{\infty} \dfrac{n}{(-3)^n (2n^2 - 1)} (x+4)^n$

d) $\sum_{n=2}^{\infty} \dfrac{e^n}{\log(n)} (x-e)^n$

e) $\sum_{n=1}^{\infty} \dfrac{\cos\left(\frac{\pi}{2n} \right)}{(2n-1)^2 (x-1)^n}$

f) $\sum_{n=1}^{\infty} \dfrac{(2x^2 - 5)^n}{n \, 3^{n+1}}$

3.6. Solved Exercises

g) $\sum_{n=1}^{\infty} \frac{\sqrt{n+1} - \sqrt{n}}{\sqrt{n+1} + \sqrt{n}} (x+2)^n$

h) $\sum_{n=1}^{\infty} \frac{\log^2(n)}{n^2} \left(\frac{2x}{x^2+1} \right)^n$

i) $\sum_{n=1}^{\infty} \frac{1}{2^n \sqrt{n+1}} \cdot \frac{1}{x^n}$

j) $\sum_{n=1}^{\infty} \left(\frac{2}{3} + \frac{1}{2n} \right)^n (2x-3)^n$

k) $\sum_{n=2}^{\infty} \frac{2^{3n} ((n-1)!)^2}{(2n)!} x^{2n}$

l) $\sum_{n=0}^{\infty} \left(\frac{2n-1}{3n+1} \right)^{2n} (x-3)^n$

m) $\sum_{n=1}^{\infty} \frac{\sin\left(\frac{\pi}{n}\right)}{4^n (n+1)} (x-1)^{2n}$

16. Find the values of x for which the series $\sum_{n=1}^{\infty} \frac{x^{2n}}{\sqrt{n}} \log\left(1 + \frac{x^2}{\sqrt{n}}\right)$ is convergent. Justify the answer.

17. Compute the radius of convergence of the series $\sum_{n=1}^{\infty} \left(\cos\left(\frac{\pi}{n}\right) + 2 - (-1)^n \right)^n x^{2n}$ and indicate the largest open interval on which the series converges.

18. Consider the power series $\sum_{n=1}^{\infty} \frac{4^n}{2^n + 5^n} x^n$. Knowing that the radius of convergence of this series is $\frac{5}{4}$, indicate, without resorting to additional calculations, whether the series converges or diverges at points 1 and 2. Investigate the convergence of the series at the point $\frac{5}{4}$.

19. Find the real values of x for which the series $\sum_{n=0}^{\infty} a_n x^n$ is absolutely convergent, conditionally convergent, and divergent, where

$$a_n = \begin{cases} 1, & \text{if } n \text{ is even} \\ 2, & \text{if } n \text{ is odd} \end{cases}$$

Justify the answer.

20. Assume that the series $\sum_{n=1}^{\infty} a_n 5^n$ converges. Is the series $\sum_{n=1}^{\infty} a_n (-2)^n$ convergent? Provide a justification for the answer.

21. Can the power series $\sum_{n=0}^{\infty} a_n (x-3)^n$ converge at the point $x = 0$ and diverge at the point $x = 1$?

22. Let $\sum_{n=0}^{\infty} b_n x^n$ be a power series whose radius of convergence is $R > 1$. Let $(f_n)_{n \in \mathbb{N}}$ be a sequence of functions such that

$$|f_n(x)| \leq |b_n|, \ \forall x \in \mathbb{R}.$$

What can be said about the uniform convergence of the series of functions $\sum_{n=0}^{\infty} f_n$?

23. Let (a_n) be a sequence of positive terms such that the series $\sum_{n=1}^{\infty} a_n (x-1)^n$ is conditionally convergent at $x = -1$.

a) What is the radius of convergence of the power series $\sum_{n=1}^{\infty} a_n (x-1)^n$? Justify.

b) Show that $x = -1$ is the only point at which the series is conditionally convergent.

24. Consider the series $\sum_{n=1}^{\infty} \frac{1}{n \, 5^n} (x-2)^n$.

a) Find the real values of x for which the series is absolutely convergent, conditionally convergent, and divergent. Justify.

b) Let f be the sum of the series. Find the set of points where f is differentiable and determine the analytical expression of f'. Justify the answer given.

25. Consider the series $\sum_{n=0}^{\infty} \frac{(-1)^n}{n+1} \left(\log(x) \right)^{n+1}$.

a) Find the real values of x for which the series is absolutely convergent, conditionally convergent, and divergent. Justify.

b) Using the series of derivatives, calculate, for each x, the sum of the series.

26. Consider the real-valued function f of a real variable, which is defined by

$$f(x) = \sum_{n=0}^{\infty} \frac{(-1)^n}{2^{2n}(n!)^2} x^{2n}.$$

a) Determine the domain of f.

b) Show that $x^2 f''(x) + x f'(x) + x^2 f(x) = 0$. Justify the answer given.

27. Let (a_n) be a sequence of real numbers such that the series $\sum_{n=1}^{\infty} a_n 2^n$ is conditionally convergent.

a) Test for convergence the series $\sum_{n=1}^{\infty} a_n$ and $\sum_{n=1}^{\infty} a_n 3^n$. Give a reason for the answer.

b) Calculate $\lim \frac{\sin(a_n)}{a_n}$. Justify.

28. Let f be a real-valued function of a real variable, defined by

$$f(x) = \sum_{n=0}^{\infty} a_n x^n,$$

such that $f'(x) = f(x)$, $\forall x \in \mathbb{R}$, and $f'(0) = 1$.

a) For each $n \in \mathbb{N}$, compute a_n.

Hint: Use the method of Mathematical Induction.

b) Using item a), identify the function f.

29. Develop the following functions in a Maclaurin series and indicate the interval on which the expansion is valid. Explain the findings for each question and provide a reason for the answers.

a) $\dfrac{5x-1}{x^2-x-2}$

b) $\displaystyle\int_0^x e^{t^2}\,dt$

c) $\displaystyle\int_0^1 \dfrac{1-e^{-tx}}{t}\,dt$

d) $\displaystyle\int_0^x \arctan(t)\,dt$

e) $\dfrac{1}{(2x+5)^2}$

f) $\dfrac{2}{3+2x} + e^{2x}$

g) $\displaystyle\int_0^x \cos(t^2)\,dt$

h) $\dfrac{x+x^2}{(1-x)^2}$

i) $x\arctan(x)$

j) $\dfrac{2x}{(1+x^2)^2}$

k) $\log(2+x^3)$

l) $\log\left(\sqrt{\dfrac{1+x}{1-x}}\right)$

m) $\log(4-x^2)$

30. a) Determine Maclaurin series of the function

$$f(x) = \frac{x}{(1-x)^2}.$$

Indicate for which values of x the development is valid. Justify the answer.

Hint: Use term by term differentiation.

b) Using item a), calculate the sum of the series

$$\sum_{n=0}^{\infty} \frac{n}{3^n}.$$

31. Let f be a real-valued function of a real variable defined by

$$f(x) = \begin{cases} \dfrac{\sin(x)}{x}, & \text{if } x \neq 0 \\ 1, & \text{if } x = 0. \end{cases}$$

a) Expand the function f in a Maclaurin series. Indicate for which values of x the development is valid. Explain the reasoning behind the response.

b) Use item a) to obtain the value of $f^{(n)}(0)$, for all $n \in \mathbb{N}$.

32. Consider the real-valued function f of a real variable defined by

$$f(x) = e^{x^2} + \cos(2x).$$

3.6. Solved Exercises

a) Compute the Maclaurin series of f, indicating the values of x for which the development is valid.

b) Using the expansion obtained in item a), indicate the value of $f^{(17)}(0)$.

33. Using only the Maclaurin expansions of the functions $f(x) = \sin(x)$, $g(x) = e^x$, and $h(x) = \dfrac{1}{1-x}$, find the sum of the following series:

a) $\displaystyle\sum_{n=0}^{\infty} (-1)^n \dfrac{1}{(2n+1)!}$

b) $\displaystyle\sum_{n=0}^{\infty} \dfrac{2^n}{n!}$

c) $\displaystyle\sum_{n=1}^{\infty} (-1)^n \dfrac{1}{2^n n}$

34. Consider the function $f : \mathbb{R} \to \mathbb{R}$ defined by
$$f(x) = \log(x^2 + 1) + \cos(2x).$$

a) Find the Maclaurin series of f and indicate the values of x for which the development is valid.

b) Using item a) compute the sum of the series
$$1 + \sum_{n=1}^{\infty} (-1)^{n-1} \dfrac{\pi^{2n}}{4^{2n}} \left(\dfrac{1}{n} - \dfrac{2^{2n}}{(2n)!} \right).$$

35. Consider the function $f : \mathbb{R} \to \mathbb{R}$ defined by
$$f(x) = \sin(2x) + \dfrac{2x}{(1 - x^2)^2}.$$

a) Express f in a power series of x indicating the values of x for which the development is valid.

b) Use item a) to calculate the sum of the series
$$\sum_{n=0}^{\infty} \left(\dfrac{(-1)^n}{(2n+1)!} + \dfrac{n+1}{2^{2n}} \right).$$

36. Represent in a power series of $x - 3$ the real-valued function f of a real variable defined by
$$f(x) = \dfrac{1}{4-x} + \log(4-x).$$

Determine the values of x for which the development is valid. Give a clear explanation of the answer.

37. Expand in a power series of $x - 3$ the real-valued function f of a real variable defined by
$$f(x) = \dfrac{1}{x^2}.$$

Determine the values of x for which the development is valid. Provide a clear explanation of the answer.

Hint: Use term by term differentiation.

38. Develop in a power series of $x + 1$ the real-valued function of a real variable f defined by
$$f(x) = \dfrac{1}{x^2 + x - 6}.$$

Indicate the values of x for which the expansion is valid. Justify.

39. Consider the function
$$f(x) = \int_1^x \dfrac{\log(t) - t + 1}{t - 1} \, dt.$$

Find the power series of $x - 1$ of f, indicating the values of x for which the development is valid.

40. Consider the function
$$f(x) = \int_2^x \dfrac{\arctan(t-2) - t + 2}{(t-2)^2} \, dt.$$

Expand f in a power series of $x - 2$, indicating the values of x for which the development is valid.

41. Let f be a function of class C^{∞} in a neighborhood of the origin that satisfies the following conditions:
$$f(0) = 1 \text{ and } f'(x) = f(x) + x, \ \forall x \in \mathbb{R}.$$

Find the Maclaurin development of f.

42. Find the Fourier series relative to the periodic functions of period 2π, defined on $]-\pi, \pi[$ as follows:

a) $\begin{cases} \pi, & \text{if } -\pi < x \leq 0 \\ 0, & \text{if } 0 < x < \pi \end{cases}$

b) $\begin{cases} 0, & \text{if } -\pi < x \leq 0 \\ e^x, & \text{if } 0 < x < \pi \end{cases}$

c) $\begin{cases} \cos(x), & \text{if } -\pi < x \leq 0 \\ \sin(x), & \text{if } 0 < x < \pi \end{cases}$

d) x^3

e) $4 - |x|$

43. Considering that the following functions, defined on the indicated intervals, are periodic with period 2π, find their respective Fourier series and their sums.

   a) x, $0 < x < 2\pi$

   b) x^2, $0 < x < 2\pi$

   c) $|x|$, $-\dfrac{\pi}{2} < x \leq \dfrac{3\pi}{2}$

   d) $\begin{cases} x, & \text{if } -\dfrac{\pi}{2} \leq x \leq \dfrac{\pi}{2} \\ \pi - x, & \text{if } \dfrac{\pi}{2} < x \leq \dfrac{3\pi}{2} \end{cases}$

   e) $\begin{cases} x^2, & \text{if } -\dfrac{\pi}{2} \leq x \leq \dfrac{\pi}{2} \\ \dfrac{\pi^2}{4}, & \text{if } \dfrac{\pi}{2} < x \leq \dfrac{3\pi}{2} \end{cases}$

44. Consider that the following functions are periodic and defined on the interval corresponding to one period. Determine the Fourier series and the respective sum for each function.

   a) $\pi \sin(\pi x)$, $0 < x < 1$

   b) $2 - x^2$, $0 < x < 1$

   c) $x - 3$, $0 < x \leq 4$

   d) $\begin{cases} 0, & \text{if } -2 \leq x \leq -1 \\ x, & \text{if } -1 < x < 1 \\ 1, & \text{if } 1 \leq x < 2 \end{cases}$

   e) $\begin{cases} -\dfrac{x}{3}, & \text{if } -3 < x \leq 0 \\ 2x - \dfrac{x^2}{3}, & \text{if } 0 < x \leq 3 \end{cases}$

45. Let $a \in \mathbb{R} \setminus \{0\}$ and $b \in \,]0, \tfrac{\pi}{2}[$. Determine the Fourier series corresponding to the periodic function of period 2π defined on the interval $[0, 2\pi]$ as follows:

$$f(x) = \begin{cases} \dfrac{a}{b}x, & \text{if } 0 \leq x \leq b \\ a, & \text{if } b < x \leq \pi - b \\ \dfrac{a}{b}(\pi - x), & \text{if } \pi - b < x \leq \pi + b \\ -a, & \text{if } \pi + b < x \leq 2\pi - b \\ \dfrac{a}{b}(x - 2\pi), & \text{if } 2\pi - b < x \leq 2\pi. \end{cases}$$

46. Let $f : [-\pi, \pi] \to \mathbb{R}$ be a piecewise C^1 function and

$$\dfrac{a_0}{2} + \sum_{n=1}^{\infty} (a_n \cos(nx) + b_n \sin(nx))$$

be its Fourier series. Consider the functions

$$g(x) = \dfrac{f(x) + f(-x)}{2} \text{ and } h(x) = \dfrac{f(x) - f(-x)}{2}.$$

Determine the Fourier series of g and h.

47. Show that if f is a periodic function with period 2π and such that $f(x + \pi) = -f(x)$, $\forall x \in \mathbb{R}$, then the Fourier series of f is

$$\sum_{n=1}^{\infty} (a_{2n-1} \cos((2n-1)x) + b_{2n-1} \sin((2n-1)x)).$$

48. Consider the real-valued function of a real variable,

$$f(x) = \begin{cases} 0, & \text{if } 0 \leq x < 2 \\ x - 2, & \text{if } 2 \leq x < 4. \end{cases}$$

   a) Find the analytical expression of the odd extension, $\tilde{f}$, of f to the interval $\,]-4, 4[$ and sketch its graph.

   b) Determine the Fourier series of $\tilde{f}$ and sketch, on $\mathbb{R}$, the graph of its sum.

49. Obtain the function

$$f(x) = \begin{cases} 1, & \text{if } 0 \leq x < \pi \\ 0, & \text{if } \pi < x \leq 2\pi \end{cases}$$

as the sum of a Fourier cosine series.

3.6. Solved Exercises

50. Consider the function $f(x) = \pi - x$.

    a) Develop the function into a Fourier sine series on $]0, \pi[$. Use the result to calculate the sum of the series
    $$\sum_{n=1}^{\infty} \frac{\sin(nx)}{n}, \quad 0 < x < 2\pi.$$

    b) Using the result from the previous item, calculate the sum of the series
    $$\sum_{n=1}^{\infty} \frac{\cos(nx)}{n^2}, \quad 0 < x < 2\pi.$$

    c) Develop the function into a Fourier cosine series on $]0, \pi[$.

    d) Calculate the sum of the series $\sum_{n=1}^{\infty} \frac{(-1)^n}{n^2}$.

51. Consider the function $f(x) = x$.

    a) Develop the function into a Fourier sine series on $]0, \pi[$. Use the result to calculate the sum of the series
    $$\sum_{n=1}^{\infty} (-1)^{n+1} \frac{\sin(nx)}{n}, \quad -\pi < x < \pi.$$

    b) Develop the function into a Fourier cosine series on $]0, \pi[$.

52. Develop into a Fourier series the periodic function of period 2π, defined on $]-\pi, \pi[$ by $f(x) = |\sin(x)|$, and use the result to show that
    $$\sum_{n=1}^{\infty} \frac{1}{4n^2 - 1} = \frac{1}{2} \quad \text{and} \quad \sum_{n=1}^{\infty} \frac{(-1)^n}{4n^2 - 1} = \frac{1}{2} - \frac{\pi}{4}.$$
    Hint: $\sin(a)\cos(b) = \frac{1}{2}\big(\sin(a+b) + \sin(a-b)\big)$.

53. Show that on the interval $]0, \pi[$ we have
    $$x^2(\pi - x)^2 = \frac{\pi^4}{30} - 3\sum_{n=1}^{\infty} \frac{\cos(2nx)}{n^4},$$
    and from this result, we obtain
    $$\sum_{n=1}^{\infty} \frac{1}{n^4} = \frac{\pi^4}{90} \quad \text{and} \quad \sum_{n=0}^{\infty} \frac{1}{(2n+1)^4} = \frac{\pi^4}{96}.$$

54. Consider the Fourier series of the periodic function of period 2π, $\tilde{f}$, defined on the interval $[-\pi, \pi]$ by $f(x) = |x|$,
    $$\frac{\pi}{2} + \frac{2}{\pi} \sum_{n=1}^{\infty} \frac{(-1)^n - 1}{n^2} \cos(nx).$$

    a) Show that the series $\sum_{n=1}^{\infty} n\, a_n \sin(nx)$, where a_n are the Fourier coefficients of the original series, is convergent on $\mathbb{R}$.

    b) Let $g(x) = -\sum_{n=1}^{\infty} n\, a_n \sin(nx)$. Sketch the graph of g on the interval $[-2\pi, 2\pi]$.

    c) Calculate the sum of the series
    $$\sum_{n=1}^{\infty} \frac{1}{(2n-1)^2} \quad \text{and} \quad \sum_{n=1}^{\infty} \frac{(-1)^n}{(2n-1)^3}.$$

55. Determine, using Fourier series, the function f such that
    $$\begin{cases} f''(x) + 3f(x) = 3x, & 0 < x < 1 \\ f(0) = f(1) = 0. \end{cases}$$

56. Consider the periodic function with period 2 defined by $f(x) = x^2$ on the interval $]0, 2]$.

    a) Determine the Fourier series of f.

    b) Show that the term by term differentiation of the series obtained in item a) does not converge to f'.

57. Consider the periodic function with period 6 defined on $]-3, 3[$ by $f(x) = x - x^2$.

    a) Find the Fourier series of f.

    b) Let $S(x)$ be the sum of the series
    $$\frac{a_0 x}{2} + \sum_{n=1}^{\infty} \frac{3}{n\pi}\left(a_n \sin\left(\frac{n\pi x}{3}\right) - b_n \cos\left(\frac{n\pi x}{3}\right)\right),$$
    where a_0, a_n, and b_n are the Fourier coefficients of f. Determine the expression of $S(x)$, $x \in]-3, 3[$.

SOLUTIONS

1. a) The terms of the series $\sum_{n=1}^{\infty} \frac{\sin(n^2 x)}{\sqrt[5]{n^7+3}}$ are functions of domain $\mathbb{R}$. The function f is the sum of the series; therefore, its domain is the subset of $\mathbb{R}$ where the series converges. The general term satisfies

$$\left|\frac{\sin(n^2 x)}{\sqrt[5]{n^7+3}}\right| \leq \frac{1}{\sqrt[5]{n^7+3}}, \quad \forall x \in \mathbb{R}, \ \forall n \in \mathbb{N}, \tag{3.13}$$

since $|\sin(x)| \leq 1$, $\forall x \in \mathbb{R}$.

We test the convergence of the series of positive terms, $\sum_{n=1}^{\infty} \frac{1}{\sqrt[5]{n^7+3}}$, by comparison with the series $\sum_{n=1}^{\infty} \frac{1}{n^{7/5}}$, which we know to be convergent. As the limit

$$\lim \frac{\frac{1}{\sqrt[5]{n^7+3}}}{\frac{1}{n^{7/5}}} = \lim \sqrt[5]{\frac{n^7}{n^7+3}} = 1$$

is finite and different from zero, we conclude, by Corollary 2 of the General Comparison Test, that the series $\sum_{n=1}^{\infty} \frac{1}{\sqrt[5]{n^7+3}}$ is convergent. By inequality (3.13), the series $\sum_{n=1}^{\infty} \left|\frac{\sin(n^2 x)}{\sqrt[5]{n^7+3}}\right|$ converges, $\forall x \in \mathbb{R}$. Then, the series $\sum_{n=1}^{\infty} \frac{\sin(n^2 x)}{\sqrt[5]{n^7+3}}$ is convergent, $\forall x \in \mathbb{R}$, that is, the domain of f is $\mathbb{R}$.

b) To conclude that f is continuous on $\mathbb{R}$, it is sufficient (see Theorem 3.2.3) that the following two conditions are satisfied:

(i) The functions $f_n(x) = \frac{\sin(n^2 x)}{\sqrt[5]{n^7+3}}$ are continuous on $\mathbb{R}$.

(ii) The series $\sum_{n=1}^{\infty} \frac{\sin(n^2 x)}{\sqrt[5]{n^7+3}}$ is uniformly convergent to f on $\mathbb{R}$.

The first condition is immediately satisfied because the functions $h_n(x) = \sin(n^2 x)$ are continuous on $\mathbb{R}$. The second is a consequence of the previous item: Inequality (3.13) and the convergence of the series $\sum_{n=1}^{\infty} \frac{1}{\sqrt[5]{n^7+3}}$ are sufficient to conclude the uniform convergence of the series $\sum_{n=1}^{\infty} \frac{\sin(n^2 x)}{\sqrt[5]{n^7+3}}$ by the Weierstrass' test. As the series converges to f, this function is continuous.

2. a) The terms of the series $\sum_{n=0}^{\infty} \frac{1}{1+(3x)^n}$ are functions with domain $\mathbb{R}$, if n is even, and $\mathbb{R} \setminus \left\{-\frac{1}{3}\right\}$, if n is odd. The function f is the sum of the series, so its domain is the subset of $\mathbb{R} \setminus \left\{-\frac{1}{3}\right\}$ where the series converges. For the series to converge, its general term must be a null sequence. But

$$\lim \frac{1}{1+(3x)^n} = \begin{cases} 1, & \text{if } |3x| < 1 \\ 0, & \text{if } |3x| > 1 \\ \frac{1}{2}, & \text{if } 3x = 1; \end{cases}$$

3.6. Solved Exercises

therefore, if $|x| \leq \dfrac{1}{3}$, $x \neq -\dfrac{1}{3}$, the series diverges.

Let x be such that $|3x| > 1$. We study the series of positive terms $\displaystyle\sum_{n=0}^{\infty} \left|\dfrac{1}{1+(3x)^n}\right|$ using D'Alembert's test:

$$\lim \dfrac{\left|\dfrac{1}{1+(3x)^{n+1}}\right|}{\left|\dfrac{1}{1+(3x)^n}\right|} = \lim \left|\dfrac{1+(3x)^n}{1+(3x)^{n+1}}\right| = \lim \left|\dfrac{\dfrac{1}{(3x)^{n+1}} + \dfrac{1}{3x}}{\dfrac{1}{(3x)^{n+1}} + 1}\right| = \dfrac{1}{|3x|}.$$

We conclude that the series converges because $\dfrac{1}{|3x|} < 1$, that is, $|x| > \dfrac{1}{3}$.

The domain of f is $\left]-\infty, -\dfrac{1}{3}\right[\cup \left]\dfrac{1}{3}, +\infty\right[$.

b) To show that f is continuous on $[1, +\infty[$, it is sufficient (see Theorem 3.2.3) that the following two conditions are met:

(i) The functions $f_n(x) = \dfrac{1}{1+(3x)^n}$ are continuous on $[1, +\infty[$.

(ii) The series $\displaystyle\sum_{n=1}^{\infty} \dfrac{1}{1+(3x)^n}$ is uniformly convergent to f on $[1, +\infty[$.

The first condition is immediately satisfied. The second condition follows from the inequality

$$\left|\dfrac{1}{1+(3x)^n}\right| \leq \dfrac{1}{(3x)^n} \leq \left(\dfrac{1}{3}\right)^n, \quad \forall x \in [1, +\infty[, \; \forall n \in \mathbb{N},$$

since the series $\displaystyle\sum_{n=1}^{\infty} \left(\dfrac{1}{3}\right)^n$ is convergent, which, by the Weierstrass' test, is sufficient to conclude the uniform convergence of the series $\displaystyle\sum_{n=1}^{\infty} \dfrac{1}{1+(3x)^n}$. As the series converges to f, this function is continuous.

3. a) Let $f(x) = \displaystyle\sum_{n=1}^{\infty} \dfrac{x}{n^x}$. The terms of the series are functions with domain $\mathbb{R}$. The function f is the sum of the series; therefore, its domain is the subset of $\mathbb{R}$ where the series converges.

– If $x = 0$, all terms are zero; hence, the series converges.

– If $x \neq 0$, then $\displaystyle\sum_{n=1}^{\infty} \dfrac{x}{n^x} = x \displaystyle\sum_{n=1}^{\infty} \dfrac{1}{n^x}$. This series is a p-series; it converges if and only if $x > 1$.

Therefore, the domain of f is the set $]1, +\infty[\cup \{0\}$.

b) Let us consider the function $g(x) = \displaystyle\sum_{n=1}^{\infty} \dfrac{1}{n^x}$. To conclude that f is continuous on $]1, +\infty[$, it is sufficient to prove that g is continuous in this set since $f(x) = x\,g(x)$. If we show that g is continuous on $[a, +\infty[$, for every $a > 1$, then g is continuous on $]1, +\infty[$. For this, it is enough (see Theorem 3.2.3) that the following two conditions are met:

(i) The functions $g_n(x) = \dfrac{1}{n^x}$ are continuous on $]a, +\infty[$.

(ii) The series $\displaystyle\sum_{n=1}^{\infty} \dfrac{1}{n^x}$ is uniformly convergent to g on $]a, +\infty[$.

The functions $g_n(x) = e^{-x \log(n)}$ are continuous, since the exponential is continuous on $\mathbb{R}$. Let us look at the second condition. Let $a > 1$ be a real constant. The inequality

$$\left| \frac{1}{n^x} \right| \leq \frac{1}{n^a}, \quad \forall x \in [a, +\infty[, \; \forall n \in \mathbb{N},$$

combined with the fact that the series $\sum_{n=1}^{\infty} \frac{1}{n^a}$ is convergent, is sufficient to conclude, by Weierstrass' test, the uniform convergence of the series $\sum_{n=1}^{\infty} \frac{1}{n^x}$ on $[a, +\infty[$. As the series converges to g, this function is continuous on $[a, +\infty[$. Since a is an arbitrary constant greater than 1, we also have g continuous on $]1, +\infty[$.

4. The terms of the series $\sum_{n=1}^{\infty} \frac{1}{(nx+1)\sqrt{n^3+1}}$ are continuous functions on $[1, 2]$. As

$$\left| \frac{1}{(nx+1)\sqrt{n^3+1}} \right| = \frac{1}{|nx+1|} \cdot \frac{1}{\sqrt{n^3+1}} \leq \frac{1}{n+1} \cdot \frac{1}{\sqrt{n^3+1}} \leq \frac{1}{\sqrt{n^3}}, \quad \forall x \in [1,2], \; \forall n \in \mathbb{N},$$

and the series of positive terms $\sum_{n=1}^{\infty} \frac{1}{\sqrt{n^3}}$ is convergent because it is a p-series with $p = \frac{3}{2}$, by Weierstrass' test, the series $\sum_{n=1}^{\infty} \frac{1}{(nx+1)\sqrt{n^3+1}}$ converges uniformly to f, on $[1, 2]$. It follows from Theorem 3.2.3 that f is continuous on $[1, 2]$; by Theorem 3.2.4, the series is integrable term by term on this interval and

$$\int_1^2 f(x)\, dx = \int_1^2 \sum_{n=1}^{\infty} \frac{1}{(nx+1)\sqrt{n^3+1}}\, dx = \sum_{n=1}^{\infty} \frac{1}{\sqrt{n^3+1}} \int_1^2 \frac{1}{nx+1}\, dx$$

$$= \sum_{n=1}^{\infty} \frac{1}{\sqrt{n^3+1}} \left[\frac{\log(nx+1)}{n} \right]_1^2 = \sum_{n=1}^{\infty} \frac{\log(2n+1) - \log(n+1)}{n\sqrt{n^3+1}}$$

$$= \sum_{n=1}^{\infty} \frac{1}{n\sqrt{n^3+1}} \log\left(\frac{2n+1}{n+1} \right).$$

5. The terms of the series $\sum_{n=1}^{\infty} \frac{\cos(nx)}{n^2}$ are functions of domain $\mathbb{R}$. The function f is the sum of the series; therefore, its domain is the subset of $\mathbb{R}$ where the series converges. Since we have

$$\left| \frac{\cos(nx)}{n^2} \right| \leq \frac{1}{n^2}, \quad \forall x \in \mathbb{R}, \; \forall n \in \mathbb{N},$$

because $|\cos(nx)| \leq 1$, $\forall x \in \mathbb{R}$, $\forall n \in \mathbb{N}$, and the series of positive terms $\sum_{n=1}^{\infty} \frac{1}{n^2}$ is convergent as it is a p-series with $p = 2$, by Weierstrass' test, the series $\sum_{n=1}^{\infty} \frac{\cos(nx)}{n^2}$ converges uniformly to f, on $\mathbb{R}$. Consequently, by Theorem 3.2.3, given that the functions $\frac{\cos(nx)}{n^2}$ are continuous on $\mathbb{R}$, f is continuous on $\mathbb{R}$; by Theorem 3.2.4,

3.6. Solved Exercises

the series is integrable term by term on every closed bounded interval of $\mathbb{R}$. In particular, f is integrable on $\left[0, \frac{\pi}{2}\right]$, and

$$\int_0^{\pi/2} f(x)\, dx = \int_0^{\pi/2} \sum_{n=1}^{\infty} \frac{\cos(nx)}{n^2}\, dx = \sum_{n=1}^{\infty} \frac{1}{n^2} \int_0^{\pi/2} \cos(nx)\, dx$$

$$= \sum_{n=1}^{\infty} \frac{1}{n^2} \left[\frac{\sin(nx)}{n}\right]_0^{\frac{\pi}{2}} = \sum_{n=1}^{\infty} \frac{\sin(\frac{n\pi}{2})}{n^3} = \sum_{n=1}^{\infty} \frac{(-1)^{n-1}}{(2n-1)^3} = \sum_{n=0}^{\infty} \frac{(-1)^n}{(2n+1)^3},$$

because

$$\sin\left(\frac{n\pi}{2}\right) = \begin{cases} 0, & \text{if } n \text{ is even} \\ -1, & \text{if } n = 3, 7, 11, \ldots \\ 1, & \text{if } n = 1, 5, 9, \ldots \end{cases}$$

6. The terms of the series $f(x) = \sum_{n=0}^{\infty} \frac{e^{-nx}}{n^4 + 5}$ are functions of domain $\mathbb{R}$. As

$$\left|\frac{e^{-nx}}{n^4 + 5}\right| \leq \frac{1}{n^4 + 5} \leq \frac{1}{n^4}, \quad \forall x \in [0, 1], \forall n \in \mathbb{N}_0,$$

since $|e^{-nx}| = e^{-nx} \leq e^0 = 1$, $\forall x \in [0, 1]$, $\forall n \in \mathbb{N}_0$, and the series of positive terms $\sum_{n=1}^{\infty} \frac{1}{n^4}$ is convergent as it is a p-series with $p = 4$. By Weierstrass' test, the series $\sum_{n=0}^{\infty} \frac{e^{-nx}}{n^4 + 5}$ converges uniformly to f, on $[0, 1]$. Therefore, by Theorem 3.2.3 since the functions $f_n(x) = \frac{e^{-nx}}{n^4 + 5}$ are continuous on $[0, 1]$, f is continuous on $[0, 1]$. Theorem 3.2.4 shows that the series is integrable term by term on this interval. We have

$$\int_0^1 f(x)\, dx = \int_0^1 \left(\frac{1}{5} + \sum_{n=1}^{\infty} \frac{e^{-nx}}{n^4 + 5}\right) dx = \frac{1}{5} + \sum_{n=1}^{\infty} \frac{1}{n^4 + 5} \int_0^1 e^{-nx}\, dx$$

$$= \frac{1}{5} + \sum_{n=1}^{\infty} \frac{1}{n^4 + 5} \left[-\frac{e^{-nx}}{n}\right]_0^1 = \frac{1}{5} + \sum_{n=1}^{\infty} \frac{1 - e^{-n}}{n(n^4 + 5)}.$$

7. a) The series of functions defined, for $x \geq 0$, by

$$\sum_{n=1}^{\infty} \left(x + \frac{1}{n}\right)^{n + \frac{x}{n}}$$

is of positive terms. Let us study this series by applying Cauchy's Root Test:

$$\lim \sqrt[n]{\left(x + \frac{1}{n}\right)^{n + \frac{x}{n}}} = \lim \sqrt[n]{\left(x + \frac{1}{n}\right)^n \left(x + \frac{1}{n}\right)^{\frac{x}{n}}}$$

$$= \lim \left(x + \frac{1}{n}\right) \sqrt[n]{\left(x + \frac{1}{n}\right)^{\frac{x}{n}}} = \lim \left(x + \frac{1}{n}\right) \left(x + \frac{1}{n}\right)^{\frac{x}{n^2}} = x.$$

By Cauchy's Root Test, the series converges if $x < 1$ and diverges if $x > 1$. It remains to see what happens if $x = 1$. In this case, we obtain the numerical series $\sum_{n=1}^{\infty} \left(1 + \frac{1}{n}\right)^{n + \frac{1}{n}}$, which is divergent because its general term is not a null sequence. In fact,

$$\lim \left(1 + \frac{1}{n}\right)^{n + \frac{1}{n}} = \lim \left(1 + \frac{1}{n}\right)^n \left(1 + \frac{1}{n}\right)^{\frac{1}{n}} = e.$$

b) Let $0 < \alpha < 1$ and $x \in [0, \alpha]$.

$$\left|\left(x + \frac{1}{n}\right)^{n + \frac{x}{n}}\right| = \left(x + \frac{1}{n}\right)^{n + \frac{x}{n}} \leq \left(\alpha + \frac{1}{n}\right)^{n + \frac{\alpha}{n}}, \quad \forall n \in \mathbb{N}.$$

As shown in item a), the numerical series $\sum_{n=1}^{\infty} \left(\alpha + \frac{1}{n}\right)^{n + \frac{\alpha}{n}}$ is convergent. By Weierstrass' test, the original series converges uniformly on $[0, \alpha]$.

8. We will use Weierstrass' test to prove that the series of functions $\sum_{n=1}^{\infty} a_n f_n(x)$ converges uniformly on $\mathbb{R}$. Since $|f_n(x)| \leq \frac{1}{n}$, $\forall n \in \mathbb{N}$, $\forall x \in \mathbb{R}$, and (a_n) is a sequence of positive terms, we have

$$|a_n f_n(x)| \leq \frac{a_n}{n}, \quad \forall n \in \mathbb{N}, \ \forall x \in \mathbb{R}.$$

We prove that the series of positive terms $\sum_{n=1}^{\infty} \frac{a_n}{n}$ is convergent by applying D'Alembert's test:

$$\lim \frac{\frac{a_{n+1}}{n+1}}{\frac{a_n}{n}} = \lim \frac{n\, a_{n+1}}{(n+1)\, a_n} = \lim \frac{n}{n+1} \cdot \frac{a_{n+1}}{a_n} = \frac{1}{2}$$

because, by hypothesis, $\lim \frac{a_{n+1}}{a_n} = \frac{1}{2}$. As the limit is less than 1, the series is convergent. By Weierstrass' test, the series of functions $\sum_{n=1}^{\infty} a_n f_n(x)$ converges uniformly on $\mathbb{R}$.

9. Since the functions $f_n : [0, 1] \to \mathbb{R}$ are continuous, if the convergence was uniform, we would have, by Theorem 3.2.4,

$$\int_0^1 \sum_{n=1}^{\infty} f_n(x)\, dx = \sum_{n=1}^{\infty} \int_0^1 f_n(x)\, dx.$$

But

$$\int_0^1 \sum_{n=1}^{\infty} f_n(x)\, dx = \int_0^1 f(x)\, dx = 0$$

and

$$\sum_{n=1}^{\infty} \int_0^1 f_n(x)\, dx = \sum_{n=1}^{\infty} \left(\frac{1}{n} - \frac{1}{n+1}\right) = 1.$$

Therefore, the series of functions is not uniformly convergent on $[0, 1]$.

3.6. Solved Exercises

10. a) The terms of the series $\sum_{n=1}^{\infty} \dfrac{\arccot(nx)}{\sqrt{n^5}}$ are functions with domain $\mathbb{R}$. The function f is the sum of the series; therefore, its domain is the subset of $\mathbb{R}$ where the series converges. The terms of the series satisfy

$$\left| \frac{\arccot(nx)}{\sqrt{n^5}} \right| \leq \frac{\pi}{\sqrt{n^5}}, \quad \forall x \in \mathbb{R}, \ \forall n \in \mathbb{N}, \tag{3.14}$$

since $|\arccot(nx)| \leq \pi$, $\forall x \in \mathbb{R}$. The series of positive terms $\sum_{n=1}^{\infty} \dfrac{1}{\sqrt{n^5}}$ is convergent because it is a p-series with $p = \dfrac{5}{2}$. Therefore, the series $\sum_{n=1}^{\infty} \dfrac{\pi}{\sqrt{n^5}}$ is also convergent. By inequality (3.14), the series $\sum_{n=1}^{\infty} \left| \dfrac{\arccot(nx)}{\sqrt{n^5}} \right|$ converges, $\forall x \in \mathbb{R}$. Then the series $\sum_{n=1}^{\infty} \dfrac{\arccot(nx)}{\sqrt{n^5}}$ is convergent, $\forall x \in \mathbb{R}$, that is, the domain of f is $\mathbb{R}$.

b) To conclude that f is continuous on $\mathbb{R}$, it is sufficient, according to Theorem 3.2.3, that the following two conditions are satisfied:

   (i) The functions $f_n(x) = \dfrac{\arccot(nx)}{\sqrt{n^5}}$ are continuous on $\mathbb{R}$.

   (ii) The series $\sum_{n=1}^{\infty} \dfrac{\arccot(nx)}{\sqrt{n^5}}$ is uniformly convergent to f on $\mathbb{R}$.

The functions $h_n(x) = \arccot(nx)$ are continuous on $\mathbb{R}$, which implies the first condition. The second condition is a consequence of the previous item: Inequality (3.14) and the fact that the series $\sum_{n=1}^{\infty} \dfrac{\pi}{\sqrt{n^5}}$ is convergent are sufficient, by Weierstrass' test, to conclude the uniform convergence of the series $\sum_{n=1}^{\infty} \dfrac{\arccot(nx)}{\sqrt{n^5}}$. As the series converges to f, this function is continuous.

c) To conclude that f is differentiable on $\mathbb{R}$, it is sufficient, by Corollary 2 of Theorem 3.2.4, that the following conditions are fulfilled:

   (i) The functions $f_n'(x)$ are continuous on $\mathbb{R}$.

   (ii) The series $\sum_{n=1}^{\infty} \dfrac{\arccot(nx)}{\sqrt{n^5}}$ is pointwise convergent to f on $\mathbb{R}$.

   (iii) The series $\sum_{n=1}^{\infty} \left(\dfrac{\arccot(nx)}{\sqrt{n^5}} \right)' = \sum_{n=1}^{\infty} -\dfrac{n}{(1+n^2x^2)\sqrt{n^5}}$ is uniformly convergent on $\mathbb{R}$.

The first condition is satisfied because the functions $(\arccot(nx))' = -\dfrac{n}{1+n^2x^2}$ are continuous on $\mathbb{R}$. We proved the second condition in the previous item. The third is again an application of Weierstrass' test. We have the inequality

$$\left| -\frac{n}{(1+n^2x^2)\sqrt{n^5}} \right| \leq \frac{n}{\sqrt{n^5}} = \frac{1}{\sqrt{n^3}}, \quad \forall x \in \mathbb{R}, \ \forall n \in \mathbb{N},$$

and the series $\sum_{n=1}^{\infty} \dfrac{1}{\sqrt{n^3}}$ is convergent because it is a p-series with $p = \dfrac{3}{2}$, so the series $\sum_{n=1}^{\infty} \left(\dfrac{\arccot(nx)}{\sqrt{n^5}} \right)'$ is uniformly convergent. We can conclude that the series converges to f', that is,

$$f'(x) = \sum_{n=1}^{\infty} \left(\frac{\operatorname{arccot}(nx)}{\sqrt{n^5}}\right)' = \sum_{n=1}^{\infty} -\frac{n}{(1+n^2x^2)\sqrt{n^5}},$$

which means that f is differentiable on $\mathbb{R}$.

11. If the series $\sum_{n=0}^{\infty} a_n \cos(nx)$ converges uniformly on $\mathbb{R}$, then it converges pointwise on $\mathbb{R}$. In particular, the series is convergent at $x = 0$. At this point, we obtain the numerical series $\sum_{n=0}^{\infty} a_n$, which is convergent.

Conversely, suppose that the series $\sum_{n=0}^{\infty} a_n$ is convergent. We have

$$|a_n \cos(nx)| \leq |a_n| = a_n, \quad \forall x \in \mathbb{R}, \forall n \in \mathbb{N}.$$

By Weierstrass' test the series $\sum_{n=0}^{\infty} a_n \cos(nx)$ is uniformly convergent on $\mathbb{R}$.

12. We know that the series $\sum_{n=1}^{+\infty}(b_n - b_{n+1})$ converges, that is, the sequence $S_n = \sum_{k=1}^{n}(b_k - b_{k+1}) = b_1 - b_{n+1}$ converges. Therefore, there exists $b = \lim b_n = b_1 - \lim S_n$, so the sequence (b_n) is bounded, that is, there exists $M > 0$ such that $|b_n| \leq M, \forall n \in \mathbb{N}$. Based on this fact, we have the following inequalities:

$$|b_n f_n(x)| = |b_n| \, |f_n(x)| \leq |b_n| \, a_n \leq M \, a_n, \quad \forall x \in \mathbb{R}, \ \forall n \in \mathbb{N}.$$

Thus, by Weierstrass' test, the series $\sum_{n=1}^{\infty} b_n f_n(x)$ is uniformly convergent on $\mathbb{R}$, defining a function f of domain $\mathbb{R}$. Since, by hypothesis, the functions f_n are continuous on $\mathbb{R}$, $\forall n \in \mathbb{N}$, the function f is continuous on its domain.

13. a) Consider the series $\sum_{n=0}^{\infty}(-1)^n(1-x)\, x^n$, $x \in [0,1]$.

   – If $x = 1$, all terms are zero. Therefore the series converges.

   – If $x \neq 1$, as the series $\sum_{n=0}^{\infty}(1-x)\,(-x)^n$ is geometric with ratio $r = -x$, it converges because $x \in [0, 1[$. In this case,

$$\sum_{n=0}^{\infty}(1-x)\,(-x)^n = (1-x) \cdot \frac{1}{1-(-x)} = \frac{1-x}{1+x}.$$

Then the series $\sum_{n=0}^{\infty}(-1)^n(1-x)\, x^n$ converges pointwise to the function $f(x) = \dfrac{1-x}{1+x}$ on $[0,1]$. This convergence will be uniform if

$$\lim_{n \to +\infty} \sup_{x \in [0,1]} |f(x) - S_n(x)| = 0,$$

where S_n denotes the sequence of partial sums of the original series.

   – If $x = 1$, $|f(1) - S_n(1)| = 0$.

3.6. Solved Exercises

- If $x \neq 1$, $S_n(x) = (1-x) \cdot \dfrac{1-(-x)^{n+1}}{1-(-x)} = \dfrac{1-x}{1+x} \cdot (1-(-x)^{n+1})$, and therefore,

$$\begin{aligned} |f(x) - S_n(x)| &= \left| \frac{1-x}{1+x} - \frac{1-x}{1+x} \cdot (1-(-x)^{n+1}) \right| \\ &= \left| \frac{1-x}{1+x} \cdot \left(1 - (1-(-x)^{n+1})\right) \right| \\ &= \left| \frac{1-x}{1+x} \cdot (-x)^{n+1} \right| = \frac{1-x}{1+x} \cdot x^{n+1}. \end{aligned}$$

Let $g(x) = |f(x) - S_n(x)| = \dfrac{1-x}{1+x} \cdot x^{n+1}$.

The function g is continuous on $[0,1]$; therefore, by Weierstrass' Theorem, it has a maximum and a minimum on this interval. As g is nonnegative and $g(0) = g(1) = 0$, the maximum is attained at some point within the interior of the interval. The function g is differentiable on $]0,1[$, so this point corresponds to a zero of the first derivative.

$$g'(x) = \left(\frac{1-x}{1+x} \cdot x^{n+1} \right)' = -\frac{x^n}{(1+x)^2} \cdot \left((n+1)x^2 + 2x - (n+1) \right),$$

and on the interval $]0,1[$,

$$g'(x) = 0 \Leftrightarrow (n+1)x^2 + 2x - (n+1) = 0 \Leftrightarrow x = \frac{-1+\sqrt{1+(n+1)^2}}{n+1}.$$

Then

$$\sup_{x \in [0,1]} |f(x) - S_n(x)| = g\left(\frac{-1+\sqrt{1+(n+1)^2}}{n+1} \right) \leq \frac{1 - \dfrac{-1+\sqrt{1+(n+1)^2}}{n+1}}{1 + \dfrac{-1+\sqrt{1+(n+1)^2}}{n+1}},$$

and, as

$$\lim_{n \to +\infty} \frac{1 - \dfrac{-1+\sqrt{1+(n+1)^2}}{n+1}}{1 + \dfrac{-1+\sqrt{1+(n+1)^2}}{n+1}} = 0,$$

we can conclude that

$$\lim_{n \to +\infty} \sup_{x \in [0,1]} |f(x) - S_n(x)| = 0,$$

that is, the series $\sum_{n=0}^{\infty} (-1)^n (1-x)\, x^n$ converges uniformly to the function f, on the interval $[0,1]$.

b) Let us consider the series $\sum_{n=0}^{\infty} (1-x)\, x^n$, $x \in [0,1]$.

- If $x = 1$, all terms are zero. Hence, the series converges.
- If $x \neq 1$, as the series $\sum_{n=0}^{\infty} (1-x)\, x^n$ is geometric with ratio $r = x$, the series converges because $x \in [0,1[$. In this case,

$$\sum_{n=0}^{\infty} (1-x)\, x^n = (1-x) \cdot \frac{1}{1-x} = 1.$$

Then, the series $\sum_{n=0}^{\infty}(1-x)\,x^n$, $x \in [0,1]$, converges pointwise to the function

$$f(x) = \begin{cases} 0, & \text{if } x = 1 \\ 1, & \text{if } x \in [0,1[. \end{cases}$$

Let $f_n(x) = (1-x)x^n$. The functions f_n are continuous on $[0,1]$, $\forall n \in \mathbb{N}_0$, because they are polynomial functions. If the series was uniformly convergent on the interval $[0,1]$, f would be continuous on this interval. Since f is not continuous at $x=1$ because $\lim_{x \to 1^-} f(x) = 1$ and $f(1) = 0$, the series $\sum_{n=0}^{\infty}(1-x)\,x^n$ does not converge uniformly to f on $[0,1]$.

14. a) Let us study the power series of x, $\sum_{n=2}^{\infty} \dfrac{\pi^n x^n}{n \log^2(n)}$. Setting $a_n = \dfrac{\pi^n}{n \log^2(n)}$, the radius of convergence of this series is, by Corollary 1 of Theorem 3.3.1

$$r = \lim \left| \frac{a_n}{a_{n+1}} \right| = \lim \left| \frac{\dfrac{\pi^n}{n \log^2(n)}}{\dfrac{\pi^{n+1}}{(n+1)\log^2(n+1)}} \right| = \lim \frac{n+1}{\pi n} \cdot \frac{\log^2(n+1)}{\log^2(n)} = \frac{1}{\pi},$$

which implies, by the same theorem, that the series is absolutely convergent if $x \in \left]-\dfrac{1}{\pi}, \dfrac{1}{\pi}\right[$ and divergent if $x \in \left]-\infty, -\dfrac{1}{\pi}\right[\cup \left]\dfrac{1}{\pi}, +\infty\right[$.

At $x = \pm \dfrac{1}{\pi}$, we have to study the corresponding numerical series.

– If $x = \dfrac{1}{\pi}$, we obtain the series $\sum_{n=2}^{\infty} \dfrac{\pi^n}{n \log^2(n)} \left(\dfrac{1}{\pi}\right)^n = \sum_{n=2}^{\infty} \dfrac{1}{n \log^2(n)}$. Let us study the convergence of this series using the Integral Test.

Let $f(x) = \dfrac{1}{x \log^2(x)}$. We have:

(i) $f(x) > 0$, $\forall x \geq 2$.

(ii) f is continuous on $[2, +\infty[$.

(iii) f is decreasing on $[2, +\infty[$ because $x \log^2(x)$ is positive and increasing.

So the numerical series $\sum_{n=2}^{\infty} \dfrac{1}{n \log^2(n)}$ converges if and only if the improper integral $\int_2^{+\infty} \dfrac{1}{x \log^2(x)}\,dx$ converges. We have

$$\int_2^{+\infty} \frac{1}{x \log^2(x)}\,dx = \lim_{x \to +\infty} \int_2^x \frac{1}{t \log^2(t)}\,dt = \lim_{x \to +\infty} \int_2^x \frac{1}{t}\bigl(\log(t)\bigr)^{-2}\,dt$$

$$= \lim_{x \to +\infty} \left[-\frac{1}{\log(t)}\right]_2^x = \lim_{x \to +\infty} \left(-\frac{1}{\log(x)} + \frac{1}{\log(2)}\right) = \frac{1}{\log(2)},$$

that is, the improper integral is convergent, so the same happens to the series. As it is a series of positive terms, it converges absolutely.

3.6. Solved Exercises

- If $x = -\dfrac{1}{\pi}$, we obtain the series $\displaystyle\sum_{n=2}^{\infty} \dfrac{\pi^n}{n \log^2(n)} \left(-\dfrac{1}{\pi}\right)^n = \sum_{n=2}^{\infty} (-1)^n \dfrac{1}{n \log^2(n)}$, which is alternating. As we just saw, the series of modules is convergent. Then the series is absolutely convergent.

Conclusion: The original series diverges if $x \in \left]-\infty, -\dfrac{1}{\pi}\right[\cup \left]\dfrac{1}{\pi}, +\infty\right[$ and is absolutely convergent if $x \in \left[-\dfrac{1}{\pi}, \dfrac{1}{\pi}\right]$.

Remark: To calculate the $\lim \dfrac{\log^2(n+1)}{\log^2(n)} = \left(\lim \dfrac{\log(n+1)}{\log(n)}\right)^2$, we proceed as follows: Consider the function $g(x) = \dfrac{\log(x+1)}{\log(x)}$; we use L'Hôpital's Rule applied to the calculation of the limit $\displaystyle\lim_{x \to +\infty} \dfrac{\log(x+1)}{\log(x)}$, where the indeterminate form of type $\dfrac{\infty}{\infty}$ arises:

$$\lim_{x \to +\infty} \dfrac{(\log(x+1))'}{(\log(x))'} = \lim_{x \to +\infty} \dfrac{\dfrac{1}{x+1}}{\dfrac{1}{x}} = \lim_{x \to +\infty} \dfrac{x}{x+1} = 1.$$

Therefore, $\displaystyle\lim_{x \to +\infty} \dfrac{\log(x+1)}{\log(x)} = 1$, which implies that $\lim \dfrac{\log^2(n+1)}{\log^2(n)} = 1$.

b) Let us study the power series of x, $\displaystyle\sum_{n=1}^{\infty} \dfrac{2^n n^n}{(n+1)^{n+1}} x^n$. Setting $a_n = \dfrac{2^n n^n}{(n+1)^{n+1}}$, the radius of convergence is by Corollary 1 of Theorem 3.3.1

$$r = \lim \left|\dfrac{a_n}{a_{n+1}}\right| = \lim \left|\dfrac{\dfrac{2^n n^n}{(n+1)^{n+1}}}{\dfrac{2^{n+1}(n+1)^{n+1}}{(n+2)^{n+2}}}\right| = \lim \dfrac{2^n n^n (n+2)^{n+2}}{(n+1)^{n+1} 2^{n+1}(n+1)^{n+1}}$$

$$= \lim \dfrac{1}{2}\left(\dfrac{n}{n+1}\right)^n \cdot \left(\dfrac{n+2}{n+1}\right)^{n+2} = \dfrac{1}{2} \cdot \dfrac{1}{e} \cdot e = \dfrac{1}{2}.$$

Then, the series is absolutely convergent if $x \in \left]-\dfrac{1}{2}, \dfrac{1}{2}\right[$ and divergent if $x \in \left]-\infty, -\dfrac{1}{2}\right[\cup \left]\dfrac{1}{2}, +\infty\right[$.

Let us study the convergence of the numerical series corresponding to the endpoints of the interval of convergence.

- If $x = \dfrac{1}{2}$, we obtain the series of positive terms $\displaystyle\sum_{n=1}^{\infty} \dfrac{2^n n^n}{(n+1)^{n+1}} \cdot \left(\dfrac{1}{2}\right)^n = \sum_{n=1}^{\infty} \dfrac{n^n}{(n+1)^{n+1}}$.

We will study this series by comparing it with the harmonic series, which is divergent. The limit

$$\lim \dfrac{\dfrac{n^n}{(n+1)^{n+1}}}{\dfrac{1}{n}} = \lim \left(\dfrac{n}{n+1}\right)^{n+1} = \lim \left(\dfrac{n}{n+1}\right)^n \cdot \dfrac{n}{n+1} = \dfrac{1}{e}$$

is finite and different from zero; therefore, by Corollary 2 of the General Comparison Test, the series is divergent.

— If $x = -\frac{1}{2}$, we obtain the series $\sum_{n=1}^{\infty} \frac{2^n n^n}{(n+1)^{n+1}} \cdot \left(-\frac{1}{2}\right)^n = \sum_{n=1}^{\infty} (-1)^n \frac{n^n}{(n+1)^{n+1}}$, which is alternating, since $\frac{n^n}{(n+1)^{n+1}} > 0$, $\forall n \in \mathbb{N}$. We proved in the previous case that the series of modules is divergent. Therefore, we need to verify whether the following conditions hold to apply Leibniz's test:

(i) $\lim \frac{n^n}{(n+1)^{n+1}} = \lim \left(\frac{n}{n+1}\right)^n \cdot \frac{1}{n+1} = \frac{1}{e} \times 0 = 0.$

(ii) $\frac{n^n}{(n+1)^{n+1}} > 0$, $\forall n \geq 1$.

(iii) The sequence of general term $\frac{n^n}{(n+1)^{n+1}} = \left(\frac{n}{n+1}\right)^n \cdot \frac{1}{n+1}$ is decreasing, as it is the product of two decreasing positive sequences.

Therefore, the series is convergent. It is conditionally convergent since the series of modules diverges.

Conclusion: The power series of x is absolutely convergent if $x \in \left]-\frac{1}{2}, \frac{1}{2}\right[$, conditionally convergent if $x = -\frac{1}{2}$, and divergent if $x \in \left]-\infty, -\frac{1}{2}\right[\cup \left[\frac{1}{2}, +\infty\right[$.

c) By Corollary 1 of Theorem 3.3.1, the power series of x, $\sum_{n=1}^{\infty} a_n x^n$, with $a_n = \frac{3^n \log(n)}{7^n n^2}$, has a radius of convergence

$$r = \lim \left|\frac{a_n}{a_{n+1}}\right| = \lim \left|\frac{\frac{3^n \log(n)}{7^n n^2}}{\frac{3^{n+1} \log(n+1)}{7^{n+1} (n+1)^2}}\right| = \lim \frac{7(n+1)^2}{3n^2} \cdot \frac{\log(n)}{\log(n+1)} = \frac{7}{3}.$$

Then, the series converges absolutely if $x \in \left]-\frac{7}{3}, \frac{7}{3}\right[$ and diverges if $x \in \left]-\infty, -\frac{7}{3}\right[\cup \left]\frac{7}{3}, +\infty\right[$.

At the endpoints of the convergence interval, we obtain two numerical series.

— If $x = -\frac{7}{3}$, we have the series $\sum_{n=1}^{\infty} \frac{3^n \log(n)}{7^n n^2} \cdot \left(-\frac{7}{3}\right)^n = \sum_{n=1}^{\infty} (-1)^n \frac{\log(n)}{n^2}$ which is alternating. Let us start by studying the series of modules, $\sum_{n=1}^{\infty} \frac{\log(n)}{n^2}$, by comparison with the series $\sum_{n=1}^{\infty} \frac{1}{n^{3/2}}$, which is convergent because it is a p-series with $p = \frac{3}{2}$. Since

$$\lim \frac{\frac{\log(n)}{n^2}}{\frac{1}{n^{3/2}}} = \lim \frac{\log(n)}{n^{1/2}} = 0,$$

by Corollary 3 of the General Comparison Test, the series $\sum_{n=1}^{\infty} \frac{\log(n)}{n^2}$ is convergent. As it is the series of modules of $\sum_{n=1}^{\infty} (-1)^n \frac{\log(n)}{n^2}$, this series is absolutely convergent.

3.6. Solved Exercises

– If $x = \dfrac{7}{3}$, we obtain the series $\displaystyle\sum_{n=1}^{\infty} \dfrac{\log(n)}{n^2}$, which we proved in the previous case to be convergent. As it is of positive terms, it is absolutely convergent.

Conclusion: The series converges absolutely if $x \in \left[-\dfrac{7}{3}, \dfrac{7}{3}\right]$ and diverges if $x \in \left]-\infty, -\dfrac{7}{3}\right[\cup \left]\dfrac{7}{3}, +\infty\right[$.

d) Let us study the power series of x, $\displaystyle\sum_{n=1}^{\infty} \dfrac{2 \times 5 \times \cdots \times (3n-1)}{(n+2)!} x^n$. Setting $a_n = \dfrac{2 \times 5 \times \cdots \times (3n-1)}{(n+2)!}$, the radius of convergence is by Corollary 1 of Theorem 3.3.1

$$r = \lim \left|\dfrac{a_n}{a_{n+1}}\right| = \lim \left|\dfrac{\dfrac{2 \times 5 \times \cdots \times (3n-1)}{(n+2)!}}{\dfrac{2 \times 5 \times \cdots \times (3n-1)(3n+2)}{(n+3)!}}\right| = \lim \dfrac{n+3}{3n+2} = \dfrac{1}{3}.$$

Then, the series converges absolutely if $x \in \left]-\dfrac{1}{3}, \dfrac{1}{3}\right[$ and diverges if $x \in \left]-\infty, -\dfrac{1}{3}\right[\cup \left]\dfrac{1}{3}, +\infty\right[$.

Let us study the behavior of the numerical series at the endpoints of the convergence interval.

– If $x = -\dfrac{1}{3}$, we obtain the alternating series

$$\sum_{n=1}^{\infty} \dfrac{2 \times 5 \times \cdots \times (3n-1)}{(n+2)!} \cdot \left(-\dfrac{1}{3}\right)^n = \sum_{n=1}^{\infty} (-1)^n \dfrac{2 \times 5 \times \cdots \times (3n-1)}{3^n (n+2)!}.$$

Let us begin by studying the series of modules

$$\sum_{n=1}^{\infty} \dfrac{2 \times 5 \times \cdots \times (3n-1)}{3^n (n+2)!} = \sum_{n=1}^{\infty} b_n,$$

using Raabe's test,

$$\lim n \left(\dfrac{b_n}{b_{n+1}} - 1\right) = \lim n \left(\dfrac{\dfrac{2 \times 5 \times \cdots \times (3n-1)}{3^n (n+2)!}}{\dfrac{2 \times 5 \times \cdots \times (3n-1)(3n+2)}{3^{n+1}(n+3)!}} - 1\right)$$

$$= \lim n \left(\dfrac{3n+9}{3n+2} - 1\right) = \lim \dfrac{7n}{3n+2} = \dfrac{7}{3} > 1,$$

showing the series is convergent. Then, $\displaystyle\sum_{n=1}^{\infty}(-1)^n b_n$ is absolutely convergent.

– If $x = \dfrac{1}{3}$, we obtain $\displaystyle\sum_{n=1}^{\infty} \dfrac{2 \times 5 \times \cdots \times (3n-1)}{3^n (n+2)!}$, which is the series of modules of the previous series that we have seen to be convergent. As it is of positive terms, it is absolutely convergent.

Conclusion: The series converges absolutely if $x \in \left[-\dfrac{1}{3}, \dfrac{1}{3}\right]$ and diverges if $x \in \left]-\infty, -\dfrac{1}{3}\right[\cup \left]\dfrac{1}{3}, +\infty\right[$.

e) Let us study the power series of x, $\sum_{n=1}^{\infty} \frac{\log(n)}{n} x^n = \sum_{n=2}^{\infty} \frac{\log(n)}{n} x^n$. Setting $a_n = \frac{\log(n)}{n}$, the radius of convergence of this series is by Corollary 1 of Theorem 3.3.1

$$r = \lim \left| \frac{a_n}{a_{n+1}} \right| = \lim \left| \frac{\frac{\log(n)}{n}}{\frac{\log(n+1)}{n+1}} \right| = \lim \frac{n+1}{n} \cdot \frac{\log(n)}{\log(n+1)} = 1;$$

therefore, the series converges absolutely if $x \in\,]-1, 1[$ and diverges if $x \in\,]-\infty, -1[\, \cup\,]1, +\infty[$. Regarding the endpoints of the convergence interval, we need to study the respective numerical series.

– If $x = -1$, we obtain the alternating series $\sum_{n=1}^{\infty} (-1)^n \frac{\log(n)}{n}$. Let us start by studying the series of modules, $\sum_{n=1}^{\infty} \frac{\log(n)}{n}$, using the Integral Test.

Let $f(x) = \frac{\log(x)}{x}$, $x \geq 1$:

(i) If $x > 1$, then $\log(x) > 0$, therefore, $f(x) > 0$, $\forall x > 1$.

(ii) f is continuous, $\forall x \geq 1$.

(iii) We can study the monotonicity of f by analyzing its derivative. We have

$$f'(x) = \frac{x \cdot \frac{1}{x} - \log(x)}{x^2} = \frac{1 - \log(x)}{x^2},$$

and thus, $f'(x) < 0$, $\forall x > e$; therefore, f is decreasing on $]e, +\infty[$.

The conditions of the Integral Test are fulfilled on the interval $[3, +\infty[$. Then the numerical series $\sum_{n=3}^{\infty} \frac{\log(n)}{n}$ and the improper integral $\int_{3}^{+\infty} \frac{\log(x)}{x} dx$ are both convergent or both divergent. Since

$$\int_{3}^{+\infty} \frac{\log(x)}{x} dx = \lim_{x \to +\infty} \int_{3}^{x} \frac{\log(t)}{t} dt = \lim_{x \to +\infty} \int_{3}^{x} \frac{1}{t} \log(t) \, dt$$

$$= \lim_{x \to +\infty} \left[\frac{\log^2(t)}{2} \right]_{3}^{x}$$

$$= \lim_{x \to +\infty} \left(\frac{\log^2(x)}{2} - \frac{\log^2(3)}{2} \right) = +\infty,$$

the improper integral is divergent; the same happens to the series.

Because the series of modules is divergent, let us apply Leibniz's test to the alternating series:

(i) $\lim \frac{\log(n)}{n} = 0$ (just note that $\lim_{x \to +\infty} \frac{\log(x)}{x}$ is an indeterminate form of type $\frac{\infty}{\infty}$ that can be lifted using L'Hôpital's Rule: $\lim_{x \to +\infty} \frac{(\log(x))'}{(x)'} = \lim_{x \to +\infty} \frac{1}{x} = 0$).

(ii) $\frac{\log(n)}{n} > 0$, $\forall n > 1$.

(iii) We saw that the function $f(x) = \frac{\log(x)}{x}$ is decreasing on $]e, +\infty[$, which implies that $\left(\frac{\log(n)}{n} \right)$ is decreasing for $n \geq 3$.

3.6. Solved Exercises

Then, the series is convergent. As the series of modules is divergent, the original series is conditionally convergent.

– If $x = 1$, we obtain the series $\sum_{n=1}^{\infty} \dfrac{\log(n)}{n}$ which, as we saw before, is divergent.

Conclusion: The power series is absolutely convergent if $x \in\ ]-1,1[$, conditionally convergent if $x = -1$, and divergent if $x \in\ ]-\infty, -1[\ \cup\ [1, +\infty[$.

Remark: The series $\sum_{n=1}^{\infty} \dfrac{\log(n)}{n}$ and $\sum_{n=3}^{\infty} \dfrac{\log(n)}{n}$ are both divergent because they only differ in a finite number of terms.

f) Let us study the power series of x, $\sum_{n=0}^{\infty} \dfrac{3n^2}{4^n(n^3+2)} x^n$. Setting $a_n = \dfrac{3n^2}{4^n(n^3+2)}$, the radius of convergence of this series is, by Corollary 1 of Theorem 3.3.1,

$$r = \lim \left|\dfrac{a_n}{a_{n+1}}\right| = \lim \left|\dfrac{\dfrac{3n^2}{4^n(n^3+2)}}{\dfrac{3(n+1)^2}{4^{n+1}((n+1)^3+2)}}\right| = \lim \dfrac{3n^2 4^{n+1}((n+1)^3+2)}{3(n+1)^2 4^n(n^3+2)}$$

$$= \lim 4\left(\dfrac{n}{n+1}\right)^2 \dfrac{(n+1)^3+2}{n^3+2} = 4.$$

The series converges absolutely if $x \in\ ]-4,4[$ and diverges if $x \in\ ]-\infty, -4[\ \cup\ ]4, +\infty[$.

Regarding the endpoints of the convergence interval, we have to study the respective numerical series.

– If $x = 4$, we obtain the series $\sum_{n=0}^{\infty} \dfrac{3n^2}{4^n(n^3+2)} 4^n = \sum_{n=0}^{\infty} \dfrac{3n^2}{n^3+2}$. We will study this series by comparing it with the harmonic series, which is divergent. The limit

$$\lim \dfrac{\dfrac{3n^2}{n^3+2}}{\dfrac{1}{n}} = \lim \dfrac{3n^3}{n^3+2} = 3$$

is finite and different from zero; therefore, the series is divergent by Corollary 2 of the General Comparison Test.

– If $x = -4$, we obtain the series $\sum_{n=0}^{\infty} \dfrac{3n^2}{4^n(n^3+2)} (-4)^n = \sum_{n=0}^{\infty} (-1)^n \dfrac{3n^2}{n^3+2}$, which is alternating since $\dfrac{3n^2}{n^3+2} > 0$, $\forall n \in \mathbb{N}$. The series of modules is the series we studied for the case $x = 4$, which we proved to be divergent. Being an alternating series, we apply Leibniz's test:

(i) $\lim \dfrac{3n^2}{n^3+2} = 0$.

(ii) $\dfrac{3n^2}{n^3+2} > 0$, $\forall n \geq 1$.

(iii) The sequence $\left(\dfrac{3n^2}{n^3+2}\right)$ is decreasing, because

$$\dfrac{3(n+1)^2}{(n+1)^3+2} - \dfrac{3n^2}{n^3+2} = 3\dfrac{-n^4-2n^3-n^2+4n+2}{((n+1)^3+2)(n^3+2)} < 0, \ \forall n \geq 2.$$

Then, the series is convergent. As the series of modules is divergent, the series is conditionally convergent.

Conclusion: The power series is absolutely convergent if $x \in \,]-4, 4[$, conditionally convergent if $x = -4$, and divergent if $x \in \,]-\infty, -4[\,\cup\, [4, +\infty[$.

g) Let $a_n = \dfrac{1 \times 4 \times \cdots \times n^2}{(2n)!} = \dfrac{(n!)^2}{(2n)!}$. By Corollary 1 of Theorem 3.3.1, the power series of x, $\sum\limits_{n=1}^{\infty} \dfrac{(n!)^2}{(2n)!} x^n$, has radius of convergence

$$r = \lim \left| \frac{a_n}{a_{n+1}} \right| = \lim \left| \frac{\dfrac{(n!)^2}{(2n)!}}{\dfrac{((n+1)!)^2}{(2n+2)!}} \right| = \lim \frac{(2n+2)(2n+1)}{(n+1)^2} = 4.$$

The series converges absolutely if $x \in \,]-4, 4[$ and diverges if $x \in \,]-\infty, -4[\,\cup\,]4, +\infty[$.

Let us study the numerical series corresponding to the endpoints of the interval.

– If $x = 4$, we obtain the series $\sum\limits_{n=1}^{\infty} \dfrac{(n!)^2}{(2n)!} 4^n = \sum\limits_{n=1}^{\infty} b_n$. Let us study this series using D'Alembert's test[1]:

$$\lim \frac{b_{n+1}}{b_n} = \lim \frac{\dfrac{((n+1)!)^2}{(2n+2)!} 4^{n+1}}{\dfrac{(n!)^2}{(2n)!} 4^n} = \lim \frac{2n+2}{2n+1} = 1.$$

Since the limit is 1, but for values greater than 1, taking into account the Note after D'Alembert's test, we can conclude that the series is divergent.

– If $x = -4$, we obtain the alternating series $\sum\limits_{n=1}^{\infty} \dfrac{(n!)^2}{(2n)!} (-4)^n = \sum\limits_{n=1}^{\infty} (-1)^n \dfrac{(n!)^2}{(2n)!} 4^n$. The series of modules $\sum\limits_{n=1}^{\infty} \dfrac{(n!)^2}{(2n)!} 4^n$ is divergent. Since we establish the divergence of the series of modules by D'Alembert's test, we can affirm that the alternating series is divergent (refer to the Note on page 105 for further details).

Conclusion: The power series converges absolutely if $x \in \,]-4, 4[$ and diverges if $x \in \,]-\infty, -4] \,\cup\, [4, +\infty[$.

h) Let us study the power series of x, $\sum\limits_{n=1}^{\infty} n^n \left(\sin\left(\dfrac{1}{n^2}\right) \right)^n x^n$. Let $a_n = n^n \left(\sin\left(\dfrac{1}{n^2}\right) \right)^n$. The radius of convergence of this series is, by Theorem 3.3.1,

$$r = \frac{1}{\overline{\lim} \sqrt[n]{|a_n|}} = \frac{1}{\overline{\lim} \sqrt[n]{n^n \left(\sin\left(\dfrac{1}{n^2}\right) \right)^n}} = \frac{1}{\lim n \sin\left(\dfrac{1}{n^2}\right)} = +\infty.$$

Moreover, the series is absolutely convergent on $\mathbb{R}$.

---

[1] In general, this test is not useful for studying the endpoints of a power series because the limit is always 1.

3.6. Solved Exercises

i) Let us study the power series of x, $\sum_{n=1}^{\infty} \dfrac{1}{2^n\, n\bigl(\arctan(n) - \frac{\pi}{2}\bigr)} x^n$. Let $a_n = \dfrac{1}{2^n\, n\bigl(\arctan(n) - \frac{\pi}{2}\bigr)}$. The radius of convergence of this series is, by Corollary 1 of Theorem 3.3.1,

$$r = \lim \left|\dfrac{a_n}{a_{n+1}}\right| = \lim \left| \dfrac{\dfrac{1}{2^n n\bigl(\arctan(n) - \frac{\pi}{2}\bigr)}}{\dfrac{1}{2^{n+1}(n+1)\bigl(\arctan(n+1) - \frac{\pi}{2}\bigr)}} \right| = \lim \dfrac{2(n+1)\bigl(\frac{\pi}{2} - \arctan(n+1)\bigr)}{n\bigl(\frac{\pi}{2} - \arctan(n)\bigr)} = 2.$$

Then, the series is absolutely convergent if $x \in\,]-2, 2[$ and diverges if $x \in\,]-\infty, -2[\, \cup\,]2, +\infty[$.
At the endpoints of the convergence interval, we obtain two numerical series.

— If $x = 2$, we have the series $\sum_{n=1}^{\infty} \dfrac{1}{2^n\, n\bigl(\arctan(n) - \frac{\pi}{2}\bigr)} 2^n = \sum_{n=1}^{\infty} \dfrac{1}{n\bigl(\arctan(n) - \frac{\pi}{2}\bigr)}$. Let us calculate the limit of its general term:

$$\lim \dfrac{1}{n\bigl(\arctan(n) - \frac{\pi}{2}\bigr)} = -1.$$

Since the general term of the series is not a null sequence, by Theorem 2.2.1, the series is divergent.

— If $x = -2$, we obtain the series $\sum_{n=1}^{\infty} \dfrac{(-2)^n}{2^n\, n\bigl(\arctan(n) - \frac{\pi}{2}\bigr)} = \sum_{n=1}^{\infty} (-1)^{n+1} \dfrac{1}{n\bigl(\frac{\pi}{2} - \arctan(n)\bigr)}$, which is alternating because $a_n = \dfrac{1}{n\bigl(\frac{\pi}{2} - \arctan(n)\bigr)} > 0$, $\forall n \in \mathbb{N}$. The limit of the general term does not exist, as the subsequence of even order terms has limit -1 and the subsequence of odd order terms has limit 1. Then, the series is divergent.

Conclusion: The power series converges absolutely if $x \in\,]-2, 2[$ and diverges if $x \in\,]-\infty, -2]\, \cup\, [2, +\infty[$.

j) Consider the power series $\sum_{n=1}^{\infty} \dfrac{3 \times 6 \times \cdots \times 3n}{(n+3)!} x^n = \sum_{n=1}^{\infty} \dfrac{3^n\, n!}{(n+3)!} x^n$. If $a_n = \dfrac{3^n\, n!}{(n+3)!}$, then the radius of convergence of this series is, by Corollary 1 of Theorem 3.3.1,

$$r = \lim \left|\dfrac{a_n}{a_{n+1}}\right| = \lim \dfrac{\dfrac{3^n\, n!}{(n+3)!}}{\dfrac{3^{n+1}(n+1)!}{(n+4)!}} = \lim \dfrac{n+4}{3(n+1)} = \dfrac{1}{3};$$

therefore, the series converges absolutely if $x \in\, \left]-\dfrac{1}{3}, \dfrac{1}{3}\right[$ and diverges if $x \in\, \left]-\infty, -\dfrac{1}{3}\right[\, \cup\, \left]\dfrac{1}{3}, +\infty\right[$.
Let us study the numerical series at the endpoints of the convergence interval.

— If $x = \dfrac{1}{3}$, we obtain the series $\sum_{n=1}^{\infty} \dfrac{n!}{(n+3)!} = \sum_{n=1}^{\infty} \dfrac{1}{(n+3)(n+2)(n+1)}$. Let us study this series by comparing it with $\sum_{n=1}^{\infty} \dfrac{1}{n^3}$ which we know to be convergent (p-series with $p = 3$):

$$\lim \dfrac{\dfrac{1}{(n+3)(n+2)(n+1)}}{\dfrac{1}{n^3}} = \lim \dfrac{n^3}{(n+3)(n+2)(n+1)} = 1.$$

Since this value is finite and different from zero, the series is convergent by Corollary 2 of the General Comparison Test. As it is a series of positive terms, it is absolutely convergent.

– If $x = -\dfrac{1}{3}$, we obtain the alternating series $\sum_{n=1}^{\infty}(-1)^n \dfrac{n!}{(n+3)!}$. As the series of modules is convergent, the alternating series is absolutely convergent.

Conclusion: The series converges absolutely if $x \in \left[-\dfrac{1}{3}, \dfrac{1}{3}\right]$ and diverges if $x \in \left]-\infty, -\dfrac{1}{3}\right[\cup \left]\dfrac{1}{3}, +\infty\right[$.

k) Let us study the power series of x, $\sum_{n=0}^{\infty}(-1)^n \dfrac{n+1}{\sqrt{e^{2n}+5}} x^n$. Let $a_n = (-1)^n \dfrac{n+1}{\sqrt{e^{2n}+5}}$. The radius of convergence of this series is, by Corollary 1 of Theorem 3.3.1,

$$r = \lim \left|\dfrac{a_n}{a_{n+1}}\right| = \lim \left|\dfrac{(-1)^n \dfrac{n+1}{\sqrt{e^{2n}+5}}}{(-1)^{n+1} \dfrac{n+2}{\sqrt{e^{2n+2}+5}}}\right| = \lim \dfrac{(n+1)\sqrt{e^{2n+2}+5}}{(n+2)\sqrt{e^{2n}+5}} = \lim \dfrac{n+1}{n+2} \cdot \sqrt{\dfrac{e^{2n+2}+5}{e^{2n}+5}} = e;$$

then, the series is absolutely convergent if $x \in]-e, e[$ and divergent if $x \in]-\infty, -e[\cup]e, +\infty[$.

At the endpoints of the convergence interval, we obtain two numerical series that we will proceed to study.

– If $x = -e$, we obtain the series $\sum_{n=0}^{\infty}(-1)^n \dfrac{n+1}{\sqrt{e^{2n}+5}}(-e)^n = \sum_{n=0}^{\infty} \dfrac{(n+1)e^n}{\sqrt{e^{2n}+5}}$. As

$$\lim \dfrac{(n+1)e^n}{\sqrt{e^{2n}+5}} = +\infty,$$

the general term is not a null sequence; therefore, by Theorem 2.2.1, the series diverges.

– If $x = e$, we obtain the series $\sum_{n=0}^{\infty}(-1)^n \dfrac{n+1}{\sqrt{e^{2n}+5}} e^n = \sum_{n=0}^{\infty}(-1)^n \dfrac{(n+1)e^n}{\sqrt{e^{2n}+5}}$. As the limit of the general term does not exist, the series is divergent.

Conclusion: The power series converges absolutely if $x \in]-e, e[$ and diverges if $x \in]-\infty, -e] \cup [e, +\infty[$.

l) Let us study the power series of x, $\sum_{n=1}^{\infty} \dfrac{1 \times 3 \times 5 \times \cdots \times (2n-1)}{2 \times 4 \times 6 \times \cdots \times 2n} \cdot \dfrac{1}{n} x^n$. Setting

$$a_n = \dfrac{1 \times 3 \times 5 \times \cdots \times (2n-1)}{2 \times 4 \times 6 \times \cdots \times 2n} \cdot \dfrac{1}{n},$$

the radius of convergence of this series is, by Corollary 1 of Theorem 3.3.1,

$$r = \lim \left|\dfrac{a_n}{a_{n+1}}\right| = \lim \left|\dfrac{\dfrac{1 \times 3 \times 5 \times \cdots \times (2n-1)}{2 \times 4 \times 6 \times \cdots \times 2n} \cdot \dfrac{1}{n}}{\dfrac{1 \times 3 \times 5 \times \cdots \times (2n-1)(2n+1)}{2 \times 4 \times 6 \times \cdots \times 2n(2n+2)} \cdot \dfrac{1}{n+1}}\right| = \lim \dfrac{(n+1)(2n+2)}{n(2n+1)} = 1.$$

The series converges absolutely if $x \in]-1, 1[$ and diverges if $x \in]-\infty, -1[\cup]1, +\infty[$.

Let us study the numerical series obtained at the endpoints of the convergence interval.

– If $x = -1$, we obtain the alternating series

$$\sum_{n=1}^{\infty}(-1)^n \dfrac{1 \times 3 \times 5 \times \cdots \times (2n-1)}{2 \times 4 \times 6 \times \cdots \times 2n} \cdot \dfrac{1}{n}.$$

3.6. Solved Exercises

Let us start by studying the series of modules,

$$\sum_{n=1}^{\infty} \frac{1 \times 3 \times 5 \times \cdots \times (2n-1)}{2 \times 4 \times 6 \times \cdots \times 2n} \cdot \frac{1}{n} = \sum_{n=1}^{\infty} a_n,$$

using Raabe's test

$$\lim n \left(\frac{a_n}{a_{n+1}} - 1 \right) = \lim n \left(\frac{(n+1)(2n+2)}{n(2n+1)} - 1 \right) = \lim \frac{3n+2}{2n+1} = \frac{3}{2} > 1,$$

which shows that the series is convergent. As it is the series of modules of $\sum_{n=1}^{\infty}(-1)^n a_n$, this series is absolutely convergent.

– If $x = 1$, we obtain the series $\sum_{n=1}^{\infty} \frac{1 \times 3 \times 5 \times \cdots \times (2n-1)}{2 \times 4 \times 6 \times \cdots \times 2n} \cdot \frac{1}{n}$, which we have seen to be convergent. As it is of positive terms, it is absolutely convergent.

Conclusion: The power series converges absolutely if $x \in [-1, 1]$ and diverges if $x \in \,]-\infty, -1[\, \cup \,]1, +\infty[$.

15. a) Let us study the power series of $x - 1$, $\sum_{n=1}^{\infty} \left(\frac{n}{2n+1} \right)^{2n} (x-1)^n$. Let $y = x - 1$, and consider the series $\sum_{n=1}^{\infty} \left(\frac{n}{2n+1} \right)^{2n} y^n$. Setting $a_n = \left(\frac{n}{2n+1} \right)^{2n}$, the radius of convergence of this series is, by Theorem 3.3.1,

$$r = \frac{1}{\overline{\lim} \sqrt[n]{|a_n|}} = \frac{1}{\overline{\lim} \sqrt[n]{\left(\frac{n}{2n+1} \right)^{2n}}} = \frac{1}{\lim \left(\frac{n}{2n+1} \right)^2} = \frac{1}{\lim \frac{n^2}{4n^2 + 4n + 1}} = 4.$$

The series converges absolutely if $y \in \,]-4, 4[$ and diverges if $y \in \,]-\infty, -4[\,\cup\,]4, +\infty[$. As $y = x - 1$, we can conclude that the series $\sum_{n=1}^{\infty} \left(\frac{n}{2n+1} \right)^{2n} (x-1)^n$ is absolutely convergent if $x \in \,]-3, 5[$ and divergent if $x \in \,]-\infty, -3[\,\cup\,]5, +\infty[$.

At the endpoints of the convergence interval, we have to study the corresponding numerical series.

– If $x = -3$, we obtain the series

$$\sum_{n=1}^{\infty} \left(\frac{n}{2n+1} \right)^{2n} (-4)^n = \sum_{n=1}^{\infty} (-1)^n \, 4^n \left(\frac{n}{2n+1} \right)^{2n} = \sum_{n=1}^{\infty} (-1)^n \left(\frac{2n}{2n+1} \right)^{2n}.$$

As

$$\lim \left(\frac{2n}{2n+1} \right)^{2n} = \frac{1}{e},$$

the general term does not have limit; therefore, by Theorem 2.2.1, the series is divergent.

– If $x = 5$, we obtain the series

$$\sum_{n=1}^{\infty} \left(\frac{n}{2n+1} \right)^{2n} 4^n = \sum_{n=1}^{\infty} \left(\frac{2n}{2n+1} \right)^{2n},$$

whose general term does not tend to zero, as we saw earlier; therefore, the series is divergent.

In summary, the power series of $x - 1$ is divergent if $x \in]-\infty, -3] \cup [5, +\infty[$ and absolutely convergent if $x \in]-3, 5[$.

b) Let us study the power series of $x - 2$, $\sum_{n=2}^{\infty} (-1)^n \frac{(x-2)^n}{n \log(n)}$. Let $y = x - 2$, and consider the series $\sum_{n=2}^{\infty} (-1)^n \frac{y^n}{n \log(n)}$. Setting $a_n = \frac{(-1)^n}{n \log(n)}$, the radius of convergence of this series is, by Corollary 1 of Theorem 3.3.1,

$$r = \lim \left| \frac{a_n}{a_{n+1}} \right| = \lim \left| \frac{\frac{(-1)^n}{n \log(n)}}{\frac{(-1)^{n+1}}{(n+1) \log(n+1)}} \right| = \lim \left(\frac{n+1}{n} \cdot \frac{\log(n+1)}{\log(n)} \right) = 1;$$

therefore, the series converges absolutely if $y \in]-1, 1[$ and diverges if $y \in]-\infty, -1[\cup]1, +\infty[$. As $y = x - 2$, the series $\sum_{n=2}^{\infty} (-1)^n \frac{(x-2)^n}{n \log(n)}$ is absolutely convergent if $x \in]1, 3[$ and divergent if $x \in]-\infty, 1[\cup]3, +\infty[$.

Let us study the numerical series obtained at the endpoints of the convergence interval.

– If $x = 1$, we obtain the series $\sum_{n=2}^{\infty} (-1)^n \frac{(-1)^n}{n \log(n)} = \sum_{n=2}^{\infty} \frac{1}{n \log(n)}$. Let us study it using the Integral Test.

Let $f(x) = \frac{1}{x \log(x)}$:

(i) As $x \geq 2 \Rightarrow \log(x) > 0$, we have $f(x) > 0$, $\forall x \geq 2$.

(ii) f is continuous on $[2, +\infty[$.

(iii) f is decreasing on $[2, +\infty[$ because $x \log(x)$ is increasing and positive.

Then the numerical series $\sum_{n=2}^{\infty} \frac{1}{n \log(n)}$ and the improper integral $\int_{2}^{+\infty} \frac{1}{x \log(x)} dx$ are both convergent or both divergent. Since

$$\int_{2}^{+\infty} \frac{1}{x \log(x)} dx = \lim_{x \to +\infty} \int_{2}^{x} \frac{1}{t \log(t)} dt = \lim_{x \to +\infty} \int_{2}^{x} \frac{\frac{1}{t}}{\log(t)} dt$$

$$= \lim_{x \to +\infty} \left[\log(\log(t)) \right]_{2}^{x} = \lim_{x \to +\infty} \left(\log(\log(x)) - \log(\log(2)) \right) = +\infty,$$

the improper integral is divergent, the same happening to the series.

– If $x = 3$, we obtain the alternating series $\sum_{n=2}^{\infty} (-1)^n \frac{1}{n \log(n)}$. We apply Leibniz's test since we already know that the series of modules is divergent:

(i) $\lim \frac{1}{n \log(n)} = 0$.

(ii) $\frac{1}{n \log(n)} > 0$, $\forall n \geq 2$.

(iii) We saw earlier that the function $f(x) = \frac{1}{x \log(x)}$ is decreasing on $[2, +\infty[$, which implies that $\left(\frac{1}{n \log(n)} \right)$ is decreasing for $n \geq 2$.

3.6. Solved Exercises

Then, we conclude that the series is convergent. As the series of modules is divergent, the series is conditionally convergent.

Conclusion: The power series of $x - 2$ is absolutely convergent if $x \in \,]1, 3[$, conditionally convergent if $x = 3$, and divergent if $x \in \,]-\infty, 1] \cup \,]3, +\infty[$.

c) Let us study the power series of $x + 4$, $\sum_{n=1}^{\infty} \dfrac{n}{(-3)^n(2n^2 - 1)}(x + 4)^n$. We can rewrite it in the form $\sum_{n=1}^{\infty} (-1)^n \dfrac{n}{3^n(2n^2 - 1)}(x + 4)^n$. Let $y = x + 4$, and consider the series $\sum_{n=1}^{\infty} (-1)^n \dfrac{n}{3^n(2n^2 - 1)} y^n$. If $a_n = (-1)^n \dfrac{n}{3^n(2n^2 - 1)}$, the radius of convergence of this series is, by Corollary 1 of Theorem 3.3.1,

$$r = \lim \left| \dfrac{a_n}{a_{n+1}} \right| = \lim \left| \dfrac{(-1)^n \dfrac{n}{3^n(2n^2 - 1)}}{(-1)^{n+1} \dfrac{n+1}{3^{n+1}(2(n+1)^2 - 1)}} \right| = \lim \dfrac{3n(2n^2 + 4n + 1)}{(n+1)(2n^2 - 1)} = 3.$$

The series converges absolutely if $y \in \,]-3, 3[$ and diverges if $y \in \,]-\infty, -3[\,\cup\,]3, +\infty[$. As $y = x + 4$, the series $\sum_{n=1}^{\infty} (-1)^n \dfrac{n}{3^n(2n^2 - 1)}(x + 4)^n$ is absolutely convergent if $x \in \,]-7, -1[$ and divergent if $x \in \,]-\infty, -7[\,\cup\,]-1, +\infty[$.

At the endpoints of the convergence interval, we have to study the respective numerical series.

– If $x = -7$, we obtain the series $\sum_{n=1}^{\infty} (-1)^n \dfrac{n}{3^n(2n^2 - 1)}(-3)^n = \sum_{n=1}^{\infty} \dfrac{n}{2n^2 - 1}$.

It is of positive terms, and we can compare it with the harmonic series, which is divergent. Since the limit

$$\lim \dfrac{\dfrac{n}{2n^2 - 1}}{\dfrac{1}{n}} = \lim \dfrac{n^2}{2n^2 - 1} = \dfrac{1}{2}$$

is finite and different from zero, the series is divergent by Corollary 2 of the General Comparison Test.

– If $x = -1$, we obtain the alternating series $\sum_{n=1}^{\infty} (-1)^n \dfrac{n}{3^n(2n^2 - 1)} 3^n = \sum_{n=1}^{\infty} (-1)^n \dfrac{n}{2n^2 - 1}$. We apply Leibniz's test since the series of modules is divergent:

(i) $\lim \dfrac{n}{2n^2 - 1} = 0$.

(ii) $\dfrac{n}{2n^2 - 1} > 0$, $\forall n \geq 1$.

(iii) $\dfrac{n+1}{2(n+1)^2 - 1} - \dfrac{n}{2n^2 - 1} = -\dfrac{2n^2 + 2n + 1}{(2n^2 + 4n + 1)(2n^2 - 1)} < 0$, $\forall n \in \mathbb{N}$, which implies that $\left(\dfrac{n}{2n^2 - 1} \right)$ is decreasing.

We can conclude that the series is convergent. Since the series of modules is divergent, the series is conditionally convergent.

Conclusion: The power series of $x + 4$ is absolutely convergent if $x \in \,]-7, -1[$, conditionally convergent if $x = -1$ and divergent if $x \in \,]-\infty, -7] \cup \,]-1, +\infty[$.

d) Let us study the power series of $x - e$, $\sum_{n=2}^{\infty} \frac{e^n}{\log(n)}(x-e)^n$. Let $y = x - e$, and consider the series $\sum_{n=2}^{\infty} \frac{e^n}{\log(n)} y^n$. If $a_n = \frac{e^n}{\log(n)}$, the radius of convergence of this series is, by Corollary 1 of Theorem 3.3.1,

$$r = \lim \left|\frac{a_n}{a_{n+1}}\right| = \lim \frac{\frac{e^n}{\log(n)}}{\frac{e^{n+1}}{\log(n+1)}} = \lim \frac{1}{e} \cdot \frac{\log(n+1)}{\log(n)} = \frac{1}{e},$$

so the series is absolutely convergent if $y \in \left]-\frac{1}{e}, \frac{1}{e}\right[$ and divergent if $y \in \left]-\infty, -\frac{1}{e}\right[\cup \left]\frac{1}{e}, +\infty\right[$. As $y = x - e$, the series $\sum_{n=2}^{\infty} \frac{e^n}{\log(n)}(x-e)^n$ is absolutely convergent if $x \in \left]e - \frac{1}{e}, e + \frac{1}{e}\right[$ and divergent if $x \in \left]-\infty, e - \frac{1}{e}\right[\cup \left]e + \frac{1}{e}, +\infty\right[$.

Let us study the numerical series corresponding to the endpoints of the interval.

– If $x = e - \frac{1}{e}$, we get the series $\sum_{n=2}^{\infty} \frac{e^n}{\log(n)} \left(-\frac{1}{e}\right)^n = \sum_{n=2}^{\infty} \frac{(-1)^n}{\log(n)}$. It is an alternating series because $a_n = \frac{1}{\log(n)} > 0$, $\forall n > 1$. Let us consider the series of modules, $\sum_{n=2}^{\infty} \frac{1}{\log(n)}$, and study it by comparing it with the harmonic series. As the limit

$$\lim \frac{\frac{1}{\log(n)}}{\frac{1}{n}} = \lim \frac{n}{\log(n)} = +\infty$$

by Corollary 4 of the General Comparison Test, the series $\sum_{n=2}^{\infty} \frac{1}{\log(n)}$ is divergent. Let us apply Leibniz's test:

(i) $\lim \frac{1}{\log(n)} = 0$.

(ii) $\frac{1}{\log(n)} > 0$, $\forall n \geq 2$.

(iii) The sequence $(\log(n))$ is increasing; therefore, $\left(\frac{1}{\log(n)}\right)$ is a decreasing sequence for $n \geq 2$.

Then, the series is convergent. Since the series of modules is divergent, the alternating series is conditionally convergent.

– If $x = e + \frac{1}{e}$, we get the series $\sum_{n=2}^{\infty} \frac{e^n}{\log(n)} \left(\frac{1}{e}\right)^n = \sum_{n=2}^{\infty} \frac{1}{\log(n)}$ which, as we have seen, is divergent.

Conclusion: The power series of $x - e$ is absolutely convergent if $x \in \left]e - \frac{1}{e}, e + \frac{1}{e}\right[$, conditionally convergent if $x = e - \frac{1}{e}$ and divergent if $x \in \left]-\infty, e - \frac{1}{e}\right[\cup \left[e + \frac{1}{e}, +\infty\right[$.

3.6. Solved Exercises

e) Let us consider the series

$$\sum_{n=1}^{\infty} \frac{\cos\left(\frac{\pi}{2n}\right)}{(2n-1)^2(x-1)^n} = \sum_{n=1}^{\infty} \frac{\cos\left(\frac{\pi}{2n}\right)}{(2n-1)^2}\left(\frac{1}{x-1}\right)^n = \sum_{n=2}^{\infty} \frac{\cos\left(\frac{\pi}{2n}\right)}{(2n-1)^2}\left(\frac{1}{x-1}\right)^n,$$

and let $y = \frac{1}{x-1}$. We study the power series of y, $\sum_{n=2}^{\infty} \frac{\cos\left(\frac{\pi}{2n}\right)}{(2n-1)^2} y^n$. Note that if $n > 1$, then $0 < \frac{\pi}{2n} < \frac{\pi}{2}$; therefore, $\cos\left(\frac{\pi}{2n}\right) > 0$. Let $a_n = \frac{\cos\left(\frac{\pi}{2n}\right)}{(2n-1)^2}$. The radius of convergence of this series is, by Corollary 1 of Theorem 3.3.1,

$$r = \lim\left|\frac{a_n}{a_{n+1}}\right| = \lim\left|\frac{\frac{\cos\left(\frac{\pi}{2n}\right)}{(2n-1)^2}}{\frac{\cos\left(\frac{\pi}{2(n+1)}\right)}{(2n+1)^2}}\right| = \lim\frac{\cos\left(\frac{\pi}{2n}\right)(2n+1)^2}{\cos\left(\frac{\pi}{2(n+1)}\right)(2n-1)^2} = 1.$$

Therefore, the series converges absolutely if $y \in\,]-1,1[$ and diverges if $y \in\,]-\infty,-1[\,\cup\,]1,+\infty[$. If $y = \pm 1$, we obtain two numerical series we will study.

- If $y = 1$, we obtain the series $\sum_{n=2}^{\infty} \frac{\cos\left(\frac{\pi}{2n}\right)}{(2n-1)^2}$, which is of positive terms. We are going to compare it with $\sum_{n=1}^{\infty} \frac{1}{n^2}$, which we know to be convergent, as it is a p-series with $p = 2$. The limit

$$\lim \frac{\frac{\cos\left(\frac{\pi}{2n}\right)}{(2n-1)^2}}{\frac{1}{n^2}} = \lim \frac{n^2 \cos\left(\frac{\pi}{2n}\right)}{(2n-1)^2} = \frac{1}{4}$$

is finite and different from zero; therefore, by Corollary 2 of the General Comparison Test, the series is convergent. As it is of positive terms, it is absolutely convergent.

- If $y = -1$, we obtain the alternating series $\sum_{n=2}^{\infty} (-1)^n \frac{\cos\left(\frac{\pi}{2n}\right)}{(2n-1)^2}$, whose series of modules is the one we studied for the case $y = 1$, which we proved to be convergent. Therefore, the series is absolutely convergent.

Conclusion: The power series of y is divergent if $y \in\,]-\infty,-1[\,\cup\,]1,+\infty[$ and absolutely convergent if $y \in [-1, 1]$.

Concerning the original series, we have, because $y = \frac{1}{x-1}$,

$$|y| \leq 1 \Leftrightarrow \left|\frac{1}{x-1}\right| \leq 1 \Leftrightarrow |x-1| \geq 1 \Leftrightarrow x \geq 2 \vee x \leq 0,$$

which implies that the series converges absolutely if $x \in\,]-\infty, 0]\,\cup\,[2, +\infty[$ and diverges if $x \in\,]0, 2[\,\setminus\{1\}$, because 1 does not belong to the domain of $\frac{1}{x-1}$.

f) Let us study the power series of $2x^2 - 5$, $\sum_{n=1}^{\infty} \frac{(2x^2-5)^n}{n\,3^{n+1}}$. Let $y = 2x^2 - 5$ and consider the series $\sum_{n=1}^{\infty} \frac{1}{n\,3^{n+1}} y^n$.

Setting $a_n = \frac{1}{n\,3^{n+1}}$, the radius of convergence of this series is, by Corollary 1 of Theorem 3.3.1,

$$r = \lim \left|\frac{a_n}{a_{n+1}}\right| = \lim \frac{\frac{1}{n\,3^{n+1}}}{\frac{1}{(n+1)\,3^{n+2}}} = \lim \frac{3(n+1)}{n} = 3;$$

therefore, the series converges absolutely if $y \in]-3,3[$ and diverges if $y \in]-\infty, -3[\cup]3, +\infty[$.
At the endpoints of the interval of convergence, it is necessary to study the respective numerical series.

- If $y = -3$, we obtain the series $\sum_{n=1}^{\infty} \frac{1}{n\,3^{n+1}}(-3)^n = \sum_{n=1}^{\infty} \frac{(-1)^n}{3n} = \frac{1}{3}\sum_{n=1}^{\infty} \frac{(-1)^n}{n}$, which is the alternating harmonic series that we know to be conditionally convergent.

- If $y = 3$, we obtain the series $\sum_{n=1}^{\infty} \frac{1}{3n} = \frac{1}{3}\sum_{n=1}^{\infty} \frac{1}{n}$, which is divergent.

But
$$|y| < 3 \;\;\Leftrightarrow\;\; |2x^2 - 5| < 3 \Leftrightarrow -3 < 2x^2 - 5 < 3 \Leftrightarrow 1 < x^2 < 4$$
$$\Leftrightarrow\;\; x \in]-2,2[\,\cap\,(]-\infty,-1[\cup]1,+\infty[) \Leftrightarrow x \in]-2,-1[\cup]1,2[.$$

In addition,
$$2x^2 - 5 = 3 \Leftrightarrow x = -2 \vee x = 2$$
and
$$2x^2 - 5 = -3 \Leftrightarrow x = -1 \vee x = 1.$$

Conclusion: The original series is absolutely convergent if $x \in]-2,-1[\cup]1,2[$, conditionally convergent if $x = -1$ or $x = 1$ and divergent if $x \in]-\infty,-2]\cup]-1,1[\cup[2,+\infty[$.

g) Let us study the power series of $x+2$, $\sum_{n=1}^{\infty} \frac{\sqrt{n+1}-\sqrt{n}}{\sqrt{n+1}+\sqrt{n}}(x+2)^n$. We can rewrite the series in the form $\sum_{n=1}^{\infty} \frac{1}{2n+1+2\sqrt{n(n+1)}}(x+2)^n$. Let $y = x+2$ and consider the series $\sum_{n=1}^{\infty} \frac{1}{2n+1+2\sqrt{n(n+1)}} y^n$.

Let $a_n = \frac{1}{2n+1+2\sqrt{n(n+1)}}$. The radius of convergence of this series is, by Corollary 1 of Theorem 3.3.1,

$$r = \lim \left|\frac{a_n}{a_{n+1}}\right| = \lim \frac{\frac{1}{2n+1+2\sqrt{n(n+1)}}}{\frac{1}{2n+3+2\sqrt{(n+1)(n+2)}}} = \lim \frac{2n+3+2\sqrt{(n+1)(n+2)}}{2n+1+2\sqrt{n(n+1)}} = 1;$$

therefore, the series absolutely converges if $y \in]-1,1[$ and diverges if $y \in]-\infty,-1[\cup]1,+\infty[$. As $y = x+2$, we can conclude that the series $\sum_{n=1}^{\infty} \frac{1}{2n+1+2\sqrt{n(n+1)}}(x+2)^n$ is absolutely convergent if $x \in]-3,-1[$ and divergent if $x \in]-\infty,-3[\cup]-1,+\infty[$.

Let us study the numerical series at the endpoints of the interval of convergence.

3.6. Solved Exercises

– If $x = -1$, we obtain the series $\sum_{n=1}^{\infty} \dfrac{1}{2n+1+2\sqrt{n(n+1)}}$. As it is of positive terms and the limit

$$\lim \frac{\dfrac{1}{2n+1+2\sqrt{n(n+1)}}}{\dfrac{1}{n}} = \lim \frac{n}{2n+1+2\sqrt{n(n+1)}} = \frac{1}{4}$$

is finite and different from zero, the series is divergent because the harmonic series is divergent (see Corollary 2 of the General Comparison Test).

– If $x = -3$, we obtain the series $\sum_{n=1}^{\infty} \dfrac{(-1)^n}{2n+1+2\sqrt{n(n+1)}}$. It is an alternating series, and we apply Leibniz's test since the series of modules is divergent:

(i) $\lim \dfrac{1}{2n+1+2\sqrt{n(n+1)}} = 0$.

(ii) $\dfrac{1}{2n+1+2\sqrt{n(n+1)}} > 0$, $\forall n \geq 1$.

(iii) $\dfrac{1}{2n+3+2\sqrt{(n+1)(n+2)}} - \dfrac{1}{2n+1+2\sqrt{n(n+1)}} < 0$, which implies that the sequence is decreasing.

Then, the alternating series is convergent. It is conditionally convergent because the series of modules is divergent.

Conclusion: The power series of $x+2$ is absolutely convergent if $x \in \,]-3, -1[$, conditionally convergent if $x = -3$ and divergent if $x \in \,]-\infty, -3[\, \cup \,[-1, +\infty[$.

h) Let us consider the series $\sum_{n=1}^{\infty} \dfrac{\log^2(n)}{n^2} \cdot \left(\dfrac{2x}{x^2+1}\right)^n$ and set $y = \dfrac{2x}{x^2+1}$. We are going to study the power series of y, $\sum_{n=1}^{\infty} \dfrac{\log^2(n)}{n^2} y^n$. The radius of convergence of this series is, by Corollary 1 of Theorem 3.3.1, with $a_n = \dfrac{\log^2(n)}{n^2}$,

$$r = \lim \left|\frac{a_n}{a_{n+1}}\right| = \lim \left|\frac{\dfrac{\log^2(n)}{n^2}}{\dfrac{\log^2(n+1)}{(n+1)^2}}\right| = \lim \frac{(n+1)^2 \log^2(n)}{n^2 \log^2(n+1)} = 1.$$

The series is absolutely convergent if $y \in \,]-1, 1[$ and diverges if $y \in \,]-\infty, -1[\, \cup \,]1, +\infty[$.
About the original series, we have

$$|y| < 1 \Leftrightarrow \left|\frac{2x}{x^2+1}\right| < 1 \Leftrightarrow -1 < \frac{2x}{x^2+1} < 1 \Leftrightarrow x \neq -1 \wedge x \neq 1.$$

Let us study the numerical series at $x = -1$ and $x = 1$.

- If $x = 1$, we obtain the series $\sum_{n=1}^{\infty} \frac{\log^2(n)}{n^2} = \sum_{n=2}^{\infty} \frac{\log^2(n)}{n^2}$, which is of positive terms, and we will study it by comparing it with the convergent series $\sum_{n=1}^{\infty} \frac{1}{n^{3/2}}$. The limit

$$\lim \frac{\frac{\log^2(n)}{n^2}}{\frac{1}{n^{3/2}}} = \lim \frac{\log^2(n)}{\sqrt{n}} = 0$$

allows us to conclude, by Corollary 3 of the General Comparison Test, that the series is convergent. As it is of positive terms, it is absolutely convergent.

- If $x = -1$, we obtain the series $\sum_{n=2}^{\infty} (-1)^n \frac{\log^2(n)}{n^2}$, which is alternating. As we have seen, the series of modules is convergent, and the alternating series is absolutely convergent.

Conclusion: The original power series is absolutely convergent on $\mathbb{R}$.

i) Let us study the power series of $\frac{1}{x}$, $\sum_{n=1}^{\infty} \frac{1}{2^n \sqrt{n+1}} \cdot \frac{1}{x^n}$. It is important to note that the set where the series converges does not include the point $x = 0$ since the functions are not defined at that point. Rather, we can set $y = \frac{1}{x}$, and consider the series $\sum_{n=1}^{\infty} \frac{1}{2^n \sqrt{n+1}} y^n$. By setting $a_n = \frac{1}{2^n \sqrt{n+1}}$, we can apply Corollary 1 of Theorem 3.3.1 to determine the radius of convergence of this series:

$$r = \lim \left| \frac{a_n}{a_{n+1}} \right| = \lim \frac{\frac{1}{2^n \sqrt{n+1}}}{\frac{1}{2^{n+1} \sqrt{n+2}}} = \lim \frac{2\sqrt{n+2}}{\sqrt{n+1}} = 2.$$

The series converges absolutely if $y \in]-2, 2[$ and diverges if $y \in]-\infty, -2[\cup]2, +\infty[$. Since $y = \frac{1}{x}$, we conclude that the series $\sum_{n=1}^{\infty} \frac{1}{2^n \sqrt{n+1}} \cdot \frac{1}{x^n}$ converges absolutely if $\frac{1}{x} \in]-2, 2[$ and diverges if $\frac{1}{x} \in]-\infty, -2[\cup]2, +\infty[$, that is, it is absolutely convergent if $x \in \left]-\infty, -\frac{1}{2}\right[\cup \left]\frac{1}{2}, +\infty\right[$ and divergent if $x \in \left]-\frac{1}{2}, \frac{1}{2}\right[\setminus \{0\}$.

Let us study the numerical series obtained at $x = -\frac{1}{2}$ and $x = \frac{1}{2}$.

- If $x = \frac{1}{2}$, we obtain the divergent series $\sum_{n=1}^{\infty} \frac{1}{\sqrt{n+1}} = \sum_{n=2}^{\infty} \frac{1}{\sqrt{n}}$.

- If $x = -\frac{1}{2}$, we obtain the alternating series $\sum_{n=1}^{\infty} (-1)^n \frac{1}{\sqrt{n+1}}$. We will apply Leibniz's test since the series of modules is divergent:

 (i) $\lim \frac{1}{\sqrt{n+1}} = 0.$
 (ii) $\frac{1}{\sqrt{n+1}} > 0, \forall n \geq 1.$

3.6. Solved Exercises

(iii) $\dfrac{1}{\sqrt{n+1}} - \dfrac{1}{\sqrt{n}} = \dfrac{\sqrt{n}-\sqrt{n+1}}{\sqrt{n}\sqrt{n+1}} < 0, \forall n \in \mathbb{N}$, which implies that $\left(\dfrac{1}{\sqrt{n}}\right)$ is decreasing.

It follows that the series is convergent. As the series of the modules is divergent, the series is conditionally convergent.

Conclusion: The power series is absolutely convergent if $x \in \left]-\infty, -\dfrac{1}{2}\right[\cup \left]\dfrac{1}{2}, +\infty\right[$, conditionally convergent if $x = -\dfrac{1}{2}$, and divergent if $x \in \left]-\dfrac{1}{2}, \dfrac{1}{2}\right] \setminus \{0\}$.

j) Let us study the power series $\displaystyle\sum_{n=1}^{\infty} \left(\dfrac{2}{3} + \dfrac{1}{2n}\right)^n (2x-3)^n$. Let $y = 2x - 3$, and consider the series $\displaystyle\sum_{n=1}^{\infty} \left(\dfrac{2}{3} + \dfrac{1}{2n}\right)^n y^n$. If $a_n = \left(\dfrac{2}{3} + \dfrac{1}{2n}\right)^n$, the radius of convergence of this series is, by Theorem 3.3.1,

$$r = \dfrac{1}{\overline{\lim}\sqrt[n]{|a_n|}} = \dfrac{1}{\overline{\lim}\sqrt[n]{\left(\dfrac{2}{3}+\dfrac{1}{2n}\right)^n}} = \dfrac{1}{\lim\left(\dfrac{2}{3}+\dfrac{1}{2n}\right)} = \dfrac{3}{2}.$$

The series converges absolutely if $y \in \left]-\dfrac{3}{2}, \dfrac{3}{2}\right[$ and diverges if $y \in \left]-\infty, -\dfrac{3}{2}\right[\cup \left]\dfrac{3}{2}, +\infty\right[$. However, $y = 2x - 3$, so the series $\displaystyle\sum_{n=1}^{\infty} \left(\dfrac{2}{3} + \dfrac{1}{2n}\right)^n (2x-3)^n$ is absolutely convergent if $x \in \left]\dfrac{3}{4}, \dfrac{9}{4}\right[$ and divergent if $x \in \left]-\infty, \dfrac{3}{4}\right[\cup \left]\dfrac{9}{4}, +\infty\right[$.

Regarding the endpoints of the interval of convergence, we have to study the respective numerical series.

– If $x = \dfrac{9}{4}$, we obtain the series $\displaystyle\sum_{n=1}^{\infty} \left(\dfrac{2}{3}+\dfrac{1}{2n}\right)^n \left(\dfrac{3}{2}\right)^n = \sum_{n=1}^{\infty}\left(1+\dfrac{3}{4n}\right)^n$. As

$$\lim\left(1+\dfrac{3}{4n}\right)^n = \lim\left(\left(1+\dfrac{3}{4n}\right)^{4n}\right)^{\frac{1}{4}} = e^{\frac{3}{4}},$$

the general term is not a null sequence; therefore, by Theorem 2.2.1, the series diverges.

– If $x = \dfrac{3}{4}$, we obtain the series $\displaystyle\sum_{n=1}^{\infty}\left(\dfrac{2}{3}+\dfrac{1}{2n}\right)^n\left(-\dfrac{3}{2}\right)^n = \sum_{n=1}^{\infty}(-1)^n\left(1+\dfrac{3}{4n}\right)^n$. As the limit of the general term does not exist, the series is divergent.

Conclusion: The power series of x is divergent if $x \in \left]-\infty, \dfrac{3}{4}\right] \cup \left[\dfrac{9}{4}, +\infty\right[$ and absolutely convergent if $x \in \left]\dfrac{3}{4}, \dfrac{9}{4}\right[$.

k) Let us study the power series of x^2, $\displaystyle\sum_{n=2}^{\infty} \dfrac{2^{3n}\left((n-1)!\right)^2}{(2n)!} x^{2n}$. Let $y = x^2$. The radius of convergence of the series $\sum_{n=2}^{\infty} \dfrac{2^{3n} ((n-1)!)^2}{(2n)!} y^n$ is, by Corollary 1 of Theorem 3.3.1, and setting $a_n = \dfrac{2^{3n} ((n-1)!)^2}{(2n)!}$,

$$r = \lim \left|\dfrac{a_n}{a_{n+1}}\right| = \lim \dfrac{\left|\dfrac{2^{3n}((n-1)!)^2}{(2n)!}\right|}{\left|\dfrac{2^{3n+3}(n!)^2}{(2n+2)!}\right|} = \lim \dfrac{(2n+2)(2n+1)}{2^3 n^2} = \dfrac{1}{2}.$$

The series is absolutely convergent if $y \in \left]-\dfrac{1}{2}, \dfrac{1}{2}\right[$ and diverges if $y \in \left]-\infty, -\dfrac{1}{2}\right[\cup \left]\dfrac{1}{2}, +\infty\right[$. Since $y = x^2$, the series $\sum_{n=2}^{\infty} \dfrac{2^{3n}((n-1)!)^2}{(2n)!} x^{2n}$ is absolutely convergent if $x \in \left]-\dfrac{1}{\sqrt{2}}, \dfrac{1}{\sqrt{2}}\right[$ and divergent if $x \in \left]-\infty, -\dfrac{1}{\sqrt{2}}\right[\cup \left]\dfrac{1}{\sqrt{2}}, +\infty\right[$.

At the endpoints of the interval of convergence, we obtain two numerical series.

- If $x = -\dfrac{1}{\sqrt{2}}$, we obtain the series $\sum_{n=1}^{\infty} \dfrac{2^{3n}((n-1)!)^2}{(2n)!} \cdot \dfrac{1}{2^n} = \sum_{n=1}^{\infty} b_n$. Using Raabe's test,

$$\lim n\left(\dfrac{b_n}{b_{n+1}} - 1\right) = \lim n \left(\dfrac{\dfrac{2^{2n}((n-1)!)^2}{(2n)!}}{\dfrac{2^{2n+2}(n!)^2}{(2n+2)!}} - 1\right) = \lim n \left(\dfrac{(2n+2)(2n+1)}{4n^2} - 1\right) = \dfrac{3}{2} > 1,$$

so the series is convergent. As it is of positive terms, it is absolutely convergent.

- If $x = \dfrac{1}{\sqrt{2}}$, we again obtain the series $\sum_{n=1}^{\infty} \dfrac{2^{2n}((n-1)!)^2}{(2n)!}$, which we have seen to be absolutely convergent.

Conclusion: The series is divergent if $x \in \left]-\infty, -\dfrac{1}{\sqrt{2}}\right[\cup \left]\dfrac{1}{\sqrt{2}}, +\infty\right[$ and is absolutely convergent if $x \in \left[-\dfrac{1}{\sqrt{2}}, \dfrac{1}{\sqrt{2}}\right]$.

l) Let us study the power series of $x - 3$, $\sum_{n=0}^{\infty} \left(\dfrac{2n-1}{3n+1}\right)^{2n} (x-3)^n$. Let $y = x - 3$, and consider the series $\sum_{n=0}^{\infty} \left(\dfrac{2n-1}{3n+1}\right)^{2n} y^n$. If $a_n = \left(\dfrac{2n-1}{3n+1}\right)^{2n}$, the radius of convergence of this series is, by Theorem 3.3.1,

$$r = \dfrac{1}{\overline{\lim} \sqrt[n]{|a_n|}} = \dfrac{1}{\overline{\lim} \sqrt[n]{\left(\dfrac{2n-1}{3n+1}\right)^{2n}}} = \dfrac{1}{\lim \left(\dfrac{2n-1}{3n+1}\right)^2} = \dfrac{9}{4}.$$

The series is absolutely convergent if $y \in \left]-\dfrac{9}{4}, \dfrac{9}{4}\right[$ and diverges if $y \in \left]-\infty, -\dfrac{9}{4}\right[\cup \left]\dfrac{9}{4}, +\infty\right[$. Since $y = x - 3$, the series $\sum_{n=0}^{\infty} \left(\dfrac{2n-1}{3n+1}\right)^{2n} (x-3)^n$ is absolutely convergent if $x \in \left]\dfrac{3}{4}, \dfrac{21}{4}\right[$ and divergent if $x \in \left]-\infty, \dfrac{3}{4}\right[\cup \left]\dfrac{21}{4}, +\infty\right[$.

3.6. Solved Exercises

At the endpoints of the interval of convergence, we obtain two numerical series.

- If $x = \frac{3}{4}$, we have the series $\sum_{n=0}^{\infty} \left(\frac{2n-1}{3n+1}\right)^{2n} \left(-\frac{9}{4}\right)^n = \sum_{n=1}^{\infty} (-1)^n \left(\frac{6n-3}{6n+2}\right)^{2n}$. As

$$\lim \left(\frac{6n-3}{6n+2}\right)^{2n} = e^{-\frac{5}{3}},$$

the limit of the general term does not exist, and, by Theorem 2.2.1, the series diverges.

- If $x = \frac{21}{4}$, we obtain the series $\sum_{n=0}^{\infty} \left(\frac{2n-1}{3n+1}\right)^{2n} \left(\frac{9}{4}\right)^n = \sum_{n=1}^{\infty} \left(\frac{6n-3}{6n+2}\right)^{2n}$. The general term is not a null sequence, so the series is divergent.

Conclusion: The power series of $x - 3$ is divergent if $x \in \left]-\infty, \frac{3}{4}\right] \cup \left[\frac{21}{4}, +\infty\right[$ and absolutely convergent if $x \in \left]\frac{3}{4}, \frac{21}{4}\right[$.

m) Let us study the power series of $x-1$, $\sum_{n=1}^{\infty} \frac{\sin\left(\frac{\pi}{n}\right)}{4^n(n+1)}(x-1)^{2n} = \sum_{n=2}^{\infty} \frac{\sin\left(\frac{\pi}{n}\right)}{4^n(n+1)}(x-1)^{2n}$. Let $y = (x-1)^2$, and consider the series $\sum_{n=2}^{\infty} \frac{\sin\left(\frac{\pi}{n}\right)}{4^n(n+1)} y^n$. The radius of convergence of this series is, by Corollary 1 of Theorem 3.3.1 and setting $a_n = \frac{\sin\left(\frac{\pi}{n}\right)}{4^n(n+1)}$,

$$r = \lim \left|\frac{a_n}{a_{n+1}}\right| = \lim \left|\frac{\frac{\sin\left(\frac{\pi}{n}\right)}{4^n(n+1)}}{\frac{\sin\left(\frac{\pi}{n+1}\right)}{4^{n+1}(n+2)}}\right| = \lim \frac{\sin\left(\frac{\pi}{n}\right) 4^{n+1}(n+2)}{\sin\left(\frac{\pi}{n+1}\right) 4^n(n+1)} = 4;$$

the series is absolutely convergent if $y \in]-4, 4[$ and divergent if $y \in]-\infty, -4[\cup]4, +\infty[$.

As $y = (x-1)^2$, the series $\sum_{n=2}^{\infty} \frac{\sin\left(\frac{\pi}{n}\right)}{4^n(n+1)}(x-1)^{2n}$ is absolutely convergent if $x \in]-1, 3[$ and divergent if $x \in]-\infty, -1[\cup]3, +\infty[$.

At the endpoints of the interval of convergence, we obtain two numerical series.

- If $x = -1$, we get the series $\sum_{n=2}^{\infty} \frac{\sin\left(\frac{\pi}{n}\right)}{4^n(n+1)}(-2)^{2n} = \sum_{n=2}^{\infty} \frac{\sin\left(\frac{\pi}{n}\right)}{n+1}$. Let us study it by comparison with $\sum_{n=1}^{\infty} \frac{1}{n^2}$, which is convergent ($p$-series with $p=2$):

$$\lim \frac{\frac{\sin\left(\frac{\pi}{n}\right)}{n+1}}{\frac{1}{n^2}} = \lim \frac{n}{n+1} \cdot \frac{\sin\left(\frac{\pi}{n}\right)}{\frac{1}{n}} = \pi.$$

Since this limit is finite and different from zero, the series is convergent by Corollary 2 of the General Comparison Test. Being of positive terms, it is absolutely convergent.

- If $x = 3$, we again obtain the series $\sum_{n=2}^{\infty} \frac{\sin\left(\frac{\pi}{n}\right)}{4^n (n+1)} 2^{2n} = \sum_{n=2}^{\infty} \frac{\sin\left(\frac{\pi}{n}\right)}{n+1}$, which we have seen to be absolutely convergent.

Conclusion: The power series of $x - 1$ is divergent if $x \in \,]-\infty, -1[\,\cup\,]3, +\infty[$ and absolutely convergent if $x \in [-1, 3]$.

16. The series $\sum_{n=1}^{\infty} \frac{x^{2n}}{\sqrt{n}} \log\left(1 + \frac{x^2}{\sqrt{n}}\right)$ is of positive terms. For it to be convergent, its general term must be a null sequence (see Theorem 2.2.1). But

$$\lim \frac{x^{2n}}{\sqrt{n}} \log\left(1 + \frac{x^2}{\sqrt{n}}\right) = \begin{cases} 0, & \text{if } |x^2| \leq 1 \\ +\infty, & \text{if } |x^2| > 1; \end{cases}$$

therefore, the series can only converge if $|x| \leq 1$.

We have

$$0 \leq \frac{x^{2n}}{\sqrt{n}} \log\left(1 + \frac{x^2}{\sqrt{n}}\right) \leq \frac{x^{2n}}{\sqrt{n}} \left(1 + \frac{x^2}{\sqrt{n}}\right) = \frac{x^{2n}}{\sqrt{n}} + \frac{x^{2n+2}}{n}, \quad \forall x \in \mathbb{R};$$

let us study the power series $\sum_{n=1}^{\infty} \frac{x^{2n}}{\sqrt{n}}$ and $\sum_{n=1}^{\infty} \frac{x^{2n+2}}{n}$.

Let $a_n = \frac{1}{\sqrt{n}}$. The radius of convergence of the first series is, by Corollary 1 of Theorem 3.3.1,

$$\lim \left|\frac{a_n}{a_{n+1}}\right| = \lim \frac{\frac{1}{\sqrt{n}}}{\frac{1}{\sqrt{n+1}}} = \lim \frac{\sqrt{n+1}}{\sqrt{n}} = 1;$$

therefore, the series is absolutely convergent if $x \in \,]-1, 1[$.

Let $b_n = \frac{1}{n}$. The radius of convergence of the second series is, by Corollary 1 of Theorem 3.3.1,

$$\lim \left|\frac{b_n}{b_{n+1}}\right| = \lim \frac{\frac{1}{n}}{\frac{1}{n+1}} = \lim \frac{n+1}{n} = 1;$$

therefore, the series is absolutely convergent if $x \in \,]-1, 1[$.

We can conclude that the series $\sum_{n=1}^{\infty} \frac{x^{2n}}{\sqrt{n}} \left(1 + \frac{x^2}{\sqrt{n}}\right)$ is convergent if $x \in \,]-1, 1[$, which implies, by the General Comparison Test, that the original series is convergent if $x \in \,]-1, 1[$.

For $x^2 = 1$, that is, $x = 1$ or $x = -1$, we obtain the series $\sum_{n=1}^{\infty} \frac{1}{\sqrt{n}} \log\left(1 + \frac{1}{\sqrt{n}}\right)$. Comparing it with the harmonic series, which is divergent,

$$\lim \frac{\frac{1}{\sqrt{n}} \log\left(1 + \frac{1}{\sqrt{n}}\right)}{\frac{1}{n}} = \lim \sqrt{n} \log\left(1 + \frac{1}{\sqrt{n}}\right) = \lim \log\left(1 + \frac{1}{\sqrt{n}}\right)^{\sqrt{n}} = 1,$$

3.6. Solved Exercises

we conclude, by Corollary 2 of the General Comparison Test, that the series is divergent. The original series converges if $x \in \,]-1,1[$ and diverges if $x \in \,]-\infty,-1] \cup [1,\infty[$.

17. Let $y = x^2$. The series $\sum_{n=1}^{\infty} \left(\cos\left(\frac{\pi}{n}\right) + 2 - (-1)^n\right)^n y^n$ setting $a_n = \left(\cos\left(\frac{\pi}{n}\right) + 2 - (-1)^n\right)^n$ has radius of convergence

$$\frac{1}{\overline{\lim} \sqrt[n]{|a_n|}} = \frac{1}{\overline{\lim} \sqrt[n]{\left(\cos\left(\frac{\pi}{n}\right) + 2 - (-1)^n\right)^n}} = \frac{1}{\overline{\lim} \left(\cos\left(\frac{\pi}{n}\right) + 2 - (-1)^n\right)} = \frac{1}{4}.$$

The series is absolutely convergent if $y \in \,\left]-\frac{1}{4}, \frac{1}{4}\right[$ and diverges if $y \in \,\left]-\infty, -\frac{1}{4}\right[\cup \left]\frac{1}{4}, +\infty\right[$.

The series $\sum_{n=1}^{\infty} \left(\cos\left(\frac{\pi}{n}\right) + 2 - (-1)^n\right)^n (x^2)^n$ is absolutely convergent if $x^2 \in \,\left]-\frac{1}{4}, \frac{1}{4}\right[$, that is, $x \in \,\left]-\frac{1}{2}, \frac{1}{2}\right[$, and this is the largest open interval on which the series converges.

18. Consider the power series of x, $\sum_{n=1}^{\infty} \frac{4^n}{2^n + 5^n} x^n$. If the radius of convergence of the series is $\frac{5}{4}$, then, by Theorem 3.3.1, the series is absolutely convergent on the interval $\left]-\frac{5}{4}, \frac{5}{4}\right[$ and diverges on $\left]-\infty, -\frac{5}{4}\right[\cup \left]\frac{5}{4}, +\infty\right[$. But $1 \in \,\left]-\frac{5}{4}, \frac{5}{4}\right[$, therefore, the series is absolutely convergent at this point. The series is divergent at $x = 2$ because $2 \in \,\left]-\infty, -\frac{5}{4}\right[\cup \left]\frac{5}{4}, +\infty\right[$.

At $x = \frac{5}{4}$, the series may be convergent or divergent as it is one of the endpoints of the interval of convergence. Let us study the series $\sum_{n=1}^{\infty} \frac{4^n}{2^n + 5^n} \left(\frac{5}{4}\right)^n = \sum_{n=1}^{\infty} \frac{5^n}{2^n + 5^n}$. Its general term is not a null sequence. In fact,

$$\lim \frac{5^n}{2^n + 5^n} = \lim \frac{1}{\left(\frac{2}{5}\right)^n + 1} = 1;$$

therefore, by Theorem 2.2.1, the series is divergent.

19. The series $\sum_{n=0}^{\infty} a_n x^n$ is a power series of x. We have

$$\sqrt[n]{a_n} = \begin{cases} 1, & \text{if } n \text{ is even} \\ \sqrt[n]{2}, & \text{if } n \text{ is odd}, \end{cases}$$

which implies that $\overline{\lim} \sqrt[n]{a_n} = 1$, because $\lim \sqrt[n]{2} = 1$. Then the radius of convergence of this series is, by Theorem 3.3.1,

$$r = \frac{1}{\overline{\lim} \sqrt[n]{a_n}} = 1;$$

therefore, the series is absolutely convergent if $x \in \,]-1, 1[$ and divergent if $x \in \,]-\infty, -1[\,\cup\,]1, +\infty[$.

If $x = -1$, we obtain the series $\sum_{n=1}^{\infty}(-1)^n a_n$, whose general term does not have a limit; therefore, by Theorem 2.2.1, the series is divergent.

If $x = 1$, we obtain the series $\sum_{n=1}^{\infty} a_n$. As $\lim a_n$ does not exist, the series is divergent.

Conclusion: The power series is absolutely convergent if $x \in\,]-1,1[$ and diverges if $x \in\,]-\infty,-1] \cup [1,+\infty[$.

20. Let us consider the power series $\sum_{n=1}^{\infty} a_n x^n$. If, by hypothesis, the series $\sum_{n=1}^{\infty} a_n 5^n$ converges, then either the power series converges absolutely on $\mathbb{R}$ or there exists $r \in \mathbb{R}^+$ such that the series converges absolutely if $x \in\,]-r, r[$ and diverges if $x \in\,]-\infty, -r[\cup]r, +\infty[$. In the first case, it is evident that the series $\sum_{n=1}^{\infty} a_n(-2)^n$ is absolutely convergent. In the second case, if the series converges at $x = 5$, then $r \geq 5$. As $-2 \in\,]-5,5[\subset\,]-r,r[$, the series $\sum_{n=1}^{\infty} a_n(-2)^n$ is absolutely convergent.

21. Suppose the radius of convergence of the power series of $x - 3$ is $r > 0$. In that case, the series is absolutely convergent on the interval $]3-r, 3+r[$, which is the largest open interval where the series is convergent. The series may also be convergent at the endpoints of the interval. If the series $\sum_{n=0}^{\infty} a_n(x-3)^n$ converges at $x = 0$, then $0 \in [3-r, 3+r]$, which implies that $r \geq 3$. But in that case, $1 \in\,]3-r, 3+r[$, so the series cannot diverge at $x = 1$.

If the radius of convergence of the series is $+\infty$, then the series converges absolutely at all points of $\mathbb{R}$, with no points where it can be divergent.

22. If the power series of x, $\sum_{n=0}^{\infty} b_n x^n$, has a radius of convergence $R > 1$, then it is absolutely convergent if $x = 1$, that is, the series $\sum_{n=0}^{\infty} b_n$ is absolutely convergent. If, for each $n \in \mathbb{N}$, $|f_n(x)| \leq |b_n|$, $\forall x \in \mathbb{R}$, Weierstrass' test allows us to conclude that the series of functions $\sum_{n=0}^{\infty} f_n(x)$ is uniformly convergent on $\mathbb{R}$.

23. a) Let us consider a power series $\sum_{n=1}^{\infty} a_n (x - x_0)^n$ with a radius of convergence $r > 0$. By Theorem 3.3.1, the series is absolutely convergent if $x \in\,]x_0 - r, x_0 + r[$ and divergent if $x \in\,]-\infty, x_0 - r[\cup]x_0 + r, +\infty[$, which implies that it can only be conditionally convergent at the endpoints of the interval of convergence, that is, at $x = x_0 - r$ or $x = x_0 + r$.

If the series $\sum_{n=1}^{\infty} a_n (x-1)^n$ is conditionally convergent at $x = -1$, then $x_0 - r = -1$. Since $x_0 = 1$, we have $r = 2$.

b) From the previous item, we know that the interval of convergence of the series is $]-1, 3[$, so the series can only be conditionally convergent at $x = -1$ or $x = 3$. If $x = 3$, we get the series $\sum_{n=1}^{\infty} a_n 2^n$. As, by hypothesis, (a_n) is a sequence of positive terms, this is a series of positive terms that, if convergent, will be absolutely convergent. Therefore, the only point where the series $\sum_{n=1}^{\infty} a_n (x-1)^n$ is conditionally convergent is $x = -1$.

3.6. Solved Exercises

24. a) Let us study the power series of $x-2$, $\sum_{n=1}^{\infty} \frac{1}{n\,5^n}(x-2)^n$. Let $y = x-2$, and consider the series $\sum_{n=1}^{\infty} \frac{1}{n\,5^n} y^n$.

The radius of convergence of this series is, by Corollary 1 of Theorem 3.3.1 and setting $a_n = \frac{1}{n\,5^n}$,

$$\lim \left| \frac{a_n}{a_{n+1}} \right| = \lim \frac{\left|\frac{1}{n\,5^n}\right|}{\left|\frac{1}{(n+1)\,5^{n+1}}\right|} = \lim \frac{(n+1)\,5^{n+1}}{n\,5^n} = 5;$$

the series is absolutely convergent if $y \in\,]-5, 5[$ and diverges if $y \in\,]-\infty, -5[\,\cup\,]5, +\infty[$. As $y = x-2$, the series $\sum_{n=1}^{\infty} \frac{1}{n\,5^n}(x-2)^n$ is absolutely convergent if $x \in\,]-3, 7[$ and divergent if $x \in\,]-\infty, -3[\,\cup\,]7, +\infty[$.

If $x = 7$, we get the series $\sum_{n=1}^{\infty} \frac{1}{n\,5^n} 5^n = \sum_{n=1}^{\infty} \frac{1}{n}$, which is divergent.

If $x = -3$, we get the series $\sum_{n=1}^{\infty} \frac{1}{n\,5^n}(-5)^n = \sum_{n=1}^{\infty} \frac{(-1)^n}{n}$, which is conditionally convergent.

Conclusion: The power series of $x-2$ is absolutely convergent if $x \in\,]-3, 7[$, conditionally convergent if $x = -3$ and divergent if $x \in\,]-\infty, -3[\,\cup\,[7, +\infty[$.

b) Let $f(x) = \sum_{n=1}^{\infty} \frac{1}{n\,5^n}(x-2)^n$, $\forall x \in\,]-3, 7[$. By Theorem 3.3.3, we have

$$f'(x) = \left(\sum_{n=1}^{\infty} \frac{1}{n\,5^n}(x-2)^n \right)' = \sum_{n=1}^{\infty} \left(\frac{1}{n\,5^n}(x-2)^n \right)' = \sum_{n=1}^{\infty} \frac{1}{n\,5^n} n(x-2)^{n-1}$$

$$= \sum_{n=1}^{\infty} \frac{1}{5^n}(x-2)^{n-1} = \sum_{n=0}^{\infty} \frac{1}{5} \left(\frac{x-2}{5} \right)^n = \frac{1}{5} \sum_{n=0}^{\infty} \left(\frac{x-2}{5} \right)^n, \quad \forall x \in\,]-3, 7[.$$

This series is geometric with $r = \frac{x-2}{5}$. Then

$$\sum_{n=0}^{\infty} \left(\frac{x-2}{5} \right)^n = \frac{1}{1 - \frac{x-2}{5}} = \frac{5}{7-x}$$

and, finally,

$$f'(x) = \frac{1}{5} \cdot \frac{5}{7-x} = \frac{1}{7-x}, \quad \forall x \in\,]-3, 7[.$$

25. Let us consider the series $\sum_{n=0}^{\infty} \frac{(-1)^n}{n+1} \big(\log(x)\big)^{n+1}$.

a) Let $y = \log(x)$, and let us study the power series $\sum_{n=0}^{\infty} \frac{(-1)^n}{n+1} y^{n+1}$. The radius of convergence of this series is, by Corollary 1 of Theorem 3.3.1 and setting $a_n = \frac{(-1)^n}{n+1}$,

$$\lim \left| \frac{a_n}{a_{n+1}} \right| = \lim \frac{\left| \frac{(-1)^n}{n+1} \right|}{\left| \frac{(-1)^{n+1}}{n+2} \right|} = \lim \frac{n+2}{n+1} = 1;$$

so the series is absolutely convergent if $y \in \,]-1, 1[$ and diverges if $y \in \,]-\infty, -1[\, \cup \,]1, +\infty[$. Since $y = \log(x)$, we conclude that the series $\sum_{n=0}^{\infty} \frac{(-1)^n}{n+1} \big(\log(x)\big)^{n+1}$ is absolutely convergent if $\log(x) \in \,]-1, 1[$ and divergent if $\log(x) \in \,]-\infty, -1[\, \cup \,]1, +\infty[$, that is, it is absolutely convergent if $x \in \,]e^{-1}, e[$ and divergent if $x \in \,]0, e^{-1}[\, \cup \,]e, +\infty[$, because the domain of $\log(x)$ is $\mathbb{R}^+$.

- If $x = e^{-1}$, we obtain the series $\sum_{n=0}^{\infty} \frac{(-1)^n}{n+1}(-1)^{n+1} = \sum_{n=0}^{\infty} \frac{(-1)^{2n+1}}{n+1} = -\sum_{n=0}^{\infty} \frac{1}{n+1} = -\sum_{n=1}^{\infty} \frac{1}{n}$, which is divergent.

- If $x = e$, we obtain the series $\sum_{n=0}^{\infty} \frac{(-1)^n}{n+1} = -\sum_{n=0}^{\infty} \frac{(-1)^{n+1}}{n+1} = -\sum_{n=1}^{\infty} \frac{(-1)^n}{n}$, which is the alternating harmonic series, therefore conditionally convergent.

Conclusion: The original series is absolutely convergent if $x \in \,]e^{-1}, e[$, conditionally convergent if $x = e$ and divergent if $x \in \,]0, e^{-1}[\, \cup \,]e, +\infty[$.

b) The function $\sum_{n=0}^{\infty} \frac{(-1)^n}{n+1} \big(\log(x)\big)^{n+1}$ is the sum of a power series with interval of convergence $]e^{-1}, e[$, so it is differentiable term by term on this interval (see Theorem 3.3.3), that is, if $x \in \,]e^{-1}, e[$, we have

$$\left(\sum_{n=0}^{\infty} \frac{(-1)^n}{n+1} \big(\log(x)\big)^{n+1} \right)' = \sum_{n=0}^{\infty} \left(\frac{(-1)^n}{n+1} \big(\log(x)\big)^{n+1} \right)' = \frac{1}{x} \sum_{n=0}^{\infty} (-1)^n \big(\log(x)\big)^n.$$

But the series $\sum_{n=0}^{\infty} (-1)^n \big(\log(x)\big)^n$ is geometric with ratio $r = -\log(x)$; therefore,

$$\frac{1}{x} \sum_{n=0}^{\infty} (-1)^n \big(\log(x)\big)^n = \frac{1}{x} \cdot \frac{1}{1 + \log(x)},$$

so

$$\left(\sum_{n=0}^{\infty} \frac{(-1)^n}{n+1} \big(\log(x)\big)^{n+1} \right)' = \frac{1}{x} \cdot \frac{1}{1 + \log(x)}.$$

Then

$$\sum_{n=0}^{\infty} \frac{(-1)^n}{n+1} \big(\log(x)\big)^{n+1} = \int \frac{1}{x} \cdot \frac{1}{1 + \log(x)} \, dx = \log\big(1 + \log(x)\big) + C.$$

Let us determine the value of the constant C. For $x = 1$ (note that this value belongs to the interval of convergence), we obtain

$$0 = \sum_{n=0}^{\infty} \frac{(-1)^n}{n+1} \big(\log(1)\big)^{n+1} = \log\big(1 + \log(1)\big) + C \Leftrightarrow C = 0;$$

3.6. Solved Exercises

thus,

$$\sum_{n=0}^{\infty} \frac{(-1)^n}{n+1} \big(\log(x)\big)^{n+1} = \log\big(1+\log(x)\big).$$

26. a) The domain of f is the subset of $\mathbb{R}$ where the series is convergent. Let us consider the power series of x^2, $\sum_{n=0}^{\infty} \frac{(-1)^n}{2^{2n}\,(n!)^2}\, x^{2n}$. Let $y = x^2$. The radius of convergence of the series $\sum_{n=0}^{\infty} \frac{(-1)^n}{2^{2n}\,(n!)^2}\, y^n$ is, by Corollary 1 of Theorem 3.3.1 and setting $a_n = \dfrac{(-1)^n}{2^{2n}\,(n!)^2}$,

$$\lim \left|\frac{a_n}{a_{n+1}}\right| = \lim \frac{\left|\dfrac{(-1)^n}{2^{2n}(n!)^2}\right|}{\left|\dfrac{(-1)^{n+1}}{2^{2n+2}\big((n+1)!\big)^2}\right|} = \lim 2^2(n+1)^2 = +\infty.$$

Then, the series converges absolutely for all $y \in \mathbb{R}$. As $y = x^2$, the series $\sum_{n=0}^{\infty} \frac{(-1)^n}{2^{2n}\,(n!)^2}\, x^{2n}$ is absolutely convergent for all $x \in \mathbb{R}$; therefore the domain of f is $\mathbb{R}$.

b) As the power series is differentiable term by term on its interval of convergence (see Theorem 3.3.3), then, for all $x \in \mathbb{R}$,

$$f'(x) = \left(\sum_{n=0}^{\infty} \frac{(-1)^n}{2^{2n}(n!)^2}\, x^{2n}\right)' = \sum_{n=1}^{\infty} \frac{(-1)^n\, 2n}{2^{2n}(n!)^2}\, x^{2n-1}$$

and

$$f''(x) = \left(\sum_{n=1}^{\infty} \frac{(-1)^n\, 2n}{2^{2n}(n!)^2}\, x^{2n-1}\right)' = \sum_{n=1}^{\infty} \frac{(-1)^n\, 2n(2n-1)}{2^{2n}(n!)^2}\, x^{2n-2}.$$

Then

$$x^2\, f(x) = x^2 \sum_{n=0}^{\infty} \frac{(-1)^n}{2^{2n}(n!)^2}\, x^{2n} = \sum_{n=0}^{\infty} \frac{(-1)^n}{2^{2n}(n!)^2}\, x^{2n+2}$$

$$x\, f'(x) = x \sum_{n=1}^{\infty} \frac{(-1)^n\, 2n}{2^{2n}(n!)^2}\, x^{2n-1} = \sum_{n=1}^{\infty} \frac{(-1)^n\, 2n}{2^{2n}(n!)^2}\, x^{2n}$$

and

$$x^2\, f''(x) = x^2 \sum_{n=1}^{\infty} \frac{(-1)^n\, 2n(2n-1)}{2^{2n}(n!)^2}\, x^{2n-2} = \sum_{n=1}^{\infty} \frac{(-1)^n\, 2n(2n-1)}{2^{2n}(n!)^2}\, x^{2n}.$$

Therefore,

$$x^2 f''(x) + x f'(x) + x^2 f(x) =$$

$$= \sum_{n=1}^{\infty} \frac{(-1)^n \, 2n(2n-1)}{2^{2n}(n!)^2} x^{2n} + \sum_{n=1}^{\infty} \frac{(-1)^n \, 2n}{2^{2n}(n!)^2} x^{2n} + \sum_{n=0}^{\infty} \frac{(-1)^n}{2^{2n}(n!)^2} x^{2n+2}$$

$$= \sum_{n=1}^{\infty} \frac{(-1)^n \left(2n(2n-1) + 2n\right)}{2^{2n}(n!)^2} x^{2n} + \sum_{n=0}^{\infty} \frac{(-1)^n}{2^{2n}(n!)^2} x^{2n+2}$$

$$= \sum_{n=1}^{\infty} \frac{(-1)^n \, 4n^2}{2^{2n}(n!)^2} x^{2n} + \sum_{n=1}^{\infty} \frac{(-1)^{n-1}}{2^{2n-2}((n-1)!)^2} x^{2n}$$

$$= \sum_{n=1}^{\infty} \frac{(-1)^n \, 4n^2}{2^{2n}(n!)^2} x^{2n} - \sum_{n=1}^{\infty} \frac{(-1)^n \, 4}{2^{2n}((n-1)!)^2} x^{2n}$$

$$= \sum_{n=1}^{\infty} \frac{(-1)^n \, 4}{2^{2n}((n-1)!)^2} x^{2n} - \sum_{n=1}^{\infty} \frac{(-1)^n \, 4}{2^{2n}((n-1)!)^2} x^{2n} = 0.$$

27. Let (a_n) be a sequence of real numbers such that the series $\sum_{n=1}^{\infty} a_n \, 2^n$ is conditionally convergent.

    a) If the series $\sum_{n=1}^{\infty} a_n \, 2^n$ is conditionally convergent, then the radius of convergence of the series $\sum_{n=1}^{\infty} a_n \, x^n$ is 2, which implies that the power series of x is absolutely convergent if $x \in \,]-2, 2[$ and divergent if $x \in \,]-\infty, -2[\,\cup\,]2, +\infty[$. It follows that the series $\sum_{n=1}^{\infty} a_n$ is absolutely convergent ($x = 1$) and the series $\sum_{n=1}^{\infty} a_n \, 3^n$ is divergent ($x = 3$).

    b) The series $\sum_{n=1}^{\infty} a_n$ is convergent; therefore, by Theorem 2.2.1, $\lim a_n = 0$. Then $\lim \frac{\sin(a_n)}{a_n} = 1$.

28. Let $f(x) = \sum_{n=0}^{\infty} a_n \, x^n$.

    a) We know, by Theorem 3.4.3, that the power series of x is unique and is equal to the Maclaurin series of f, that is, $f(x) = \sum_{n=0}^{\infty} \frac{f^{(n)}(0)}{n!} x^n$, $\forall n \in \mathbb{N}_0$. Therefore, $a_n = \frac{f^{(n)}(0)}{n!}$. We will prove, by induction, that

    $$f^{(n)}(x) = f(x), \quad \forall n \in \mathbb{N}, \quad \forall x \in \mathbb{R}.$$

    By hypothesis, $f'(x) = f(x)$, $\forall x \in \mathbb{R}$, which proves the equality for $n = 1$.
    Assume that $f^{(n)}(x) = f(x)$, $\forall x \in \mathbb{R}$. We prove that $f^{(n+1)}(x) = f(x)$, $\forall x \in \mathbb{R}$.

    $$f^{(n+1)}(x) = \left(f^{(n)}(x)\right)' = \left(f(x)\right)' = f(x), \quad \forall x \in \mathbb{R}.$$

    Then $f^{(n)}(0) = f(0) = f'(0) = 1$ and $a_n = \frac{f^{(n)}(0)}{n!} = \frac{1}{n!}$, $\forall n \in \mathbb{N}$.

3.6. Solved Exercises

b) Taking into account part a), we have $f(x) = \sum_{n=0}^{\infty} \dfrac{x^n}{n!}$, $\forall x \in \mathbb{R}$. However, this series is the Maclaurin series of e^x, that is, $f(x) = e^x$.

29. a) Let $f(x) = \dfrac{5x-1}{x^2-x-2}$. As $x^2 - x - 2 = (x+1)(x-2)$, we have

$$f(x) = \frac{5x-1}{x^2-x-2} = \frac{A}{x-2} + \frac{B}{x+1} = \frac{A(x+1) + B(x-2)}{(x-2)(x+1)} = \frac{(A+B)x + A - 2B}{(x-2)(x+1)}.$$

Then

$$\begin{cases} A + B = 5 \\ A - 2B = -1 \end{cases} \Leftrightarrow \begin{cases} A = 3 \\ B = 2; \end{cases}$$

therefore,

$$f(x) = \frac{3}{x-2} + \frac{2}{x+1} = -\frac{3}{2-x} + \frac{2}{1-(-x)} = -\frac{3}{2} \cdot \frac{1}{1-\frac{x}{2}} + \frac{2}{1-(-x)}.$$

Since $\dfrac{1}{1-(-x)}$ is the sum of a geometric series with ratio $r = -x$ and the first term equal to 1 and $\dfrac{1}{1-\frac{x}{2}}$ is the sum of a geometric series with ratio $r = \dfrac{x}{2}$ and the first term equal to 1, we have

$$\frac{1}{1-(-x)} = \sum_{n=0}^{\infty} (-x)^n = \sum_{n=0}^{\infty} (-1)^n x^n$$

and

$$\frac{1}{1-\frac{x}{2}} = \sum_{n=0}^{\infty} \left(\frac{x}{2}\right)^n.$$

These equalities are valid if $|-x| < 1$ and $\left|\dfrac{x}{2}\right| < 1$, respectively, that is, $|x| < 1$ and $|x| < 2$. We can write the Maclaurin series expansion of f:

$$f(x) = -\frac{3}{2} \sum_{n=0}^{\infty} \left(\frac{x}{2}\right)^n + 2 \sum_{n=0}^{\infty} (-1)^n x^n = \sum_{n=0}^{\infty} \left(2(-1)^n - \frac{3}{2^{n+1}}\right) x^n,$$

which is valid if $|x| < 1$.

b) Let $f(x) = \displaystyle\int_0^x e^{t^2} \, dt$. The function $g(t) = e^{t^2}$ is continuous on $\mathbb{R}$, so the domain of f is $\mathbb{R}$. We know, from Example 3.4.5, that $e^x = \sum_{n=0}^{\infty} \dfrac{x^n}{n!}$, $\forall x \in \mathbb{R}$; therefore,

$$e^{t^2} = \sum_{n=0}^{\infty} \frac{(t^2)^n}{n!} = \sum_{n=0}^{\infty} \frac{t^{2n}}{n!}, \quad \forall t \in \mathbb{R}.$$

As it is a power series, by Corollary 1 of Theorem 3.3.2, it is integrable term by term on every closed bounded interval of $\mathbb{R}$. Therefore,

$$f(x) = \int_0^x e^{t^2} \, dt = \int_0^x \sum_{n=0}^{\infty} \frac{t^{2n}}{n!} \, dt = \sum_{n=0}^{\infty} \int_0^x \frac{t^{2n}}{n!} \, dt = \sum_{n=0}^{\infty} \frac{1}{n!} \int_0^x t^{2n} \, dt$$

$$= \sum_{n=0}^{\infty} \frac{1}{n!} \left[\frac{t^{2n+1}}{2n+1}\right]_0^x = \sum_{n=0}^{\infty} \frac{1}{n!} \cdot \frac{x^{2n+1}}{2n+1} = \sum_{n=0}^{\infty} \frac{x^{2n+1}}{n!(2n+1)},$$

with the expansion valid on $\mathbb{R}$.

c) We know, from Example 3.4.5, that
$$e^x = \sum_{n=0}^{\infty} \frac{x^n}{n!}, \quad \forall x \in \mathbb{R}.$$

Therefore, for all $x \in \mathbb{R}$ and all $t \in \mathbb{R}$,
$$e^{-tx} = \sum_{n=0}^{\infty} \frac{(-tx)^n}{n!} = \sum_{n=0}^{\infty} \frac{(-1)^n t^n x^n}{n!}.$$

Then
$$f(x) = \int_0^1 \frac{1-e^{-tx}}{t}\,dt = \int_0^1 \frac{1 - \sum_{n=0}^{\infty} \frac{(-1)^n t^n x^n}{n!}}{t}\,dt = \int_0^1 \frac{-\sum_{n=1}^{\infty} \frac{(-1)^n t^n x^n}{n!}}{t}\,dt$$
$$= \int_0^1 \sum_{n=1}^{\infty} -\frac{(-1)^n t^{n-1} x^n}{n!}\,dt = \int_0^1 \sum_{n=1}^{\infty} \frac{(-1)^{n+1} x^n}{n!} t^{n-1}\,dt.$$

For each $x \in \mathbb{R} \setminus \{0\}$, the power series of t has an infinite radius of convergence:
$$\lim \frac{\left|\frac{(-1)^n x^n}{n!}\right|}{\left|\frac{(-1)^{n+1} x^{n+1}}{(n+1)!}\right|} = \lim \frac{\frac{|x|^n}{n!}}{\frac{|x|^{n+1}}{(n+1)!}} = \lim \frac{n+1}{|x|} = +\infty,$$

and if $x = 0$, all terms of the series are zero.

The interval of convergence of the series is $\mathbb{R}$; therefore, by Corollary 1 of Theorem 3.3.2, the series is integrable term by term on every closed interval of $\mathbb{R}$. In particular, we have on $[0, 1]$
$$f(x) = \int_0^1 \sum_{n=1}^{\infty} \frac{(-1)^{n+1} x^n}{n!} t^{n-1}\,dt = \sum_{n=1}^{\infty} \int_0^1 \frac{(-1)^{n+1} x^n}{n!} t^{n-1}\,dt$$
$$= \sum_{n=1}^{\infty} \frac{(-1)^{n+1} x^n}{n!} \int_0^1 t^{n-1}\,dt = \sum_{n=1}^{\infty} \frac{(-1)^{n+1} x^n}{n!} \left[\frac{t^n}{n}\right]_0^1 = \sum_{n=1}^{\infty} \frac{(-1)^{n+1}}{n \cdot n!} x^n.$$

This development is valid on $\mathbb{R}$.

d) Let $f(x) = \int_0^x \arctan(t)\,dt$. By the Fundamental Theorem of Integral Calculus, $f'(x) = \arctan(x)$. Considering that $(\arctan(x))' = \frac{1}{1+x^2}$ and $\frac{1}{1+x^2}$ is the sum of a geometric series with ratio $r = -x^2$ and the first term equal to 1, we have
$$\frac{1}{1+x^2} = \sum_{n=0}^{\infty} (-x^2)^n = \sum_{n=0}^{\infty} (-1)^n x^{2n}$$

if, and only if, $|x^2| < 1$, that is, $|x| < 1$. Then
$$\arctan(x) = \int_0^x \frac{1}{1+t^2}\,dt = \int_0^x \sum_{n=0}^{\infty} (-1)^n t^{2n}\,dt = \sum_{n=0}^{\infty} \left(\int_0^x (-1)^n t^{2n}\,dt\right)$$
$$= \sum_{n=0}^{\infty} (-1)^n \left[\frac{t^{2n+1}}{2n+1}\right]_0^x = \sum_{n=0}^{\infty} (-1)^n \frac{x^{2n+1}}{2n+1}$$

3.6. Solved Exercises

because power series are integrable term by term on every interval contained on its interval of convergence (see Corollary 1 of Theorem 3.3.2).

By the same theorem, we have

$$\int_0^x \arctan(t)\, dt = \int_0^x \left(\sum_{n=0}^{\infty}(-1)^n \frac{t^{2n+1}}{2n+1}\right) dt = \sum_{n=0}^{\infty}\left(\int_0^x (-1)^n \frac{t^{2n+1}}{2n+1}\, dt\right)$$

$$= \sum_{n=0}^{\infty} \frac{(-1)^n}{2n+1}\left[\frac{t^{2n+2}}{2n+2}\right]_0^x = \sum_{n=0}^{\infty} \frac{(-1)^n}{2n+1} \cdot \frac{x^{2n+2}}{2n+2}$$

$$= \sum_{n=0}^{\infty} \frac{(-1)^n}{(2n+1)(2n+2)} x^{2n+2}.$$

Therefore,

$$f(x) = \sum_{n=0}^{\infty} \frac{(-1)^n}{(2n+1)(2n+2)} x^{2n+2};$$

this development is valid on $]-1, 1[$.

e) Taking into account that $\left(\dfrac{1}{2x+5}\right)' = -\dfrac{2}{(2x+5)^2}$, we can write

$$f(x) = \frac{1}{(2x+5)^2} = -\frac{1}{2}\left(\frac{1}{2x+5}\right)'.$$

We know that

$$\frac{1}{2x+5} = \frac{\frac{1}{5}}{1+\frac{2}{5}x} = \frac{1}{5}\sum_{n=0}^{\infty}\left(-\frac{2}{5}x\right)^n = \sum_{n=0}^{\infty}(-1)^n \frac{2^n}{5^{n+1}} x^n,$$

and this development is valid if $\left|-\dfrac{2}{5}x\right| < 1$, that is, $x \in \left]-\dfrac{5}{2}, \dfrac{5}{2}\right[$. It is a power series, so it is differentiable term by term on its interval of convergence (see Theorem 3.3.3), that is,

$$\left(\frac{1}{2x+5}\right)' = \left(\sum_{n=0}^{\infty}(-1)^n \frac{2^n}{5^{n+1}} x^n\right)' = \sum_{n=0}^{\infty}\left((-1)^n \frac{2^n}{5^{n+1}} x^n\right)' = \sum_{n=1}^{\infty}(-1)^n \frac{n 2^n}{5^{n+1}} x^{n-1}, \ \forall x \in \left]-\frac{5}{2}, \frac{5}{2}\right[.$$

Then

$$f(x) = \frac{1}{(2x+5)^2} = -\frac{1}{2}\sum_{n=1}^{\infty}(-1)^n \frac{n 2^n}{5^{n+1}} x^{n-1} = \sum_{n=1}^{\infty}(-1)^{n+1} \frac{n 2^{n-1}}{5^{n+1}} x^{n-1}$$

$$= \sum_{n=0}^{\infty}(-1)^n \frac{2^n (n+1)}{5^{n+2}} x^n, \quad \forall x \in \left]-\frac{5}{2}, \frac{5}{2}\right[.$$

f) Let $f(x) = \dfrac{2}{3+2x} + e^{2x}$. We know that

$$\frac{2}{3+2x} = \frac{2}{3} \cdot \frac{1}{1+\frac{2x}{3}} = \frac{2}{3}\sum_{n=0}^{\infty}\left(-\frac{2x}{3}\right)^n = \sum_{n=0}^{\infty}(-1)^n \frac{2^{n+1} x^n}{3^{n+1}},$$

and this development is valid if $\left|-\dfrac{2x}{3}\right| < 1$, that is, $x \in \left]-\dfrac{3}{2}, \dfrac{3}{2}\right[$.

We also know that
$$e^{2x} = \sum_{n=0}^{\infty} \frac{(2x)^n}{n!} = \sum_{n=0}^{\infty} \frac{2^n \, x^n}{n!}, \quad \forall x \in \mathbb{R}.$$

Then
$$f(x) = \frac{2}{3+2x} + e^{2x} = \sum_{n=0}^{\infty} (-1)^n \frac{2^{n+1} \, x^n}{3^{n+1}} + \sum_{n=0}^{\infty} \frac{2^n \, x^n}{n!} = \sum_{n=0}^{\infty} 2^n \left(\frac{(-1)^n \, 2}{3^{n+1}} + \frac{1}{n!} \right) x^n,$$

development valid on $\left] -\frac{3}{2}, \frac{3}{2} \right[$.

g) By Example 3.4.4, we have the Maclaurin series representation of the function $\sin(x)$:
$$\sin(x) = \sum_{n=0}^{\infty} (-1)^n \frac{x^{2n+1}}{(2n+1)!}, \quad \forall x \in \mathbb{R}.$$

Then, by Theorem 3.3.3,
$$\cos(x) = \bigl(\sin(x)\bigr)' = \sum_{n=0}^{\infty} \left((-1)^n \frac{x^{2n+1}}{(2n+1)!} \right)', \quad \forall x \in \mathbb{R},$$

that is,
$$\cos(x) = \sum_{n=0}^{\infty} (-1)^n \frac{(2n+1) \, x^{2n}}{(2n+1)!} = \sum_{n=0}^{\infty} (-1)^n \frac{x^{2n}}{(2n)!}, \quad \forall x \in \mathbb{R}.$$

Therefore, for every $t \in \mathbb{R}$,
$$\cos(t^2) = \sum_{n=0}^{\infty} (-1)^n \frac{(t^2)^{2n}}{(2n)!} = \sum_{n=0}^{\infty} (-1)^n \frac{t^{4n}}{(2n)!},$$

and
$$f(x) = \int_0^x \cos(t^2) \, dt = \int_0^x \sum_{n=0}^{\infty} (-1)^n \frac{t^{4n}}{(2n)!} \, dt.$$

The interval of convergence of the series is $\mathbb{R}$; therefore, by Corollary 1 of Theorem 3.3.2, the series is integrable term by term on every closed interval of $\mathbb{R}$. We have
$$f(x) = \int_0^x \sum_{n=0}^{\infty} (-1)^n \frac{t^{4n}}{(2n)!} \, dt = \sum_{n=0}^{\infty} \left(\frac{(-1)^n}{(2n)!} \int_0^x t^{4n} \, dt \right)$$
$$= \sum_{n=0}^{\infty} \frac{(-1)^n}{(2n)!} \left[\frac{t^{4n+1}}{4n+1} \right]_0^x = \sum_{n=0}^{\infty} \frac{(-1)^n}{(2n)!(4n+1)} x^{4n+1}.$$

Thus, the power series development of x is
$$f(x) = \sum_{n=0}^{\infty} \frac{(-1)^n}{(2n)!(4n+1)} x^{4n+1}, \quad \forall x \in \mathbb{R}.$$

3.6. Solved Exercises

h) Let us consider the following derivative: $\left(\dfrac{1}{1-x}\right)' = \dfrac{1}{(1-x)^2}$. Based on this, we can express $f(x)$ as follows:
$$f(x) = \frac{x+x^2}{(1-x)^2} = (x+x^2)\left(\frac{1}{1-x}\right)'.$$

We know that
$$\frac{1}{1-x} = \sum_{n=0}^{\infty} x^n.$$

Moreover, this development is valid on the interval $]-1,1[$. According to Theorem 3.3.3, this power series is differentiable term by term on its interval of convergence, that is,
$$\left(\frac{1}{1-x}\right)' = \left(\sum_{n=0}^{\infty} x^n\right)' = \sum_{n=0}^{\infty} (x^n)' = \sum_{n=1}^{\infty} n\, x^{n-1}, \quad \forall x \in\,]-1,1[.$$

Then
$$\begin{aligned} f(x) &= (x+x^2)\left(\frac{1}{1-x}\right)' = (x+x^2) \sum_{n=1}^{\infty} n\, x^{n-1} = x \sum_{n=1}^{\infty} n\, x^{n-1} + x^2 \sum_{n=1}^{\infty} n\, x^{n-1} \\ &= \sum_{n=1}^{\infty} n\, x^n + \sum_{n=1}^{\infty} n\, x^{n+1} = \sum_{n=0}^{\infty} (n+1)\, x^{n+1} + \sum_{n=1}^{\infty} n\, x^{n+1} \\ &= x + \sum_{n=1}^{\infty} (2n+1)\, x^{n+1} = \sum_{n=0}^{\infty} (2n+1)\, x^{n+1}, \quad \forall x \in\,]-1,1[. \end{aligned}$$

i) Let $f(x) = x\arctan(x)$. We know that $\bigl(\arctan(x)\bigr)' = \dfrac{1}{1+x^2}$, and this function is the sum of a geometric series with ratio $r=-x^2$ and the first term equal to 1. Therefore,
$$\frac{1}{1+x^2} = \sum_{n=0}^{\infty} (-x^2)^n = \sum_{n=0}^{\infty} (-1)^n x^{2n}$$

if and only if $|-x^2| < 1$, that is, if and only if $|x| < 1$. Then, if $x \in\,]-1,1[$,
$$\begin{aligned} \arctan(x) &= \int_0^x \frac{1}{1+t^2}\, dt = \int_0^x \sum_{n=0}^{\infty} (-1)^n t^{2n}\, dt = \sum_{n=0}^{\infty} \left(\int_0^x (-1)^n t^{2n}\, dt\right) \\ &= \sum_{n=0}^{\infty} (-1)^n \left[\frac{t^{2n+1}}{2n+1}\right]_0^x = \sum_{n=0}^{\infty} (-1)^n \frac{x^{2n+1}}{2n+1} \end{aligned}$$

because power series, by Corollary 1 of Theorem 3.3.2, are integrable term by term on every closed interval contained in their interval of convergence. Therefore,
$$f(x) = x \sum_{n=0}^{\infty} (-1)^n \frac{x^{2n+1}}{2n+1} = \sum_{n=0}^{\infty} (-1)^n \frac{x^{2n+2}}{2n+1},$$

and this development is valid on $]-1,1[$.

j) Taking into account that $\left(\dfrac{1}{1+x^2}\right)' = -\dfrac{2x}{(1+x^2)^2}$, we can write

$$f(x) = \frac{2x}{(1+x^2)^2} = -\left(\frac{1}{1+x^2}\right)'.$$

We know that $\dfrac{1}{1+x^2}$ is the sum of a geometric series with ratio $r = -x^2$ and the first term equal to 1. Thus,

$$\frac{1}{1+x^2} = \sum_{n=0}^{\infty} (-x^2)^n;$$

this development is valid if $|-x^2| < 1$, that is, if $x \in]-1, 1[$. This is a power series that, by Theorem 3.3.3, is differentiable term by term on its interval of convergence, that is,

$$\left(\frac{1}{1+x^2}\right)' = \left(\sum_{n=0}^{\infty}(-1)^n x^{2n}\right)' = \sum_{n=0}^{\infty} \left((-1)^n x^{2n}\right)' = \sum_{n=1}^{\infty} (-1)^n 2n\, x^{2n-1}, \quad \forall x \in]-1, 1[.$$

Then

$$f(x) = \frac{2x}{(1+x^2)^2} = -\sum_{n=1}^{\infty} (-1)^n\, 2n\, x^{2n-1} = \sum_{n=1}^{\infty} (-1)^{n+1}\, 2n\, x^{2n-1}, \quad \forall x \in]-1, 1[.$$

k) Let $f(x) = \log(2 + x^3)$. We have $f'(x) = \left(\log(2+x^3)\right)' = \dfrac{3x^2}{2+x^3}$ and

$$\frac{1}{2+x^3} = \frac{1}{2} \cdot \frac{1}{1-\left(-\dfrac{x^3}{2}\right)} = \frac{1}{2}\sum_{n=0}^{\infty}\left(-\frac{x^3}{2}\right)^n = \sum_{n=0}^{\infty} \frac{(-1)^n}{2^{n+1}} x^{3n};$$

this development is valid if, and only if, $\left|-\dfrac{x^3}{2}\right| < 1$, that is, if and only if $|x| < \sqrt[3]{2}$. Thus,

$$f'(x) = 3x^2 \sum_{n=0}^{\infty} \frac{(-1)^n}{2^{n+1}} x^{3n} = \sum_{n=0}^{\infty} \frac{3(-1)^n}{2^{n+1}} x^{3n+2},$$

if and only if $x \in]-\sqrt[3]{2}, \sqrt[3]{2}[$.

Since $\displaystyle\int_0^x f'(t)\, dt = f(x) - f(0) = f(x) - \log(2)$, we have

$$f(x) = \int_0^x f'(t)\, dt + \log(2) = \log(2) + \int_0^x \sum_{n=0}^{\infty} \frac{3(-1)^n}{2^{n+1}} t^{3n+2}\, dt$$

$$= \log(2) + \sum_{n=0}^{\infty}\left(\frac{3(-1)^n}{2^{n+1}} \int_0^x t^{3n+2}\, dt\right)$$

because, by Corollary 1 of Theorem 3.3.2, a power series is integrable term by term on its interval of convergence. Then

$$f(x) = \log(2) + \sum_{n=0}^{\infty} \frac{3(-1)^n}{2^{n+1}} \left[\frac{t^{3n+3}}{3n+3}\right]_0^x = \log(2) + \sum_{n=0}^{\infty} \frac{3(-1)^n}{2^{n+1}} \frac{x^{3n+3}}{3n+3}$$

$$= \log(2) + \sum_{n=0}^{\infty} \frac{(-1)^n}{2^{n+1}(n+1)} x^{3n+3},$$

and the development is valid on $]-\sqrt[3]{2}, \sqrt[3]{2}[$.

3.6. Solved Exercises

l) Let $f(x) = \log\left(\sqrt{\dfrac{1+x}{1-x}}\right)$, and calculate its derivative:

$$f'(x) = \left(\log\left(\sqrt{\dfrac{1+x}{1-x}}\right)\right)' = \dfrac{\left(\sqrt{\dfrac{1+x}{1-x}}\right)'}{\sqrt{\dfrac{1+x}{1-x}}} = \dfrac{\dfrac{\left(\dfrac{1+x}{1-x}\right)'}{2\sqrt{\dfrac{1+x}{1-x}}}}{\sqrt{\dfrac{1+x}{1-x}}} = \dfrac{\dfrac{2}{(1-x)^2}}{2\dfrac{1+x}{1-x}} = \dfrac{1}{1-x^2}.$$

Knowing that $\dfrac{1}{1-x^2}$ is the sum of a geometric series with ratio $r = x^2$ and the first term equal to 1, we have

$$\dfrac{1}{1-x^2} = \sum_{n=0}^{\infty} x^{2n}, \quad |x^2| < 1,$$

and we can write the power series expansion of x of f':

$$f'(x) = \sum_{n=0}^{\infty} x^{2n};$$

the development is valid on $]-1,1[$. As, by Corollary 1 of Theorem 3.3.2, a power series is term by term integrable on its interval of convergence, we have

$$f(x) - f(0) = f(x) = \int_0^x \sum_{n=0}^{\infty} t^{2n}\, dt = \sum_{n=0}^{\infty} \int_0^x t^{2n}\, dt = \sum_{n=0}^{\infty} \left[\dfrac{t^{2n+1}}{2n+1}\right]_0^x = \sum_{n=0}^{\infty} \dfrac{x^{2n+1}}{2n+1}.$$

The development of f in a power series of x is

$$f(x) = \sum_{n=0}^{\infty} \dfrac{x^{2n+1}}{2n+1},$$

which is valid on the interval $]-1,1[$.

m) Considering that $\left(\log(4-x^2)\right)' = -\dfrac{2x}{4-x^2} = -\dfrac{x}{2}\cdot\dfrac{1}{1-(\frac{x}{2})^2}$ and that the function $g(x) = \dfrac{1}{1-(\frac{x}{2})^2}$ is the sum of a geometric series with ratio $r = \left(\dfrac{x}{2}\right)^2$ and the first term equal to 1, we can write

$$f'(x) = -\dfrac{x}{2}\cdot\sum_{n=0}^{\infty}\left(\dfrac{x}{2}\right)^{2n} = -\sum_{n=0}^{\infty}\dfrac{x^{2n+1}}{2^{2n+1}};$$

this development is valid if $\left|\left(\dfrac{x}{2}\right)^2\right| < 1$, that is, $x \in\,]-2,2[$. This is a power series that, by Corollary 1 of Theorem 3.3.2, is term by term integrable on its interval of convergence, that is,

$$f(x) - \log(4) = -\int_0^x \sum_{n=0}^{\infty}\dfrac{t^{2n+1}}{2^{2n+1}}\, dt = -\sum_{n=0}^{\infty}\int_0^x \dfrac{t^{2n+1}}{2^{2n+1}}\, dt = -\sum_{n=0}^{\infty}\dfrac{1}{2^{2n+1}(2n+2)}x^{2n+2}, \quad \forall x \in\,]-2,2[.$$

Then

$$f(x) = \log(4) - \sum_{n=0}^{\infty}\dfrac{1}{2^{2n+1}(2n+2)}x^{2n+2};$$

this development is valid on $]-2,2[$.

30. a) Since $\left(\dfrac{1}{1-x}\right)' = \dfrac{1}{(1-x)^2}$, we can write

$$f(x) = \dfrac{x}{(1-x)^2} = x\left(\dfrac{1}{1-x}\right)'.$$

We know that $\dfrac{1}{1-x}$ is the sum of a geometric series with ratio $r = x$ and the first term equal to 1. Therefore,

$$\dfrac{1}{1-x} = \sum_{n=0}^{\infty} x^n;$$

this development is valid on $]-1, 1[$. It is a power series that, by Theorem 3.3.3, is term by term differentiable on its interval of convergence, that is,

$$\left(\dfrac{1}{1-x}\right)' = \left(\sum_{n=0}^{\infty} x^n\right)' = \sum_{n=0}^{\infty} (x^n)' = \sum_{n=1}^{\infty} n\,x^{n-1}, \quad \forall x \in\,]-1, 1[.$$

Then

$$f(x) = \dfrac{x}{(1-x)^2} = x\sum_{n=1}^{\infty} n\,x^{n-1} = \sum_{n=1}^{\infty} n\,x^n, \quad \forall x \in\,]-1, 1[.$$

b) Rewriting the general term of the numerical series, we get

$$\sum_{n=0}^{\infty} \dfrac{n}{3^n} = \sum_{n=1}^{\infty} n\left(\dfrac{1}{3}\right)^n.$$

By item a), with $x = \dfrac{1}{3}$,

$$\sum_{n=1}^{\infty} n\left(\dfrac{1}{3}\right)^n = f\left(\dfrac{1}{3}\right) = \dfrac{\dfrac{1}{3}}{\left(1-\dfrac{1}{3}\right)^2} = \dfrac{3}{4}.$$

31. Let $f(x) = \begin{cases} \dfrac{\sin(x)}{x}, & \text{if } x \neq 0 \\ 1, & \text{if } x = 0. \end{cases}$

a) We know that

$$\sin(x) = \sum_{n=0}^{\infty} (-1)^n \dfrac{x^{2n+1}}{(2n+1)!}, \quad \forall x \in \mathbb{R};$$

therefore,

$$\dfrac{\sin(x)}{x} = \sum_{n=0}^{\infty} (-1)^n \dfrac{x^{2n}}{(2n+1)!}, \quad \forall x \in \mathbb{R} \setminus \{0\}.$$

Observing that for $x = 0$, the previous series is convergent and has sum 1, we conclude that

$$f(x) = \sum_{n=0}^{\infty} (-1)^n \dfrac{x^{2n}}{(2n+1)!}, \quad \forall x \in \mathbb{R}.$$

3.6. Solved Exercises

b) Given the uniqueness of the power series development (see Theorem 3.4.3), the series obtained in item a) is the Maclaurin series of f, so, for each $n \in \mathbb{N}$, the coefficient a_n of x^n in the previous series is $\dfrac{f^{(n)}(0)}{n!}$, that is, if

$$a_n = \begin{cases} \dfrac{(-1)^k}{(2k+1)!}, & \text{if } n = 2k,\ k \in \mathbb{N}_0 \\ 0, & \text{if } n = 2k+1,\ k \in \mathbb{N}_0, \end{cases}$$

then

$$f^{(n)}(0) = n!\, a_n = \begin{cases} \dfrac{(-1)^k}{2k+1}, & \text{if } n = 2k,\ k \in \mathbb{N} \\ 0, & \text{if } n = 2k+1,\ k \in \mathbb{N}_0. \end{cases}$$

32. a) Let $f(x) = e^{x^2} + \cos(2x)$. Taking into account that

$$\cos(x) = \sum_{n=0}^{\infty} (-1)^n \frac{x^{2n}}{(2n)!}, \quad \forall x \in \mathbb{R},$$

we can write

$$\cos(2x) = \sum_{n=0}^{\infty} (-1)^n \frac{(2x)^{2n}}{(2n)!}, \quad \forall x \in \mathbb{R}.$$

Knowing that

$$e^x = \sum_{n=0}^{\infty} \frac{x^n}{n!}, \quad \forall x \in \mathbb{R},$$

we have

$$e^{x^2} = \sum_{n=0}^{\infty} \frac{(x^2)^n}{n!}, \quad \forall x \in \mathbb{R}.$$

Therefore, the power series development of f is

$$f(x) = \sum_{n=0}^{\infty} \frac{(x^2)^n}{n!} + \sum_{n=0}^{\infty} (-1)^n \frac{(2x)^{2n}}{(2n)!} = \sum_{n=0}^{\infty} \left(\frac{1}{n!} + (-1)^n \frac{2^{2n}}{(2n)!} \right) x^{2n},$$

which is valid on $\mathbb{R}$.

b) By Theorem 3.4.3, the power series of x of a function f is unique. Therefore, the series we obtained in the previous item is the Maclaurin series of f, that is,

$$\sum_{n=0}^{\infty} \frac{f^{(n)}(0)}{n!} x^n = \sum_{n=0}^{\infty} \left(\frac{1}{n!} + (-1)^n \frac{2^{2n}}{(2n)!} \right) x^{2n}.$$

Considering that in this series, all terms of odd order are zero, all derivatives of odd order of f at $x = 0$ are zero. In particular, $f^{(17)}(0) = 0$.

33. a) We know that $\sin(x) = \sum_{n=0}^{\infty} (-1)^n \dfrac{x^{2n+1}}{(2n+1)!}$, $\forall x \in \mathbb{R}$. If $x = 1$, we obtain $\sum_{n=0}^{\infty} (-1)^n \dfrac{1}{(2n+1)!} = \sin(1)$.

b) Upon substituting $x = 2$ in the equality $e^x = \sum_{n=0}^{\infty} \dfrac{x^n}{n!}$, which holds true on $\mathbb{R}$, we obtain $e^2 = \sum_{n=0}^{\infty} \dfrac{2^n}{n!}$.

c) We have the following equality, valid on $]-1,1[$: $\dfrac{1}{1-x} = \sum_{n=0}^{\infty} x^n$. By Corollary 1 of Theorem 3.3.2, we can integrate this series term by term obtaining

$$-\log(1-x) + \log(1) = \sum_{n=0}^{\infty} \frac{x^{n+1}}{n+1} = \sum_{n=1}^{\infty} \frac{x^n}{n}.$$

If $x = -\dfrac{1}{2}$, we arrive at the following result:

$$-\log\left(\frac{3}{2}\right) = \sum_{n=1}^{\infty} \frac{(-\frac{1}{2})^n}{n} = \sum_{n=1}^{\infty} (-1)^n \frac{1}{2^n\, n}.$$

34. a) Let $f(x) = \log(x^2+1) + \cos(2x)$, and calculate its derivative:

$$f'(x) = \frac{2x}{x^2+1} - 2\sin(2x).$$

Taking into account that

$$\sin(x) = \sum_{n=0}^{\infty} (-1)^n \frac{x^{2n+1}}{(2n+1)!}, \quad \forall x \in \mathbb{R},$$

we can write

$$\sin(2x) = \sum_{n=0}^{\infty} (-1)^n \frac{(2x)^{2n+1}}{(2n+1)!}, \quad \forall x \in \mathbb{R}.$$

Knowing that $\dfrac{1}{1+x^2}$ is the sum of a geometric series with ratio $r = -x^2$ and the first term equal to 1, we have

$$\frac{1}{1+x^2} = \sum_{n=0}^{\infty} (-1)^n x^{2n}, \quad |-x^2| < 1.$$

Then, the power series expansion of x of f' is

$$\begin{aligned} f'(x) &= 2x \sum_{n=0}^{\infty} (-1)^n x^{2n} - 2 \sum_{n=0}^{\infty} (-1)^n \frac{(2x)^{2n+1}}{(2n+1)!} \\ &= \sum_{n=0}^{\infty} (-1)^n 2\, x^{2n+1} - \sum_{n=0}^{\infty} (-1)^n \frac{2^{2n+2}\, x^{2n+1}}{(2n+1)!} \\ &= \sum_{n=0}^{\infty} (-1)^n \left(2 - \frac{2^{2n+2}}{(2n+1)!}\right) x^{2n+1}, \end{aligned}$$

which is valid on $]-1,1[$.

3.6. Solved Exercises

By Corollary 1 of Theorem 3.3.2, we have

$$\begin{aligned}
f(x) - f(0) = f(x) - 1 &= \int_0^x \sum_{n=0}^{\infty} (-1)^n \left(2 - \frac{2^{2n+2}}{(2n+1)!}\right) t^{2n+1}\, dt \\
&= \sum_{n=0}^{\infty} (-1)^n \left(2 - \frac{2^{2n+2}}{(2n+1)!}\right) \int_0^x t^{2n+1}\, dt \\
&= \sum_{n=0}^{\infty} (-1)^n \left(\frac{2}{2n+2} - \frac{2^{2n+2}}{(2n+2)!}\right) x^{2n+2} \\
&= \sum_{n=0}^{\infty} (-1)^n \left(\frac{1}{n+1} - \frac{2^{2n+2}}{(2n+2)!}\right) x^{2n+2}.
\end{aligned}$$

The power series expansion of f in terms of x is

$$1 + \sum_{n=0}^{\infty} (-1)^n \left(\frac{1}{n+1} - \frac{2^{2n+2}}{(2n+2)!}\right) x^{2n+2} = 1 + \sum_{n=1}^{\infty} (-1)^{n-1} \left(\frac{1}{n} - \frac{2^{2n}}{(2n)!}\right) x^{2n},$$

which is valid on the interval $\,]-1, 1[$.

b) Let $x = \dfrac{\pi}{4}$. Substituting this value into the expansion obtained in item a), we get the series

$$1 + \sum_{n=1}^{\infty} (-1)^{n-1} \frac{\pi^{2n}}{4^{2n}} \left(\frac{1}{n} - \frac{2^{2n}}{(2n)!}\right).$$

This series has sum $f\left(\dfrac{\pi}{4}\right) = \log\left(\left(\dfrac{\pi}{4}\right)^2 + 1\right) + \cos\left(2\dfrac{\pi}{4}\right) = \log\left(\left(\dfrac{\pi}{4}\right)^2 + 1\right).$

35. a) We can write

$$f(x) = \sin(2x) + \frac{2x}{(1-x^2)^2} = \sin(2x) + \left(\frac{1}{1-x^2}\right)'.$$

Considering that

$$\sin(x) = \sum_{n=0}^{\infty} (-1)^n \frac{x^{2n+1}}{(2n+1)!}, \quad \forall x \in \mathbb{R},$$

we get

$$\sin(2x) = \sum_{n=0}^{\infty} (-1)^n \frac{(2x)^{2n+1}}{(2n+1)!}, \quad \forall x \in \mathbb{R}.$$

Since $\dfrac{1}{1-x^2}$ is the sum of a geometric series with ratio $r = x^2$ and the first term equal to 1, we have

$$\frac{1}{1-x^2} = \sum_{n=0}^{\infty} x^{2n}, \quad |x^2| < 1.$$

As, by Theorem 3.3.3, power series are differentiable term by term on their interval of convergence,

$$f(x) = \sum_{n=0}^{\infty} (-1)^n \frac{(2x)^{2n+1}}{(2n+1)!} + \left(\sum_{n=0}^{\infty} x^{2n}\right)'$$

$$= \sum_{n=0}^{\infty} (-1)^n \frac{(2x)^{2n+1}}{(2n+1)!} + \sum_{n=1}^{\infty} 2nx^{2n-1}$$

$$= \sum_{n=0}^{\infty} (-1)^n \frac{(2x)^{2n+1}}{(2n+1)!} + \sum_{n=0}^{\infty} 2(n+1)x^{2n+1}$$

$$= \sum_{n=0}^{\infty} \left((-1)^n \frac{2^{2n+1}}{(2n+1)!} + 2(n+1)\right) x^{2n+1},$$

development valid on $]-1,1[$.

b) Let $x = \dfrac{1}{2}$. Substituting this value into the development obtained in item a), we get

$$\sum_{n=0}^{\infty} \left((-1)^n \frac{2^{2n+1}}{(2n+1)!} + 2(n+1)\right) \left(\frac{1}{2}\right)^{2n+1} = \sum_{n=0}^{\infty} \left(\frac{(-1)^n}{(2n+1)!} + \frac{n+1}{2^{2n}}\right);$$

therefore, the sum of this series is $f\left(\dfrac{1}{2}\right) = \sin(1) + \dfrac{1}{\left(1-\dfrac{1}{4}\right)^2} = \sin(1) + \dfrac{16}{9}$.

36. Let $f(x) = \dfrac{1}{4-x} + \log(4-x)$. Taking into account that $\log(4-x) = -\displaystyle\int_3^x \dfrac{1}{4-t}\,dt$, we can write

$$f(x) = \frac{1}{4-x} - \int_3^x \frac{1}{4-t}\,dt.$$

We know that

$$\frac{1}{4-x} = \frac{1}{1-(x-3)} = \sum_{n=0}^{\infty} (x-3)^n;$$

this development is valid if $|x-3| < 1$, that is, $x \in \,]2,4[$. It is a power series so, by Corollary 1 of Theorem 3.3.2, it is integrable term by term on its interval of convergence, that is,

$$\log(4-x) = -\int_3^x \frac{1}{4-t}\,dt = -\int_3^x \left(\sum_{n=0}^{\infty}(t-3)^n\right)dt = -\sum_{n=0}^{\infty} \int_3^x (t-3)^n\,dt = -\sum_{n=0}^{\infty} \frac{(x-3)^{n+1}}{n+1}.$$

Then

$$f(x) = \frac{1}{4-x} + \log(4-x) = \sum_{n=0}^{\infty} (x-3)^n - \sum_{n=0}^{\infty} \frac{(x-3)^{n+1}}{n+1}$$

$$= 1 + \sum_{n=1}^{\infty} (x-3)^n - \sum_{n=1}^{\infty} \frac{(x-3)^n}{n} = 1 + \sum_{n=1}^{\infty} \frac{n-1}{n}(x-3)^n, \quad \forall x \in \,]2,4[.$$

37. Let $f(x) = \dfrac{1}{x^2}$. Considering that $\dfrac{1}{x^2} = -\left(\dfrac{1}{x}\right)'$, we can write

$$f(x) = -\left(\frac{1}{x}\right)' = -\left(\frac{1}{3+x-3}\right)' = -\frac{1}{3}\left(\frac{1}{1+\frac{x-3}{3}}\right)'.$$

3.6. Solved Exercises

Since
$$\frac{1}{1+\frac{x-3}{3}} = \sum_{n=0}^{\infty} \left(-\frac{x-3}{3}\right)^n = \sum_{n=0}^{\infty} (-1)^n \frac{(x-3)^n}{3^n},$$

we also have
$$f(x) = -\frac{1}{3}\left(\sum_{n=0}^{\infty}(-1)^n \frac{(x-3)^n}{3^n}\right)' = \left(\sum_{n=0}^{\infty}(-1)^{n+1}\frac{(x-3)^n}{3^{n+1}}\right)';$$

this development is valid if $|x-3| < 3$, that is, $x \in \,]0,6[$. It is a power series that, by Theorem 3.3.3, is differentiable term by term on its interval of convergence, that is,

$$f(x) = \sum_{n=0}^{\infty}\left((-1)^{n+1}\frac{(x-3)^n}{3^{n+1}}\right)' = \sum_{n=1}^{\infty}(-1)^{n+1}\frac{n(x-3)^{n-1}}{3^{n+1}}.$$

The development is valid on the interval $]0,6[$.

38. Let $f(x) = \dfrac{1}{x^2+x-6}$. As $x^2 + x - 6 = (x-2)(x+3)$, we get

$$f(x) = \frac{1}{x^2+x-6} = \frac{A}{x-2} + \frac{B}{x+3} = \frac{A(x+3)+B(x-2)}{(x-2)(x+3)} = \frac{(A+B)x+3A-2B}{(x-2)(x+3)},$$

so
$$\begin{cases} A+B=0 \\ 3A-2B=1 \end{cases} \Leftrightarrow \begin{cases} A=\dfrac{1}{5} \\ B=-\dfrac{1}{5}; \end{cases}$$

therefore,
$$f(x) = \frac{1}{5}\cdot\frac{1}{x-2} - \frac{1}{5}\cdot\frac{1}{x+3} = -\frac{1}{5}\cdot\frac{1}{2-x} - \frac{1}{5}\cdot\frac{1}{3+x}$$
$$= -\frac{1}{5}\cdot\frac{1}{3-(x+1)} - \frac{1}{5}\cdot\frac{1}{2+(x+1)}$$
$$= -\frac{1}{15}\cdot\frac{1}{1-\dfrac{x+1}{3}} - \frac{1}{10}\cdot\frac{1}{1-\left(-\dfrac{x+1}{2}\right)}.$$

Knowing that $\dfrac{1}{1-\dfrac{x+1}{3}}$ is the sum of a geometric series with ratio $r = \dfrac{x+1}{3}$ and the first term equal to 1 and $\dfrac{1}{1-\left(-\dfrac{x+1}{2}\right)}$ is the sum of a geometric series with ratio $r = -\dfrac{x+1}{2}$ and the first term equal to 1, we have

$$\frac{1}{1-\dfrac{x+1}{3}} = \sum_{n=0}^{\infty}\left(\frac{x+1}{3}\right)^n$$

and
$$\frac{1}{1-\left(-\dfrac{x+1}{2}\right)} = \sum_{n=0}^{\infty}\left(-\frac{x+1}{2}\right)^n = \sum_{n=0}^{\infty}(-1)^n\left(\frac{x+1}{2}\right)^n,$$

equalities valid if $\left|\dfrac{x+1}{3}\right| < 1$ and $\left|-\dfrac{x+1}{2}\right| < 1$, respectively. If $x \in\,]-3,1[$, we can write the development of f in powers of $x+1$:

$$f(x) = -\dfrac{1}{15}\sum_{n=0}^{\infty}\left(\dfrac{x+1}{3}\right)^n - \dfrac{1}{10}\sum_{n=0}^{\infty}(-1)^n\left(\dfrac{x+1}{2}\right)^n = \sum_{n=0}^{\infty}\dfrac{1}{5}\left(\dfrac{(-1)^{n+1}}{2^{n+1}} - \dfrac{1}{3^{n+1}}\right)(x+1)^n.$$

39. Let $g(t) = \log(t)$. Since $\bigl(\log(t)\bigr)' = \dfrac{1}{t}$ and $\dfrac{1}{t} = \dfrac{1}{1+(t-1)}$ is the sum of a geometric series with ratio $r = -(t-1)$ and the first term equal to 1, we have

$$\dfrac{1}{1+(t-1)} = \sum_{n=0}^{\infty}\bigl(-(t-1)\bigr)^n = \sum_{n=0}^{\infty}(-1)^n(t-1)^n$$

if, and only if, $|t-1| < 1$, that is, $t \in\,]0,2[$. Then

$$\log(t) - \log(1) = \log(t) = \int_1^t \sum_{n=0}^{\infty}(-1)^n(x-1)^n\,dx = \sum_{n=0}^{\infty}\left(\int_1^t (-1)^n(x-1)^n\,dx\right)$$

$$= \sum_{n=0}^{\infty}(-1)^n\left[\dfrac{(x-1)^{n+1}}{n+1}\right]_1^t = \sum_{n=0}^{\infty}(-1)^n\dfrac{(t-1)^{n+1}}{n+1},$$

because, by Corollary 1 of Theorem 3.3.2, power series are integrable term by term on every interval contained on their interval of convergence. Again, by Corollary 1 of Theorem 3.3.2, we have

$$f(x) - f(1) = f(x) = \int_1^x \dfrac{\sum_{n=0}^{\infty}(-1)^n \dfrac{(t-1)^{n+1}}{n+1} - (t-1)}{t-1}\,dt = \int_1^x \dfrac{\sum_{n=1}^{\infty}(-1)^n \dfrac{(t-1)^{n+1}}{n+1}}{t-1}\,dt$$

$$= \int_1^x \sum_{n=1}^{\infty}(-1)^n\dfrac{(t-1)^n}{n+1}\,dt = \sum_{n=1}^{\infty}\left(\int_1^x (-1)^n\dfrac{(t-1)^n}{n+1}\,dt\right)$$

$$= \sum_{n=1}^{\infty}\dfrac{(-1)^n}{n+1}\left[\dfrac{(t-1)^{n+1}}{n+1}\right]_1^x = \sum_{n=1}^{\infty}\dfrac{(-1)^n}{n+1}\cdot\dfrac{(x-1)^{n+1}}{n+1}$$

$$= \sum_{n=1}^{\infty}\dfrac{(-1)^n}{(n+1)^2}(x-1)^{n+1}.$$

Therefore,

$$f(x) = \sum_{n=1}^{\infty}\dfrac{(-1)^n}{(n+1)^2}(x-1)^{n+1}.$$

Moreover, this development is valid on $]0,2[$.

40. Let $g(t) = \arctan(t-2)$. Taking into account that $\bigl(\arctan(t-2)\bigr)' = \dfrac{1}{1+(t-2)^2}$ and $\dfrac{1}{1+(t-2)^2}$ is the sum of a geometric series with ratio $r = -(t-2)^2$ and the first term equal to 1, we have

$$\dfrac{1}{1+(t-2)^2} = \sum_{n=0}^{\infty}\bigl(-(t-2)^2\bigr)^n = \sum_{n=0}^{\infty}(-1)^n(t-2)^{2n}$$

3.6. Solved Exercises

if, and only if, $|(t-2)^2| < 1$, that is, $t \in \,]1, 3[$. Then, by Corollary 1 of Theorem 3.3.2,

$$\arctan(t-2) - \arctan(0) = \arctan(t-2) = \int_2^t \sum_{n=0}^{\infty} (-1)^n (x-2)^{2n}\, dx = \sum_{n=0}^{\infty} \left(\int_2^t (-1)^n (x-2)^{2n}\, dx \right)$$
$$= \sum_{n=0}^{\infty} (-1)^n \left[\frac{(x-2)^{2n+1}}{2n+1} \right]_2^t = \sum_{n=0}^{\infty} (-1)^n \frac{(t-2)^{2n+1}}{2n+1}.$$

By the corollary mentioned above, we have

$$f(x) = \int_2^x \frac{\sum_{n=0}^{\infty} (-1)^n \frac{(t-2)^{2n+1}}{2n+1} - (t-2)}{(t-2)^2}\, dt = \int_2^x \frac{\sum_{n=1}^{\infty} (-1)^n \frac{(t-2)^{2n+1}}{2n+1}}{(t-2)^2}\, dt$$
$$= \int_2^x \sum_{n=1}^{\infty} (-1)^n \frac{(t-2)^{2n-1}}{2n+1}\, dt = \sum_{n=1}^{\infty} \left(\int_2^x (-1)^n \frac{(t-2)^{2n-1}}{2n+1}\, dt \right)$$
$$= \sum_{n=1}^{\infty} \frac{(-1)^n}{2n+1} \left[\frac{(t-2)^{2n}}{2n} \right]_2^x = \sum_{n=1}^{\infty} \frac{(-1)^n}{2n+1} \cdot \frac{(x-2)^{2n}}{2n}$$
$$= \sum_{n=1}^{\infty} \frac{(-1)^n}{2n(2n+1)} (x-2)^{2n}.$$

Therefore,

$$f(x) = \sum_{n=1}^{\infty} \frac{(-1)^n}{2n(2n+1)} (x-2)^{2n}.$$

Moreover, this development is valid on $]1, 3[$.

41. We know that the Maclaurin series is $\sum_{n=0}^{\infty} \frac{f^{(n)}(0)}{n!} x^n$. Therefore, we need to calculate $f^{(n)}(0)$, $\forall n \in \mathbb{N}$. We will prove, by induction, that $f^{(n)}(x) = f''(x)$, $\forall n > 2$, $\forall x \in \mathbb{R}$.

By hypothesis, $f'(x) = f(x) + x$, $\forall x \in \mathbb{R}$, which implies that $f''(x) = f'(x) + 1$ and $f'''(x) = f''(x)$, $\forall x \in \mathbb{R}$ which proves the equality for $n = 3$.

Suppose that $f^{(n)}(x) = f''(x)$, $\forall x \in \mathbb{R}$. We prove that $f^{(n+1)}(x) = f''(x)$, $\forall x \in \mathbb{R}$.

$$f^{(n+1)}(x) = \left(f^{(n)}(x) \right)' = (f''(x))' = f'''(x) = f''(x).$$

Then, $f'(0) = f(0) + 0 = 1$ and $f^{(n)}(0) = f''(0) = f'(0) + 1 = 2$, so the Maclaurin series of f is

$$\sum_{n=0}^{\infty} \frac{f^{(n)}(0)}{n!} x^n = 1 + x + \sum_{n=2}^{\infty} \frac{2}{n!} x^n.$$

42. a) Let $f(x) = \begin{cases} \pi, & \text{if } -\pi < x \leq 0 \\ 0, & \text{if } 0 < x < \pi. \end{cases}$

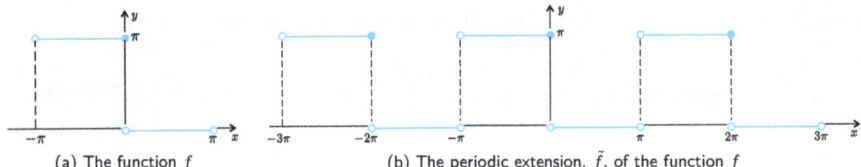

(a) The function f (b) The periodic extension, $\tilde{f}$, of the function f

Figure 3.29: Graphs of exercise 42 a)

The graph of f can be seen in Fig. 3.29. By Definition 3.5.3, the Fourier coefficients are

$$a_0 = \frac{1}{\pi}\int_{-\pi}^{\pi} f(x)\,dx = \frac{1}{\pi}\int_{-\pi}^{0} \pi\,dx = [x]_{-\pi}^{0} = \pi;$$

$$a_n = \frac{1}{\pi}\int_{-\pi}^{0} \pi \cos(nx)\,dx = \left[\frac{\sin(nx)}{n}\right]_{-\pi}^{0} = 0;$$

$$b_n = \frac{1}{\pi}\int_{-\pi}^{0} \pi \sin(nx)\,dx = \left[-\frac{\cos(nx)}{n}\right]_{-\pi}^{0} = -\frac{1}{n} + \frac{\cos(-n\pi)}{n} = \frac{(-1)^n - 1}{n} = \begin{cases} 0, & \text{if } n \text{ is even} \\ -\dfrac{2}{n}, & \text{if } n \text{ is odd.} \end{cases}$$

Thus, the Fourier series of f is

$$\frac{\pi}{2} - \sum_{n=1}^{\infty} \frac{2}{2n-1} \sin\bigl((2n-1)x\bigr).$$

Since f is a piecewise C^1 function, by Theorem 3.5.1 the series converges to $\tilde{f}(x)$ if $x \neq k\pi$, $k \in \mathbb{Z}$, and takes the value $\dfrac{\pi}{2}$, if $x = k\pi$, $k \in \mathbb{Z}$. In Fig. 3.30, we see an illustration of the sum of the series.

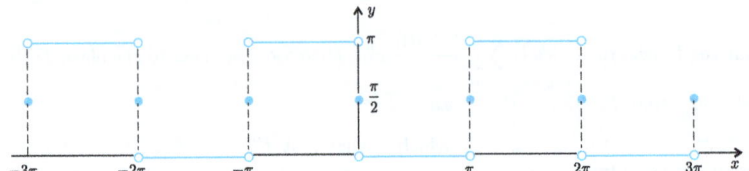

Figure 3.30: The sum of the series of exercise 42 a)

b) Let us consider the function $f(x) = \begin{cases} 0, & \text{if } -\pi < x \leq 0 \\ e^x, & \text{if } 0 < x < \pi. \end{cases}$

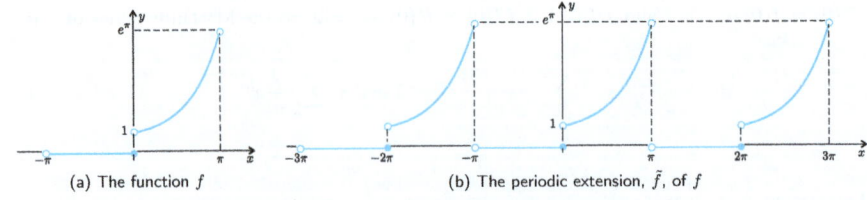

(a) The function f (b) The periodic extension, $\tilde{f}$, of f

Figure 3.31: Graphs of exercise 42 b)

3.6. Solved Exercises

The graph of f can be seen in Fig. 3.31. By Definition 3.5.3, the Fourier coefficients are

$$a_0 = \frac{1}{\pi}\int_{-\pi}^{\pi} f(x)\,dx = \frac{1}{\pi}\int_0^{\pi} e^x\,dx = \frac{1}{\pi}\left[e^x\right]_0^{\pi} = \frac{1}{\pi}(e^{\pi}-1);$$

$$a_n = \frac{1}{\pi}\int_0^{\pi} e^x \cos(nx)\,dx;$$

$$b_n = \frac{1}{\pi}\int_0^{\pi} e^x \sin(nx)\,dx = \frac{1}{\pi}\left(\left[e^x\sin(nx)\right]_0^{\pi} - \int_0^{\pi} n\,e^x \cos(nx)\,dx\right)$$

$$= -\frac{n}{\pi}\int_0^{\pi} e^x \cos(nx)\,dx.$$

We have

$$\int_0^{\pi} e^x \cos(nx)\,dx = \left[e^x \cos(nx)\right]_0^{\pi} + \int_0^{\pi} n\,e^x \sin(nx)\,dx = e^{\pi}\cos(n\pi) - e^0 + n\int_0^{\pi} e^x \sin(nx)\,dx$$

$$= (-1)^n e^{\pi} - 1 + n\left(\left[e^x \sin(nx)\right]_0^{\pi} - \int_0^{\pi} n\,e^x \cos(nx)\,dx\right)$$

$$= (-1)^n e^{\pi} - 1 - n^2 \int_0^{\pi} e^x \cos(nx)\,dx.$$

From this equality, we obtain

$$(1+n^2)\int_0^{\pi} e^x \cos(nx)\,dx = (-1)^n e^{\pi} - 1;$$

therefore,

$$\int_0^{\pi} e^x \cos(nx)\,dx = \frac{(-1)^n e^{\pi} - 1}{n^2+1}.$$

Using this equality, the coefficients are

$$a_n = \frac{1}{\pi}\int_0^{\pi} e^x \cos(nx)\,dx = \frac{(-1)^n e^{\pi} - 1}{\pi(n^2+1)};$$

$$b_n = -\frac{n}{\pi}\cdot\frac{(-1)^n e^{\pi} - 1}{n^2+1}.$$

The Fourier series of f is

$$\frac{e^{\pi}-1}{2\pi} + \sum_{n=1}^{\infty}\left(\frac{(-1)^n e^{\pi} - 1}{\pi(n^2+1)}\cos(nx) - \frac{n}{\pi}\cdot\frac{(-1)^n e^{\pi} - 1}{n^2+1}\sin(nx)\right)$$

$$= \frac{e^{\pi}-1}{2\pi} + \frac{1}{\pi}\sum_{n=1}^{\infty}\frac{(-1)^n e^{\pi} - 1}{n^2+1}\bigl(\cos(nx) - n\sin(nx)\bigr).$$

Since f is a piecewise C^1 function, by Theorem 3.5.1 the series converges pointwise to the function

$$S(x) = \begin{cases} \tilde{f}(x), & \text{if } x \neq k\pi,\ k \in \mathbb{Z} \\ \dfrac{1}{2}, & \text{if } x = 2k\pi,\ k \in \mathbb{Z} \\ \dfrac{e^{\pi}}{2}, & \text{if } x = (2k+1)\pi,\ k \in \mathbb{Z}. \end{cases}$$

Figure 3.32 displays the sum of the series.

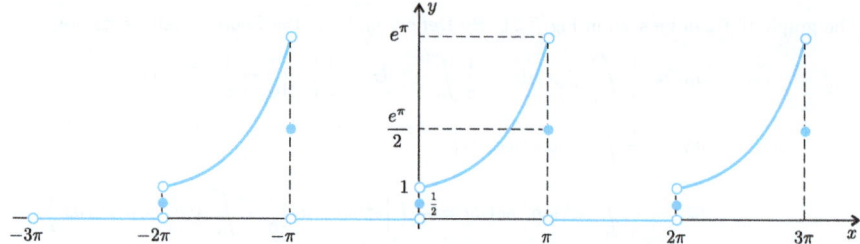

Figure 3.32: The sum of the series of exercise 42 b)

c) Let f be the function defined by $f(x) = \begin{cases} \cos(x), & \text{if } -\pi < x \leq 0 \\ \sin(x), & \text{if } 0 < x < \pi. \end{cases}$

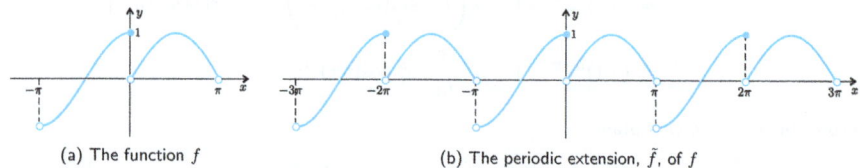

(a) The function f \qquad (b) The periodic extension, $\tilde{f}$, of f

Figure 3.33: Graphs of exercise 42 c)

The graph of f can be seen in Fig. 3.33. Using equalities (3.7), (3.8), and (3.11), we can compute the Fourier coefficients:

$$a_0 = \frac{1}{\pi}\int_{-\pi}^{\pi} f(x)\,dx = \frac{1}{\pi}\left(\int_{-\pi}^{0} \cos(x)\,dx + \int_{0}^{\pi} \sin(x)\,dx\right)$$

$$= \frac{1}{\pi}\left(\Big[\sin(x)\Big]_{-\pi}^{0} + \Big[-\cos(x)\Big]_{0}^{\pi}\right) = \frac{2}{\pi};$$

$$a_n = \frac{1}{\pi}\left(\int_{-\pi}^{0} \cos(x)\cos(nx)\,dx + \int_{0}^{\pi} \sin(x)\cos(nx)\,dx\right)$$

$$= \frac{1}{\pi}\left(\frac{1}{2}\int_{-\pi}^{0}\Big(\cos((n+1)x) + \cos((n-1)x)\Big)\,dx + \frac{1}{2}\int_{0}^{\pi}\Big(\sin((n+1)x) + \sin((1-n)x)\Big)\,dx\right)$$

$$= \begin{cases} \dfrac{1}{2\pi}\left(\left[\dfrac{\sin((n+1)x)}{n+1} + \dfrac{\sin((1-n)x)}{1-n}\right]_{-\pi}^{0} + \left[-\dfrac{\cos((n+1)x)}{n+1} - \dfrac{\cos((1-n)x)}{1-n}\right]_{0}^{\pi}\right), & \text{if } n \neq 1 \\[2ex] \dfrac{1}{2\pi}\left(\left[\dfrac{\sin(2x)}{2} + x\right]_{-\pi}^{0} + \left[-\dfrac{\cos(2x)}{2}\right]_{0}^{\pi}\right), & \text{if } n = 1 \end{cases}$$

$$= \begin{cases} \dfrac{1}{2\pi}\left(-\dfrac{\cos((n+1)\pi)}{n+1} - \dfrac{\cos((1-n)\pi)}{1-n} + \dfrac{1}{n+1} + \dfrac{1}{1-n}\right), & \text{if } n \neq 1 \\[2ex] \dfrac{1}{2}, & \text{if } n = 1 \end{cases}$$

3.6. Solved Exercises

$$= \begin{cases} \dfrac{1}{2\pi}\left(\dfrac{(-1)^n+1}{n+1} - \dfrac{(-1)^n+1}{n-1}\right) = -\dfrac{1}{\pi}\cdot\dfrac{(-1)^n+1}{n^2-1}, & \text{if } n \neq 1 \\ \dfrac{1}{2}, & \text{if } n = 1; \end{cases}$$

$$b_n = \dfrac{1}{\pi}\left(\int_{-\pi}^{0} \cos(x)\sin(nx)\,dx + \int_{0}^{\pi} \sin(x)\sin(nx)\,dx\right)$$

$$= \dfrac{1}{\pi}\left(\dfrac{1}{2}\int_{-\pi}^{0}\Big(\sin((n+1)x)+\sin((n-1)x)\Big)\,dx + \dfrac{1}{2}\int_{0}^{\pi}\Big(\cos((n-1)x)-\cos((n+1)x)\Big)\,dx\right)$$

$$= \begin{cases} \dfrac{1}{2\pi}\left(\left[-\dfrac{\cos((n+1)x)}{n+1} - \dfrac{\cos((n-1)x)}{n-1}\right]_{-\pi}^{0} + \left[\dfrac{\sin((n-1)x)}{n-1} - \dfrac{\sin((n+1)x)}{n+1}\right]_{0}^{\pi}\right), & \text{if } n \neq 1 \\ \dfrac{1}{2\pi}\left(\left[-\dfrac{\cos(2x)}{2}\right]_{-\pi}^{0} + \left[x - \dfrac{\sin(2x)}{2}\right]_{0}^{\pi}\right), & \text{if } n = 1. \end{cases}$$

$$= \begin{cases} \dfrac{1}{2\pi}\left(-\dfrac{1}{n+1} - \dfrac{1}{n-1} + \dfrac{\cos((n+1)(-\pi))}{n+1} + \dfrac{\cos((n-1)(-\pi))}{n-1}\right), & \text{if } n \neq 1 \\ \dfrac{1}{2}, & \text{if } n = 1 \end{cases}$$

$$= \begin{cases} \dfrac{1}{2\pi}\left(\dfrac{(-1)^{n+1}-1}{n+1} + \dfrac{(-1)^{n-1}-1}{n-1}\right) = -\dfrac{1}{\pi}\cdot\dfrac{n(1+(-1)^n)}{n^2-1}, & \text{if } n \neq 1 \\ \dfrac{1}{2}, & \text{if } n = 1. \end{cases}$$

The Fourier series of f is

$$\dfrac{1}{\pi} + \dfrac{1}{2}\cos(x) + \dfrac{1}{2}\sin(x) - \dfrac{1}{\pi}\sum_{n=2}^{\infty}\dfrac{(-1)^n+1}{n^2-1}\big(\cos(nx) + n\sin(nx)\big)$$

$$= \dfrac{1}{\pi} + \dfrac{1}{2}\big(\cos(x)+\sin(x)\big) - \dfrac{2}{\pi}\sum_{n=1}^{\infty}\dfrac{1}{4n^2-1}\big(\cos(2nx) + 2n\sin(2nx)\big).$$

Since f is a piecewise C^1 function, by Theorem 3.5.1 the series converges pointwise to the function

$$S(x) = \begin{cases} \tilde{f}(x), & \text{if } x \neq k\pi,\ k \in \mathbb{Z} \\ \dfrac{1}{2}, & \text{if } x = 2k\pi,\ k \in \mathbb{Z} \\ -\dfrac{1}{2}, & \text{if } x = (2k+1)\pi,\ k \in \mathbb{Z}. \end{cases}$$

The sum of the series is illustrated in Fig. 3.34.

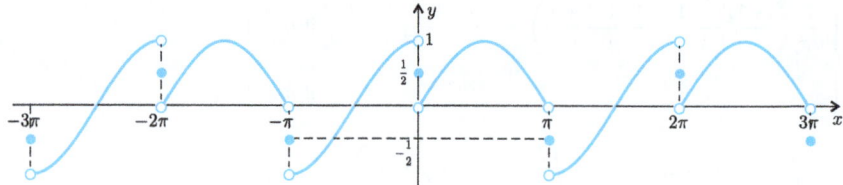

Figure 3.34: The sum of the series of exercise 42 c)

d) The function $f(x) = x^3$ defined on $]-\pi, \pi[$ is odd (see Fig. 3.35); therefore, $a_n = 0$, $\forall n \in \mathbb{N}_0$. Let us calculate the remaining Fourier coefficients.

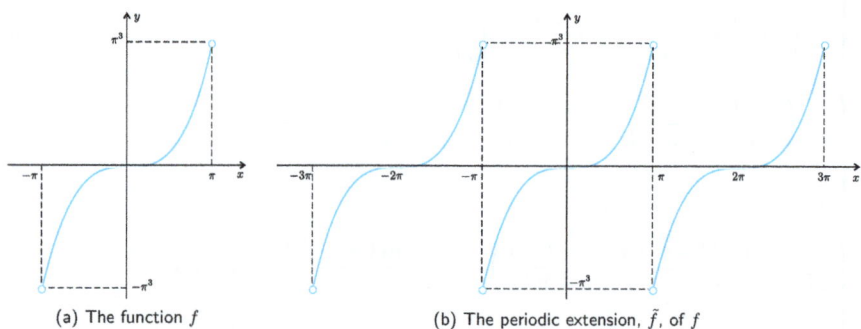

(a) The function f (b) The periodic extension, $\tilde{f}$, of f

Figure 3.35: Graphs of exercise 42 d)

$$b_n = \frac{2}{\pi}\int_0^\pi x^3 \sin(nx)\,dx = \frac{2}{\pi}\left(\left[-x^3 \frac{\cos(nx)}{n}\right]_0^\pi + \int_0^\pi 3x^2 \frac{\cos(nx)}{n}\,dx\right)$$

$$= \frac{2}{\pi}\left(-\frac{\pi^3 \cos(n\pi)}{n} + \frac{3}{n}\left(\left[x^2 \frac{\sin(nx)}{n}\right]_0^\pi - \int_0^\pi 2x \frac{\sin(nx)}{n}\,dx\right)\right)$$

$$= -\frac{2\pi^2(-1)^n}{n} - \frac{12}{n^2\pi}\int_0^\pi x \sin(nx)\,dx = -\frac{2\pi^2(-1)^n}{n} - \frac{12}{n^2\pi}\left(\left[-x\frac{\cos(nx)}{n}\right]_0^\pi + \int_0^\pi \frac{\cos(nx)}{n}\,dx\right)$$

$$= -\frac{2\pi^2(-1)^n}{n} - \frac{12}{n^2\pi}\left(-\pi\frac{\cos(n\pi)}{n} + \left[\frac{\sin(nx)}{n^2}\right]_0^\pi\right) = -\frac{2\pi^2(-1)^n}{n} + \frac{12(-1)^n}{n^3}$$

$$= 2(-1)^n\left(\frac{6}{n^3} - \frac{\pi^2}{n}\right).$$

The Fourier series of f is

$$2\sum_{n=1}^\infty (-1)^n \left(\frac{6}{n^3} - \frac{\pi^2}{n}\right)\sin(nx).$$

Since f is a piecewise C^1 function, by Theorem 3.5.1 the series converges pointwise to the function

$$S(x) = \begin{cases} \tilde{f}(x), & \text{if } x \neq (2k+1)\pi,\ k \in \mathbb{Z} \\ 0, & \text{if } x = (2k+1)\pi,\ k \in \mathbb{Z}. \end{cases}$$

3.6. Solved Exercises

The sum of the series is illustrated in Fig. 3.36.

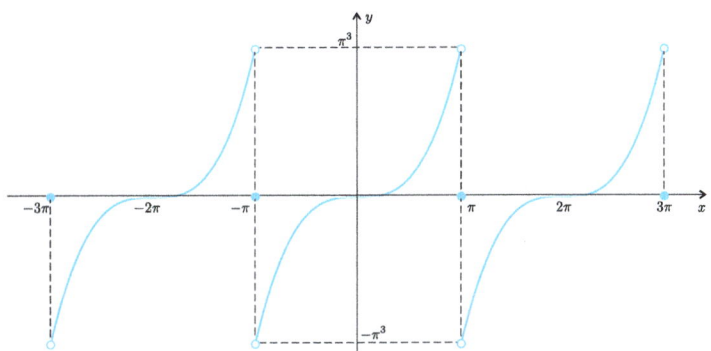

Figure 3.36: The sum of the series of exercise 42 d)

e) The function $f(x) = 4 - |x|$ defined on $]-\pi, \pi[$ is even (see Fig. 3.37); therefore, $b_n = 0$, $\forall n \in \mathbb{N}$. The Fourier coefficients are

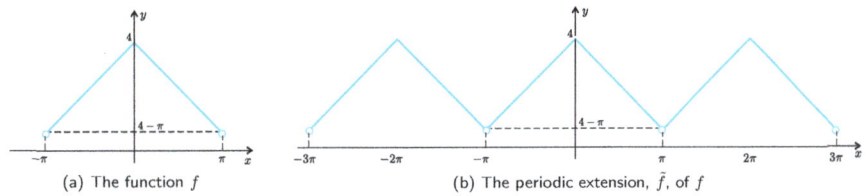

(a) The function f (b) The periodic extension, $\tilde{f}$, of f

Figure 3.37: Graphs of exercise 42 e)

$$a_0 = \frac{2}{\pi}\int_0^\pi f(x)\,dx = \frac{2}{\pi}\int_0^\pi (4-x)\,dx = \frac{2}{\pi}\left[4x - \frac{x^2}{2}\right]_0^\pi = 8 - \pi;$$

$$a_n = \frac{2}{\pi}\int_0^\pi (4-x)\cos(nx)\,dx = \frac{2}{\pi}\left(\left[\frac{4\sin(nx)}{n}\right]_0^\pi - \int_0^\pi x\cos(nx)\,dx\right)$$

$$= -\frac{2}{\pi}\left(\left[\frac{x\sin(nx)}{n}\right]_0^\pi - \int_0^\pi \frac{\sin(nx)}{n}\,dx\right) = \frac{2}{\pi}\left[-\frac{\cos(nx)}{n^2}\right]_0^\pi = \frac{2(1-(-1)^n)}{n^2\pi}.$$

The Fourier series of f is

$$4 - \frac{\pi}{2} + \frac{2}{\pi}\sum_{n=1}^\infty \frac{1-(-1)^n}{n^2}\cos(nx) = 4 - \frac{\pi}{2} + \frac{4}{\pi}\sum_{n=1}^\infty \frac{1}{(2n-1)^2}\cos((2n-1)x).$$

As $\tilde{f}$ is a piecewise C^1 function and extendable by continuity to $\mathbb{R}$, by Theorem 3.5.1 the series converges pointwise to the extension by continuity of $\tilde{f}$ on $\mathbb{R}$.

43. a) Let us consider the periodic function of period 2π defined on the interval $]0, 2\pi[$ by $f(x) = x$ (see Fig. 3.38).

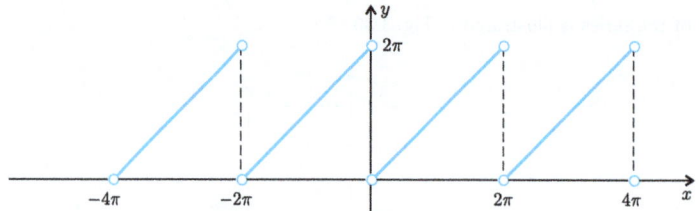

Figure 3.38: The function of exercise 43 a)

By Proposition 3.5.3, the coefficients of the Fourier series can be calculated by the formulas

$$a_0 = \frac{1}{\pi} \int_0^{2\pi} x\, dx = \frac{1}{\pi} \left[\frac{x^2}{2}\right]_0^{2\pi} = 2\pi;$$

$$a_n = \frac{1}{\pi} \int_0^{2\pi} x \cos(nx)\, dx = \frac{1}{\pi} \left(\left[x \frac{\sin(nx)}{n}\right]_0^{2\pi} - \int_0^{2\pi} \frac{\sin(nx)}{n}\, dx \right) = 0;$$

$$b_n = \frac{1}{\pi} \int_0^{2\pi} x \sin(nx)\, dx = \frac{1}{\pi} \left(\left[-x \frac{\cos(nx)}{n}\right]_0^{2\pi} + \int_0^{2\pi} \frac{\cos(nx)}{n}\, dx \right)$$

$$= \frac{1}{\pi} \left(-\frac{2\pi}{n} + \left[\frac{\sin(nx)}{n^2}\right]_0^{2\pi} \right) = -\frac{2}{n}.$$

The Fourier series of f is

$$\pi - 2 \sum_{n=1}^{\infty} \frac{1}{n} \sin(nx).$$

By Theorem 3.5.2, the series converges pointwise to the function, S, periodic of period 2π, illustrated in Fig. 3.39 and whose analytical expression is

$$S(x) = \begin{cases} x - 2k\pi, & \text{if } 2k\pi < x < (2k+2)\pi,\ k \in \mathbb{Z} \\ \pi, & \text{if } x = 2k\pi,\ k \in \mathbb{Z}. \end{cases}$$

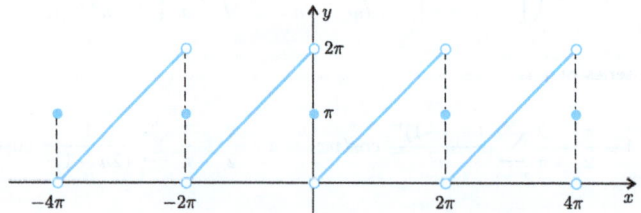

Figure 3.39: The sum of the series of exercise 43 a)

b) Let us consider the periodic function of period 2π defined on $]0, 2\pi[$ by $f(x) = x^2$ (refer to Fig. 3.40).

3.6. Solved Exercises

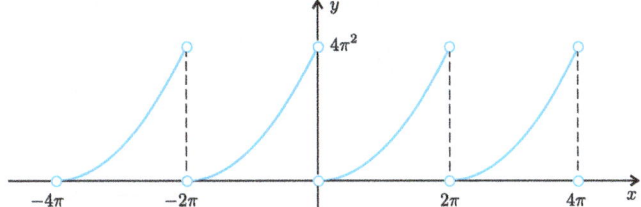

Figure 3.40: The function of exercise 43 b)

Using the results from the previous item and Proposition 3.5.3, the coefficients of the Fourier series are

$$a_0 = \frac{1}{\pi}\int_0^{2\pi} x^2\,dx = \frac{1}{\pi}\left[\frac{x^3}{3}\right]_0^{2\pi} = \frac{8\pi^2}{3};$$

$$a_n = \frac{1}{\pi}\int_0^{2\pi} x^2\cos(nx)\,dx = \frac{1}{\pi}\left(\left[x^2\frac{\sin(nx)}{n}\right]_0^{2\pi} - \int_0^{2\pi} 2x\frac{\sin(nx)}{n}\,dx\right) = \frac{4}{n^2};$$

$$b_n = \frac{1}{\pi}\int_0^{2\pi} x^2\sin(nx)\,dx = \frac{1}{\pi}\left(\left[-x^2\frac{\cos(nx)}{n}\right]_0^{2\pi} + \int_0^{2\pi} 2x\frac{\cos(nx)}{n}\,dx\right) = -\frac{4\pi}{n}.$$

The Fourier series of f is

$$\frac{4\pi^2}{3} + 4\sum_{n=1}^{\infty}\left(\frac{1}{n^2}\cos(nx) - \frac{\pi}{n}\sin(nx)\right).$$

By Theorem 3.5.2, the series converges pointwise to the function, S, periodic of period 2π, illustrated in Fig. 3.41 and whose analytical expression is

$$S(x) = \begin{cases} (x - 2k\pi)^2, & \text{if } 2k\pi < x < (2k+2)\pi,\ k\in\mathbb{Z} \\ 2\pi^2, & \text{if } x = 2k\pi,\ k\in\mathbb{Z}. \end{cases}$$

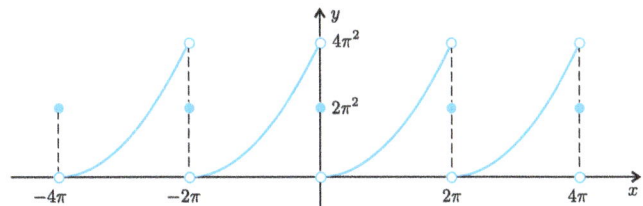

Figure 3.41: The sum of the series of exercise 43 b)

c) Let us consider the periodic function of period 2π defined on $\left]-\frac{\pi}{2}, \frac{3\pi}{2}\right]$ by $f(x) = |x|$ (see Fig. 3.42).

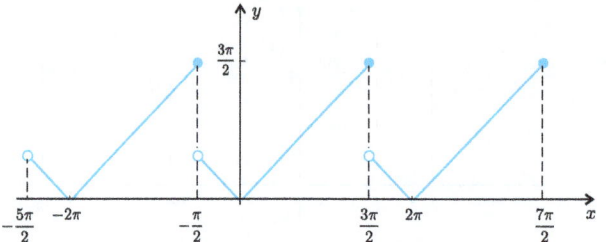

Figure 3.42: The function of exercise 43 c)

By Proposition 3.5.3, the coefficients of the Fourier series are

$$a_0 = \frac{1}{\pi}\int_{-\pi/2}^{3\pi/2} |x|\, dx = \frac{2}{\pi}\left[\frac{x^2}{2}\right]_0^{\pi/2} + \frac{1}{\pi}\left[\frac{x^2}{2}\right]_{\pi/2}^{3\pi/2} = \frac{5\pi}{4};$$

$$a_n = \frac{1}{\pi}\int_{-\pi/2}^{3\pi/2} |x|\cos(nx)\, dx = \frac{2}{\pi}\int_0^{\pi/2} x\cos(nx)\, dx + \frac{1}{\pi}\int_{\pi/2}^{3\pi/2} x\cos(nx)\, dx$$

$$= \frac{2}{\pi}\left(\left[x\frac{\sin(nx)}{n}\right]_0^{\pi/2} - \int_0^{\pi/2}\frac{\sin(nx)}{n}\, dx\right) + \frac{1}{\pi}\left(\left[x\frac{\sin(nx)}{n}\right]_{\pi/2}^{3\pi/2} - \int_{\pi/2}^{3\pi/2}\frac{\sin(nx)}{n}\, dx\right)$$

$$= -\frac{1}{n}\sin\left(\frac{n\pi}{2}\right) + \frac{2}{\pi n^2}\cos\left(\frac{n\pi}{2}\right) - \frac{2}{\pi n^2};$$

$$b_n = \frac{1}{\pi}\int_{-\pi/2}^{3\pi/2} |x|\sin(nx)\, dx = \frac{1}{\pi}\int_{\pi/2}^{3\pi/2} x\sin(nx)\, dx = \frac{1}{\pi}\left(\left[-x\frac{\cos(nx)}{n}\right]_{\pi/2}^{3\pi/2} + \int_{\pi/2}^{3\pi/2}\frac{\cos(nx)}{n}\, dx\right)$$

$$= -\frac{2}{\pi n^2}\sin\left(\frac{n\pi}{2}\right) - \frac{1}{n}\cos\left(\frac{n\pi}{2}\right).$$

The Fourier series of f is

$$\frac{5\pi}{8} - \sum_{n=1}^{\infty}\left(\frac{1}{n}\sin\left(\frac{n\pi}{2}\right) - \frac{2}{\pi n^2}\cos\left(\frac{n\pi}{2}\right) + \frac{2}{\pi n^2}\right)\cos(nx) + \left(\frac{2}{\pi n^2}\sin\left(\frac{n\pi}{2}\right) + \frac{1}{n}\cos\left(\frac{n\pi}{2}\right)\right)\sin(nx).$$

By Theorem 3.5.1, the series converges pointwise to the function, S, periodic with period 2π, illustrated in Fig. 3.43 and whose analytical expression is

$$S(x) = \begin{cases} |x - 2k\pi|, & \text{if } \left(2k - \frac{1}{2}\right)\pi < x < \left(2k + \frac{3}{2}\right)\pi,\ k \in \mathbb{Z} \\ \pi, & \text{if } x = \left(2k + \frac{3}{2}\right)\pi,\ k \in \mathbb{Z}. \end{cases}$$

d) Let f be the periodic function with period 2π defined on $\left[-\frac{\pi}{2}, \frac{3\pi}{2}\right]$ by

$$\begin{cases} x, & \text{if } -\frac{\pi}{2} \leq x \leq \frac{\pi}{2} \\ \pi - x, & \text{if } \frac{\pi}{2} < x \leq \frac{3\pi}{2}. \end{cases}$$

3.6. Solved Exercises

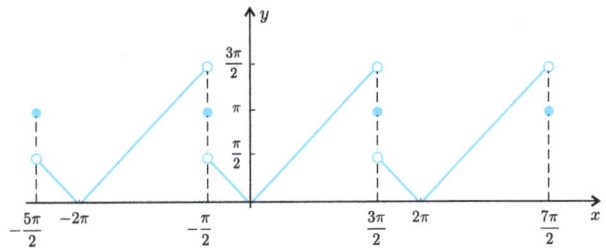

Figure 3.43: The sum of the series of exercise 43 c)

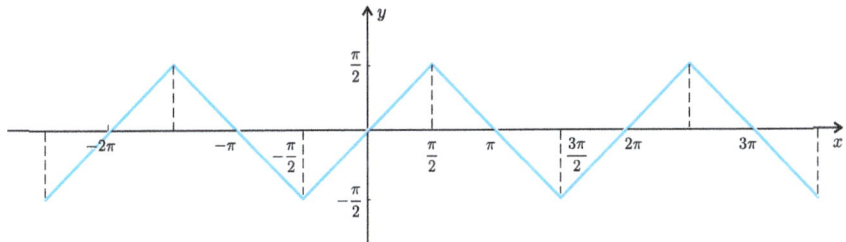

Figure 3.44: The function of exercise 43 d)

It is easy to verify that it is an odd function (see Fig. 3.44). As it is periodic with period 2π, it follows from Propositions 3.5.3 and 3.5.6 that $a_n = 0$, $\forall n \in \mathbb{N}_0$. Let us calculate the remaining coefficients.

$$b_n = \frac{2}{\pi}\left(\int_0^{\pi/2} x \sin(nx)\, dx + \int_{\pi/2}^{\pi} (\pi - x) \sin(nx)\, dx\right)$$

$$= \frac{2}{\pi}\left(\left[-x \frac{\cos(nx)}{n}\right]_0^{\pi/2} - \int_0^{\pi/2} -\frac{\cos(nx)}{n}\, dx + \pi \left[-\frac{\cos(nx)}{n}\right]_{\pi/2}^{\pi} - \int_{\pi/2}^{\pi} x \sin(nx)\, dx\right)$$

$$= \frac{2}{\pi}\left(\frac{-\pi}{2n}\cos\left(\frac{n\pi}{2}\right) + \left[\frac{\sin(nx)}{n^2}\right]_0^{\pi/2} - \frac{\pi}{n}\left(\cos(n\pi) - \cos\left(\frac{n\pi}{2}\right)\right) - \int_{\pi/2}^{\pi} x \sin(nx)\, dx\right)$$

$$= \frac{2}{\pi}\left(\frac{\pi}{2n}\cos\left(\frac{n\pi}{2}\right) + \frac{1}{n^2}\sin\left(\frac{n\pi}{2}\right) - \frac{\pi}{n}(-1)^n - \left[-x\frac{\cos(nx)}{n}\right]_{\pi/2}^{\pi} + \int_{\pi/2}^{\pi} -\frac{\cos(nx)}{n}\, dx\right)$$

$$= \frac{2}{\pi}\left(\frac{\pi}{2n}\cos\left(\frac{n\pi}{2}\right) + \frac{1}{n^2}\sin\left(\frac{n\pi}{2}\right) - \frac{\pi}{n}(-1)^n + \frac{\pi}{n}\cos(n\pi) - \frac{\pi}{2n}\cos\left(\frac{n\pi}{2}\right) - \left[\frac{\sin(nx)}{n^2}\right]_{\pi/2}^{\pi}\right)$$

$$= \frac{4}{\pi n^2}\sin\left(\frac{n\pi}{2}\right) = \begin{cases} 0, & \text{if } n \text{ is even} \\ \dfrac{4(-1)^{(n-1)/2}}{n^2 \pi}, & \text{if } n \text{ is odd.} \end{cases}$$

The Fourier series of f is

$$\frac{4}{\pi} \sum_{n=1}^{\infty} \frac{(-1)^{n+1}}{(2n-1)^2} \sin((2n-1)x).$$

Since f is a continuous function and piecewise C^1, Theorem 3.5.1 allows us to conclude that the series converges pointwise to f on $\mathbb{R}$.

e) Let f be the periodic function of period 2π defined on $\left[-\frac{\pi}{2}, \frac{3\pi}{2}\right]$ by

$$\begin{cases} x^2, & \text{if } -\frac{\pi}{2} \leq x \leq \frac{\pi}{2} \\ \frac{\pi^2}{4}, & \text{if } \frac{\pi}{2} < x \leq \frac{3\pi}{2}. \end{cases}$$

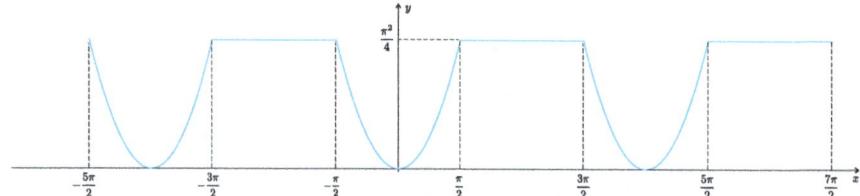

Figure 3.45: The function of exercise 43 e)

It is easy to verify that it is an even function (see Fig. 3.45). As it is periodic with period 2π, it follows from Propositions 3.5.3 and 3.5.6 that $b_n = 0$, $\forall n \in \mathbb{N}$. Let us calculate the remaining coefficients.

$$a_0 = \frac{1}{\pi} \int_{-\pi/2}^{3\pi/2} f(x)\, dx = \frac{2}{\pi} \int_0^{\pi/2} x^2\, dx + \frac{1}{\pi} \int_{\pi/2}^{3\pi/2} \frac{\pi^2}{4}\, dx = \frac{2}{\pi} \left[\frac{x^3}{3}\right]_0^{\pi/2} + \frac{\pi}{4}\left[x\right]_{\pi/2}^{3\pi/2} = \frac{\pi^2}{3};$$

$$a_n = \frac{2}{\pi} \int_0^{\pi/2} x^2 \cos(nx)\, dx + \frac{1}{\pi} \int_{\pi/2}^{3\pi/2} \frac{\pi^2}{4} \cos(nx)\, dx$$

$$= \frac{2}{\pi} \left(\left[x^2 \frac{\sin(nx)}{n}\right]_0^{\pi/2} - \int_0^{\pi/2} 2x \frac{\sin(nx)}{n}\, dx\right) + \frac{\pi}{4}\left[\frac{\sin(nx)}{n}\right]_{\pi/2}^{3\pi/2}$$

$$= \frac{2}{\pi}\left(\frac{\pi^2}{4n}\sin\left(\frac{n\pi}{2}\right) + \frac{2}{n}\left[x\frac{\cos(nx)}{n}\right]_0^{\pi/2} - \frac{2}{n}\int_0^{\pi/2}\frac{\cos(nx)}{n}\,dx\right) + \frac{\pi}{4n}\left(\sin\left(\frac{3n\pi}{2}\right) - \sin\left(\frac{n\pi}{2}\right)\right)$$

$$= \frac{2}{\pi}\left(\frac{\pi^2}{4n}\sin\left(\frac{n\pi}{2}\right) + \frac{\pi}{n^2}\cos\left(\frac{n\pi}{2}\right) - \frac{2}{n^2}\left[\frac{\sin(nx)}{n}\right]_0^{\pi/2}\right) - \frac{\pi}{2n}\sin\left(\frac{n\pi}{2}\right)$$

$$= \frac{2}{\pi}\left(\frac{\pi^2}{4n}\sin\left(\frac{n\pi}{2}\right) + \frac{\pi}{n^2}\cos\left(\frac{n\pi}{2}\right) - \frac{2}{n^3}\sin\left(\frac{n\pi}{2}\right)\right) - \frac{\pi}{2n}\sin\left(\frac{n\pi}{2}\right)$$

$$= \frac{2}{n^2}\cos\left(\frac{n\pi}{2}\right) - \frac{4}{n^3\pi}\sin\left(\frac{n\pi}{2}\right).$$

The Fourier series of f is

$$\frac{\pi^2}{6} + \sum_{n=1}^{\infty}\left(\frac{2}{n^2}\cos\left(\frac{n\pi}{2}\right) - \frac{4}{n^3\pi}\sin\left(\frac{n\pi}{2}\right)\right)\cos(nx).$$

Since f is a continuous function and piecewise C^1, by Theorem 3.5.1, the series converges pointwise to f on $\mathbb{R}$.

3.6. Solved Exercises

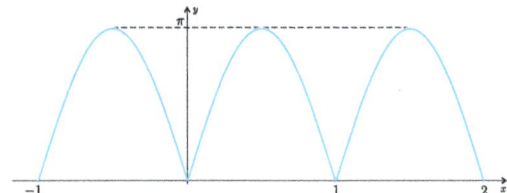

Figure 3.46: The function of exercise 44 a)

44. a) Let f be the periodic function of period 1 defined on the interval $]0,1[$ by $f(x) = \pi \sin(\pi x)$. It is easy to verify that it is an even function (see Fig. 3.46) and, therefore, $b_n = 0$, $\forall n \in \mathbb{N}$. As it is periodic with period $2l = 1$, it follows from Proposition 3.5.9 that the coefficients of its development in Fourier series are given by

$$a_0 = \frac{1}{l}\int_{-l}^{l} f(x)\,dx = 4\int_0^{1/2} \pi \sin(\pi x)\,dx = 4\bigl[-\cos(\pi x)\bigr]_0^{1/2} = 4;$$

$$a_n = \frac{1}{l}\int_{-l}^{l} f(x)\cos\left(\frac{n\pi x}{l}\right)dx = 4\int_0^{1/2} \pi \sin(\pi x)\cos(2n\pi x)\,dx.$$

Using equality (3.8), the previous integral is equal to

$$2\int_0^{1/2} \pi\Bigl(\sin\bigl((2n+1)\pi x\bigr) - \sin\bigl((2n-1)\pi x\bigr)\Bigr)dx = 2\left[-\frac{\cos\bigl((2n+1)\pi x\bigr)}{2n+1} + \frac{\cos\bigl((2n-1)\pi x\bigr)}{2n-1}\right]_0^{1/2}$$

and so

$$a_n = -\frac{4}{(2n-1)(2n+1)}.$$

The Fourier series of f is

$$2 - 4\sum_{n=1}^{\infty} \frac{1}{(2n-1)(2n+1)}\cos(2n\pi x).$$

Since f is a continuous function and piecewise C^1, Theorem 3.5.2 allows us to conclude that the series converges pointwise to f on $\mathbb{R}$.

b) Let f be the periodic function of period 1 defined on the interval $]0,1[$ by $f(x) = 2 - x^2$ (see Fig. 3.47).

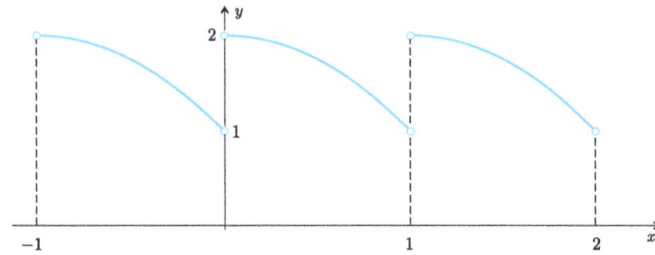

Figure 3.47: The function of exercise 44 b)

As it is periodic with period $2l = 1$, it follows from Propositions 3.5.9 and 3.5.3 that the coefficients of its Fourier series development are given by

$$a_0 = \frac{1}{l}\int_{-l}^{l} f(x)\,dx = 2\int_{-1/2}^{1/2} f(x)\,dx = 2\int_0^1 (2-x^2)\,dx = 2\left[2x - \frac{x^3}{3}\right]_0^1 = \frac{10}{3};$$

$$a_n = \frac{1}{l}\int_{-l}^{l} f(x)\cos\left(\frac{n\pi x}{l}\right)dx = 2\int_0^1 (2-x^2)\cos(2n\pi x)\,dx$$

$$= 4\left[\frac{\sin(2n\pi x)}{2n\pi}\right]_0^1 - 2\int_0^1 x^2 \cos(2n\pi x)\,dx = -2\left(\left[x^2\frac{\sin(2n\pi x)}{2n\pi}\right]_0^1 - \int_0^1 2x\frac{\sin(2n\pi x)}{2n\pi}\,dx\right)$$

$$= \frac{2}{n\pi}\int_0^1 x\sin(2n\pi x)\,dx = \frac{2}{n\pi}\left(\left[-x\frac{\cos(2n\pi x)}{2n\pi}\right]_0^1 + \int_0^1 \frac{\cos(2n\pi x)}{2n\pi}\,dx\right)$$

$$= \frac{2}{n\pi}\left(-\frac{\cos(2n\pi)}{2n\pi} + \left[\frac{\sin(2n\pi x)}{(2n\pi)^2}\right]_0^1\right) = -\frac{1}{(n\pi)^2};$$

$$b_n = \frac{1}{l}\int_{-l}^{l} f(x)\sin\left(\frac{n\pi x}{l}\right)dx = 2\int_0^1 (2-x^2)\sin(2n\pi x)\,dx$$

$$= 4\left[-\frac{\cos(2n\pi x)}{2n\pi}\right]_0^1 - 2\int_0^1 x^2 \sin(2n\pi x)\,dx = -2\left(\left[-x^2\frac{\cos(2n\pi x)}{2n\pi}\right]_0^1 + \int_0^1 2x\frac{\cos(2n\pi x)}{2n\pi}\,dx\right)$$

$$= -2\left(-\frac{1}{2n\pi} + \frac{1}{n\pi}\left[x\frac{\sin(2n\pi x)}{2n\pi}\right]_0^1 - \frac{1}{n\pi}\int_0^1 \frac{\sin(2n\pi x)}{2n\pi}\,dx\right)$$

$$= \frac{1}{n\pi} + \frac{1}{(n\pi)^2}\left[-\frac{\cos(2n\pi)}{2n\pi}\right]_0^1 = \frac{1}{n\pi}.$$

The Fourier series of f is

$$\frac{5}{3} + \sum_{n=1}^{\infty}\left(-\frac{1}{(n\pi)^2}\cos(2n\pi x) + \frac{1}{n\pi}\sin(2n\pi x)\right).$$

Since f is piecewise C^1, by Theorem 3.5.2 the series converges pointwise to the function, S, periodic with period 1, illustrated in Fig. 3.48 and analytically defined by

$$S(x) = \begin{cases} f(x), & \text{if } x \notin \mathbb{Z} \\ \dfrac{3}{2}, & \text{if } x \in \mathbb{Z}. \end{cases}$$

c) Let f be the periodic function with period 4 defined on the interval $]0,4]$ by $f(x) = x - 3$ (see Fig. 3.49). As it is periodic with period $2l = 4$, it follows from Proposition 3.5.9 that the coefficients of its development in Fourier series are given by

$$a_0 = \frac{1}{l}\int_{-l}^{l} f(x)\,dx = \frac{1}{2}\int_{-2}^{2} f(x)\,dx = \frac{1}{2}\int_0^4 (x-3)\,dx = \frac{1}{2}\left[\frac{x^2}{2} - 3x\right]_0^4 = -2;$$

$$a_n = \frac{1}{l}\int_{-l}^{l} f(x)\cos\left(\frac{n\pi x}{l}\right)dx = \frac{1}{2}\int_0^4 (x-3)\cos\left(\frac{n\pi x}{2}\right)dx$$

3.6. Solved Exercises

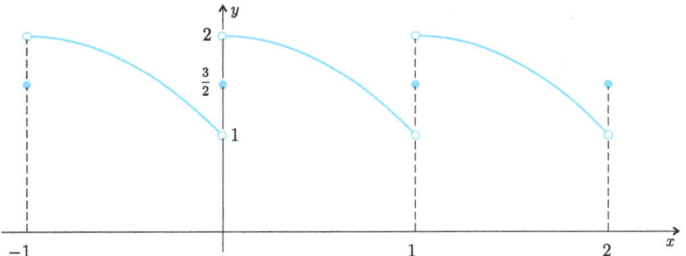

Figure 3.48: The sum of the series of exercise 44 b)

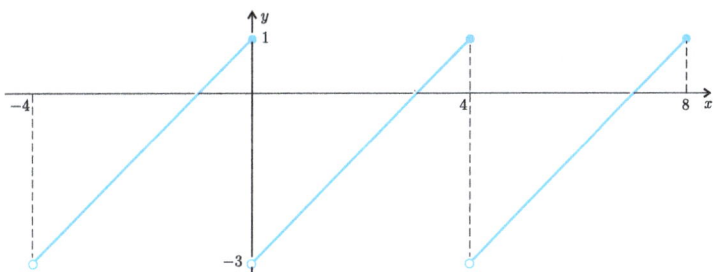

Figure 3.49: The function of exercise 44 c)

$$= \frac{1}{2}\left(\int_0^4 x\cos\left(\frac{n\pi x}{2}\right)dx - \frac{6}{n\pi}\left[\sin\left(\frac{n\pi x}{2}\right)\right]_0^4\right) = \frac{1}{n\pi}\left(\left[x\sin\left(\frac{n\pi x}{2}\right)\right]_0^4 - \int_0^4 \sin\left(\frac{n\pi x}{2}\right)dx\right)$$

$$= \frac{2}{(n\pi)^2}\left[\cos\left(\frac{n\pi x}{2}\right)\right]_0^4 = 0;$$

$$b_n = \frac{1}{l}\int_{-l}^{l} f(x)\sin\left(\frac{n\pi x}{l}\right)dx = \frac{1}{2}\int_0^4 (x-3)\sin\left(\frac{n\pi x}{2}\right)dx$$

$$= \frac{1}{2}\left(\int_0^4 x\sin\left(\frac{n\pi x}{2}\right)dx + \frac{6}{n\pi}\left[\cos\left(\frac{n\pi x}{2}\right)\right]_0^4\right) = \frac{1}{n\pi}\left(\left[-x\cos\left(\frac{n\pi x}{2}\right)\right]_0^4 + \int_0^4 \cos\left(\frac{n\pi x}{2}\right)dx\right)$$

$$= \frac{1}{n\pi}\left(-4 + \frac{2}{n\pi}\left[\sin\left(\frac{n\pi x}{2}\right)\right]_0^4\right) = -\frac{4}{n\pi}.$$

The Fourier series of f is

$$-1 - 4\sum_{n=1}^{\infty}\frac{1}{n\pi}\sin\left(\frac{n\pi x}{2}\right).$$

Since f is piecewise C^1, Theorem 3.5.2 allows us to conclude that the series converges pointwise to the function, S, periodic with period 4, illustrated in Fig. 3.50.

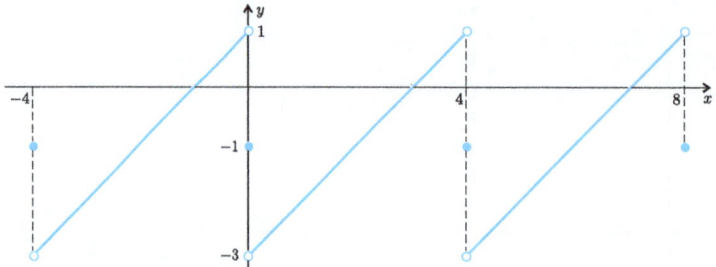

Figure 3.50: The sum of the series of exercise 44 c)

The analytical expression of the sum of the series is

$$S(x) = \begin{cases} f(x), & \text{if } x \neq 4k,\ k \in \mathbb{Z} \\ -1, & \text{if } x = 4k,\ k \in \mathbb{Z}. \end{cases}$$

d) Let us consider the function f, periodic with period $2l = 4$, defined on $[-2, 2[$ by

$$f(x) = \begin{cases} 0, & \text{if } -2 \leq x \leq -1 \\ x, & \text{if } -1 < x < 1 \\ 1, & \text{if } 1 \leq x < 2 \end{cases}$$

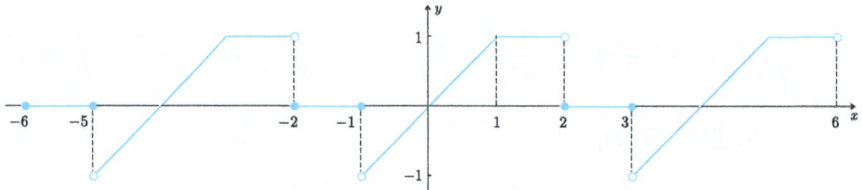

Figure 3.51: The function of exercise 44 d)

(see Fig. 3.51). As it is periodic with period $2l = 4$, it follows from Proposition 3.5.9 that the coefficients of its Fourier series development are given by

$$a_0 = \frac{1}{l} \int_{-l}^{l} f(x)\, dx = \frac{1}{2} \int_{-2}^{2} f(x)\, dx = \frac{1}{2} \left(\int_{-1}^{1} x\, dx + \int_{1}^{2} 1\, dx \right) = \frac{1}{2};$$

$$a_n = \frac{1}{l} \int_{-l}^{l} f(x) \cos\left(\frac{n\pi x}{l}\right) dx = \frac{1}{2} \left(\int_{-1}^{1} x \cos\left(\frac{n\pi x}{2}\right) dx + \int_{1}^{2} \cos\left(\frac{n\pi x}{2}\right) dx \right)$$

$$= \frac{1}{n\pi} \left[\sin\left(\frac{n\pi x}{2}\right) \right]_{1}^{2} = -\frac{1}{n\pi} \sin\left(\frac{n\pi}{2}\right);$$

3.6. Solved Exercises

$$b_n = \frac{1}{l}\int_{-l}^{l} f(x) \sin\left(\frac{n\pi x}{l}\right) dx = \frac{1}{2}\left(\int_{-1}^{1} x \sin\left(\frac{n\pi x}{2}\right) dx + \int_{1}^{2} \sin\left(\frac{n\pi x}{2}\right) dx\right)$$

$$= \frac{1}{n\pi}\left(\left[-x \cos\left(\frac{n\pi x}{2}\right)\right]_{-1}^{1} + \int_{-1}^{1} \cos\left(\frac{n\pi x}{2}\right) dx - \left[\cos\left(\frac{n\pi x}{2}\right)\right]_{1}^{2}\right)$$

$$= \frac{1}{n\pi}\left(-\cos\left(\frac{n\pi}{2}\right) - \cos\left(\frac{-n\pi}{2}\right) + \frac{2}{n\pi}\left[\sin\left(\frac{n\pi x}{2}\right)\right]_{-1}^{1} - \cos(n\pi) + \cos\left(\frac{n\pi}{2}\right)\right)$$

$$= \frac{1}{n\pi}\left(-\cos\left(\frac{n\pi}{2}\right) + \frac{4}{n\pi}\sin\left(\frac{n\pi}{2}\right) - (-1)^n\right).$$

The Fourier series of f is

$$\frac{1}{4} + \sum_{n=1}^{\infty} \left(-\frac{1}{n\pi} \sin\left(\frac{n\pi}{2}\right) \cos\left(\frac{n\pi x}{2}\right) + \frac{1}{n\pi}\left(-\cos\left(\frac{n\pi}{2}\right) + \frac{4}{n\pi}\sin\left(\frac{n\pi}{2}\right) - (-1)^n\right) \sin\left(\frac{n\pi x}{2}\right)\right).$$

As f is piecewise C^1, by Theorem 3.5.2 the series converges pointwise to the function, S, periodic with period 4, illustrated in Fig. 3.52 and whose analytical expression is

$$S(x) = \begin{cases} f(x), & \text{if } x \neq 4k+2 \wedge x \neq 4k+3, \, k \in \mathbb{Z} \\ \dfrac{1}{2}, & \text{if } x = 4k+2, \, k \in \mathbb{Z} \\ -\dfrac{1}{2}, & \text{if } x = 4k+3, \, k \in \mathbb{Z}. \end{cases}$$

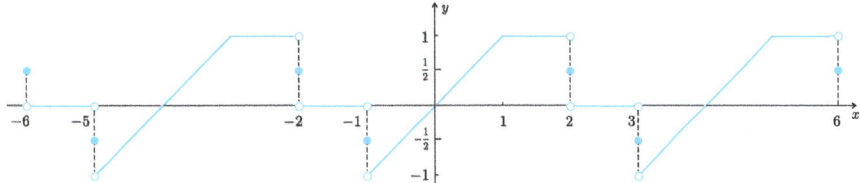

Figure 3.52: The sum of the series of exercise 44 d)

e) Let us consider the function f, periodic with period 6, defined on the interval $]-3, 3]$ by

$$f(x) = \begin{cases} -\dfrac{x}{3}, & \text{if } -3 < x \leq 0 \\ 2x - \dfrac{x^2}{3}, & \text{if } 0 < x \leq 3 \end{cases}$$

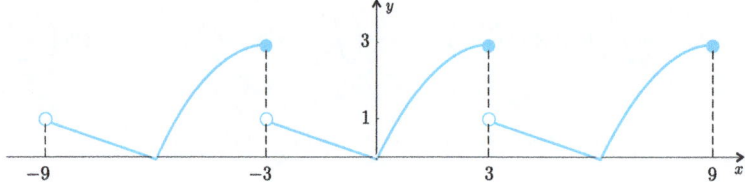

Figure 3.53: The function of exercise 44 e)

As it is periodic with period $2l = 6$ (see Fig. 3.53), it follows from Proposition 3.5.9 that the coefficients of its Fourier series development are given by

$$a_0 = \frac{1}{l}\int_{-l}^{l} f(x)\,dx = \frac{1}{3}\int_{-3}^{3} f(x)\,dx = \frac{1}{3}\left(\int_{-3}^{0} -\frac{x}{3}\,dx + \int_{0}^{3}\left(2x - \frac{x^2}{3}\right)dx\right)$$

$$= \frac{1}{3}\left(\left[-\frac{x^2}{6}\right]_{-3}^{0} + \left[x^2 - \frac{x^3}{9}\right]_{0}^{3}\right) = \frac{5}{2};$$

$$a_n = \frac{1}{l}\int_{-l}^{l} f(x)\cos\left(\frac{n\pi x}{l}\right)dx = \frac{1}{3}\left(\int_{-3}^{0} -\frac{x}{3}\cos\left(\frac{n\pi x}{3}\right)dx + \int_{0}^{3}\left(2x - \frac{x^2}{3}\right)\cos\left(\frac{n\pi x}{3}\right)dx\right)$$

$$= \frac{1}{n\pi}\left(\left[-\frac{x}{3}\sin\left(\frac{n\pi x}{3}\right)\right]_{-3}^{0} + \frac{1}{3}\int_{-3}^{0}\sin\left(\frac{n\pi x}{3}\right)dx\right) + \frac{1}{3}\int_{0}^{3}\left(2x - \frac{x^2}{3}\right)\cos\left(\frac{n\pi x}{3}\right)dx$$

$$= \frac{1}{(n\pi)^2}\left[-\cos\left(\frac{n\pi x}{3}\right)\right]_{-3}^{0} + \frac{1}{n\pi}\left(\left[\left(2x - \frac{x^2}{3}\right)\sin\left(\frac{n\pi x}{3}\right)\right]_{0}^{3} - \int_{0}^{3}\left(2 - \frac{2x}{3}\right)\sin\left(\frac{n\pi x}{3}\right)dx\right)$$

$$= \frac{1}{(n\pi)^2}(-1 + \cos(n\pi)) - \frac{1}{(n\pi)^2}\left(\left[-3\left(2 - \frac{2x}{3}\right)\cos\left(\frac{n\pi x}{3}\right)\right]_{0}^{3} - \int_{0}^{3} 2\cos\left(\frac{n\pi x}{3}\right)dx\right)$$

$$= \frac{1}{(n\pi)^2}(-1 + (-1)^n) - \frac{1}{(n\pi)^2}\left(6 - \frac{6}{n\pi}\left[\sin\left(\frac{n\pi x}{3}\right)\right]_{0}^{3}\right) = \frac{(-1)^n - 7}{(n\pi)^2};$$

$$b_n = \frac{1}{l}\int_{-l}^{l} f(x)\sin\left(\frac{n\pi x}{l}\right)dx = \frac{1}{3}\left(\int_{-3}^{0} -\frac{x}{3}\sin\left(\frac{n\pi x}{3}\right)dx + \int_{0}^{3}\left(2x - \frac{x^2}{3}\right)\sin\left(\frac{n\pi x}{3}\right)dx\right)$$

$$= \frac{1}{n\pi}\left(\left[\frac{x}{3}\cos\left(\frac{n\pi x}{3}\right)\right]_{-3}^{0} - \frac{1}{3}\int_{-3}^{0}\cos\left(\frac{n\pi x}{3}\right)dx\right) + \frac{1}{3}\int_{0}^{3}\left(2x - \frac{x^2}{3}\right)\sin\left(\frac{n\pi x}{3}\right)dx$$

$$= \frac{1}{n\pi}\left((-1)^n - \frac{1}{n\pi}\left[\sin\left(\frac{n\pi x}{3}\right)\right]_{-3}^{0} + \left[-\left(2x - \frac{x^2}{3}\right)\cos\left(\frac{n\pi x}{3}\right)\right]_{0}^{3} + \int_{0}^{3}\left(2 - \frac{2x}{3}\right)\cos\left(\frac{n\pi x}{3}\right)dx\right)$$

$$= \frac{1}{n\pi}\left((-1)^n - 3\cos(n\pi) + \frac{3}{n\pi}\left[\left(2 - \frac{2x}{3}\right)\sin\left(\frac{n\pi x}{3}\right)\right]_{0}^{3} + \frac{3}{n\pi}\int_{0}^{3}\frac{2}{3}\sin\left(\frac{n\pi x}{3}\right)dx\right)$$

$$= \frac{1}{n\pi}\left(-2(-1)^n + \frac{6}{(n\pi)^2}\left[-\cos\left(\frac{n\pi x}{3}\right)\right]_{0}^{3}\right) = \frac{6(1-(-1)^n)}{(n\pi)^3} - \frac{2(-1)^n}{n\pi}.$$

The Fourier series of f is

$$\frac{5}{4} + \sum_{n=1}^{\infty}\left(\frac{(-1)^n - 7}{(n\pi)^2}\cos\left(\frac{n\pi x}{3}\right) + \left(\frac{6(1-(-1)^n)}{(n\pi)^3} - \frac{2(-1)^n}{n\pi}\right)\sin\left(\frac{n\pi x}{3}\right)\right).$$

3.6. Solved Exercises

Since f is piecewise C^1, Theorem 3.5.2 allows us to conclude that the series converges pointwise to the function, S, periodic of period 4, illustrated in Fig. 3.54 and whose analytical expression is

$$S(x) = \begin{cases} f(x), & \text{if } x \neq 6k+3,\ k \in \mathbb{Z} \\ 2, & \text{if } x = 6k+3,\ k \in \mathbb{Z}. \end{cases}$$

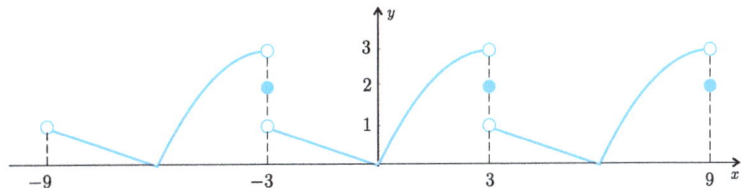

Figure 3.54: The sum of the series of exercise 44 e)

45. In Fig. 3.55 we can see the graph of the periodic function of period 2π defined on the interval $[0, 2\pi]$ in the following way:

$$f(x) = \begin{cases} \dfrac{a}{b}x, & \text{if } 0 \leq x \leq b \\ a, & \text{if } b < x \leq \pi - b \\ \dfrac{a}{b}(\pi - x), & \text{if } \pi - b < x \leq \pi + b \\ -a, & \text{if } \pi + b < x \leq 2\pi - b \\ \dfrac{a}{b}(x - 2\pi), & \text{if } 2\pi - b < x \leq 2\pi \end{cases}$$

for the case $a = b = 1$.

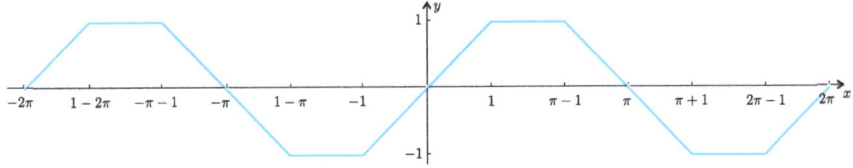

Figure 3.55: The function of exercise 45 if $a = b = 1$

It is easy to see that f is odd; therefore, $a_n = 0$, $\forall n \in \mathbb{N}_0$. By Proposition 3.5.3 the coefficients b_n can be calculated by the formulas:

$$b_n = \frac{1}{\pi}\int_0^{2\pi} f(x)\sin(nx)\,dx = \frac{1}{\pi}\int_{-\pi}^{\pi} f(x)\sin(nx)\,dx = \frac{2}{\pi}\int_0^{\pi} f(x)\sin(nx)\,dx$$

$$= \frac{2}{\pi}\left(\int_0^b \frac{a}{b}x\sin(nx)\,dx + \int_b^{\pi-b} a\sin(nx)\,dx + \int_{\pi-b}^{\pi} \frac{a}{b}(\pi - x)\sin(nx)\,dx\right)$$

$$= \frac{2}{\pi}\left(\frac{a}{b}\left[-x\frac{\cos(nx)}{n}\right]_0^b + \frac{a}{b}\int_0^b \frac{\cos(nx)}{n}\,dx - a\left[\frac{\cos(nx)}{n}\right]_b^{\pi-b} + \frac{a}{b}\int_{\pi-b}^\pi (\pi-x)\sin(nx)\,dx\right)$$

$$= \frac{2}{\pi}\left(-\frac{a}{n}\cos(nb) + \frac{a}{b}\left[\frac{\sin(nx)}{n^2}\right]_0^b - a\left[\frac{\cos(nx)}{n}\right]_b^{\pi-b} - \frac{a}{b}\left[(\pi-x)\frac{\cos(nx)}{n}\right]_{\pi-b}^\pi - \frac{a}{b}\int_{\pi-b}^\pi \frac{\cos(nx)}{n}\,dx\right)$$

$$= \frac{2}{\pi}\left(-\frac{a}{n}\cos(nb) + \frac{a}{n^2 b}\sin(nb) - a\left[\frac{\cos(nx)}{n}\right]_b^{\pi-b} + \frac{a}{n}\cos(n\pi - nb) - \frac{a}{b}\int_{\pi-b}^\pi \frac{\cos(nx)}{n}\,dx\right)$$

$$= \frac{2}{\pi}\left(-\frac{a}{n}\cos(nb) + \frac{a}{n^2 b}\sin(nb) - \frac{a}{n}(\cos(n\pi - nb) - \cos(nb)) + \frac{a}{n}\cos(n\pi - nb) - \frac{a}{b}\left[\frac{\sin(nx)}{n^2}\right]_{\pi-b}^\pi\right)$$

$$= \frac{2}{\pi}\left(\frac{a}{n^2 b}\sin(nb) + \frac{a}{n^2 b}\sin(n\pi - nb)\right) = \frac{2a}{b\,n^2\pi}(\sin(nb) + \sin(n\pi - nb)) = \frac{2a}{b\,n^2\pi}(1 - (-1)^n)\sin(nb).$$

The Fourier series of f is

$$\sum_{n=1}^\infty \frac{2a}{b\,n^2\pi}(1 - (-1)^n)\sin(nb)\sin(nx) = \frac{4a}{b\,\pi}\sum_{n=1}^\infty \frac{\sin((2n-1)b)}{(2n-1)^2}\sin((2n-1)x).$$

Since f is a continuous function and piecewise of class C^1, Theorem 3.5.1 allows us to conclude that the series converges pointwise to f.

46. Let $f:[-\pi,\pi] \to \mathbb{R}$ be piecewise C^1 and

$$\frac{a_0}{2} + \sum_{n=1}^\infty (a_n \cos(nx) + b_n \sin(nx))$$

its Fourier series.

It is easy to prove by change of variables that

$$\int_{-\pi}^\pi f(-x)\,dx = \int_{-\pi}^\pi f(x)\,dx;$$

$$\int_{-\pi}^\pi f(-x)\cos(nx)\,dx = \int_{-\pi}^\pi f(x)\cos(nx)\,dx;$$

$$\int_{-\pi}^\pi f(-x)\sin(nx)\,dx = -\int_{-\pi}^\pi f(x)\sin(nx)\,dx.$$

Let us calculate the Fourier coefficients of $g(x) = \dfrac{f(x) + f(-x)}{2}$.

$$c_0 = \frac{1}{\pi}\int_{-\pi}^\pi g(x)\,dx = \frac{1}{\pi}\int_{-\pi}^\pi \frac{f(x) + f(-x)}{2}\,dx = \frac{1}{2\pi}\left(\int_{-\pi}^\pi f(x)\,dx + \int_{-\pi}^\pi f(-x)\,dx\right)$$

$$= \frac{1}{\pi}\int_{-\pi}^\pi f(x)\,dx = a_0;$$

3.6. Solved Exercises

$$c_n = \frac{1}{\pi}\int_{-\pi}^{\pi} g(x)\cos(nx)\,dx = \frac{1}{\pi}\int_{-\pi}^{\pi}\frac{f(x)+f(-x)}{2}\cos(nx)\,dx$$

$$= \frac{1}{2\pi}\left(\int_{-\pi}^{\pi} f(x)\cos(nx)\,dx + \int_{-\pi}^{\pi} f(-x)\cos(nx)\,dx\right)$$

$$= \frac{1}{\pi}\int_{-\pi}^{\pi} f(x)\cos(nx)\,dx = a_n;$$

$$d_n = \frac{1}{\pi}\int_{-\pi}^{\pi} g(x)\sin(nx)\,dx = \frac{1}{\pi}\int_{-\pi}^{\pi}\frac{f(x)+f(-x)}{2}\sin(nx)\,dx$$

$$= \frac{1}{2\pi}\left(\int_{-\pi}^{\pi} f(x)\sin(nx)\,dx + \int_{-\pi}^{\pi} f(-x)\sin(nx)\,dx\right) = 0.$$

The Fourier series of g is

$$\frac{a_0}{2} + \sum_{n=1}^{\infty} a_n \cos(nx).$$

The coefficients of $h(x) = \dfrac{f(x) - f(-x)}{2}$ are

$$c_0^* = \frac{1}{\pi}\int_{-\pi}^{\pi} h(x)\,dx = \frac{1}{\pi}\int_{-\pi}^{\pi}\frac{f(x)-f(-x)}{2}\,dx = \frac{1}{2\pi}\left(\int_{-\pi}^{\pi} f(x)\,dx - \int_{-\pi}^{\pi} f(-x)\,dx\right) = 0;$$

$$c_n^* = \frac{1}{\pi}\int_{-\pi}^{\pi} h(x)\cos(nx)\,dx = \frac{1}{\pi}\int_{-\pi}^{\pi}\frac{f(x)-f(-x)}{2}\cos(nx)\,dx$$

$$= \frac{1}{2\pi}\left(\int_{-\pi}^{\pi} f(x)\cos(nx)\,dx - \int_{-\pi}^{\pi} f(-x)\cos(nx)\,dx\right) = 0;$$

$$d_n^* = \frac{1}{\pi}\int_{-\pi}^{\pi} h(x)\sin(nx)\,dx = \frac{1}{\pi}\int_{-\pi}^{\pi}\frac{f(x)-f(-x)}{2}\sin(nx)\,dx$$

$$= \frac{1}{2\pi}\left(\int_{-\pi}^{\pi} f(x)\sin(nx)\,dx - \int_{-\pi}^{\pi} f(-x)\sin(nx)\,dx\right)$$

$$= \frac{1}{\pi}\int_{-\pi}^{\pi} f(x)\sin(nx)\,dx = b_n.$$

The Fourier series of h is

$$\sum_{n=1}^{\infty} b_n \sin(nx).$$

47. Let $a \in \mathbb{R}$. From $f(x+\pi) = -f(x)$, $\forall x \in \mathbb{R}$, and the fact that f is periodic with period 2π, we get

$$\int_a^{a+\pi} f(x)\cos(2nx)\,dx = -\int_{a+\pi}^{a+2\pi} f(x)\cos(2nx)\,dx \tag{3.15}$$

$$\int_a^{a+\pi} f(x)\sin(2nx)\,dx = -\int_{a+\pi}^{a+2\pi} f(x)\sin(2nx)\,dx. \tag{3.16}$$

Indeed, by the variable change $t = x - \pi$, we obtain

$$\int_{a+\pi}^{a+2\pi} f(x) \cos(2nx) \, dx = \int_a^{a+\pi} f(t+\pi) \cos(2n(t+\pi)) \, dt = -\int_a^{a+\pi} f(t) \cos(2nt + 2n\pi) \, dt$$

$$\int_{a+\pi}^{a+2\pi} f(x) \sin(2nx) \, dx = \int_a^{a+\pi} f(t+\pi) \sin(2n(t+\pi)) \, dt = -\int_a^{a+\pi} f(t) \sin(2nt + 2n\pi) \, dt.$$

Moreover, the result follows from the periodicity of the two trigonometric functions. Similarly, we can show that

$$\int_a^{a+\pi} f(x) \cos((2n-1)x) \, dx = \int_{a+\pi}^{a+2\pi} f(x) \cos((2n-1)x) \, dx \tag{3.17}$$

$$\int_a^{a+\pi} f(x) \sin((2n-1)x) \, dx = \int_{a+\pi}^{a+2\pi} f(x) \sin((2n-1)x) \, dx. \tag{3.18}$$

Let us calculate the Fourier coefficients of the function f using equalities (3.15), (3.16), (3.17), and (3.18).

$$a_0 = \frac{1}{\pi} \int_{-\pi}^{\pi} f(x) \, dx = \frac{1}{\pi} \left(\int_{-\pi}^{0} f(x) \, dx + \int_0^{\pi} f(x) \, dx \right) = \frac{1}{\pi} \left(-\int_0^{\pi} f(x) \, dx + \int_0^{\pi} f(x) \, dx \right) = 0;$$

$$a_n = \frac{1}{\pi} \int_{-\pi}^{\pi} f(x) \cos(nx) \, dx = \frac{1}{\pi} \left(\int_{-\pi}^{0} f(x) \cos(nx) \, dx + \int_0^{\pi} f(x) \cos(nx) \, dx \right)$$

$$= \begin{cases} \frac{1}{\pi} \left(-\int_0^{\pi} f(x) \cos(nx) \, dx + \int_0^{\pi} f(x) \cos(nx) \, dx \right) = 0, & \text{if } n \text{ is even} \\ \frac{1}{\pi} \left(\int_0^{\pi} f(x) \cos(nx) \, dx + \int_0^{\pi} f(x) \cos(nx) \, dx \right) = \frac{2}{\pi} \int_0^{\pi} f(x) \cos(nx) \, dx, & \text{if } n \text{ is odd;} \end{cases}$$

$$b_n = \frac{1}{\pi} \int_{-\pi}^{\pi} f(x) \sin(nx) \, dx = \frac{1}{\pi} \left(\int_{-\pi}^{0} f(x) \sin(nx) \, dx + \int_0^{\pi} f(x) \sin(nx) \, dx \right)$$

$$= \begin{cases} \frac{1}{\pi} \left(-\int_0^{\pi} f(x) \sin(nx) \, dx + \int_0^{\pi} f(x) \sin(nx) \, dx \right) = 0, & \text{if } n \text{ is even} \\ \frac{1}{\pi} \left(\int_0^{\pi} f(x) \sin(nx) \, dx + \int_0^{\pi} f(x) \sin(nx) \, dx \right) = \frac{2}{\pi} \int_0^{\pi} f(x) \sin(nx) \, dx, & \text{if } n \text{ is odd.} \end{cases}$$

The Fourier series of f is

$$\sum_{n=1}^{\infty} \left(a_{2n-1} \cos((2n-1)x) + b_{2n-1} \sin((2n-1)x) \right).$$

48. a) The graph of the function

$$f(x) = \begin{cases} 0, & \text{if } 0 \leq x < 2 \\ x - 2, & \text{if } 2 \leq x < 4 \end{cases}$$

and its odd extension can be seen in Fig. 3.56.

The analytical expression of the odd extension to the interval $]-4, 4[$ is

$$\tilde{f}(x) = \begin{cases} x + 2, & \text{if } -4 < x \leq -2 \\ 0, & \text{if } -2 < x < 2 \\ x - 2, & \text{if } 2 \leq x < 4. \end{cases}$$

3.6. Solved Exercises

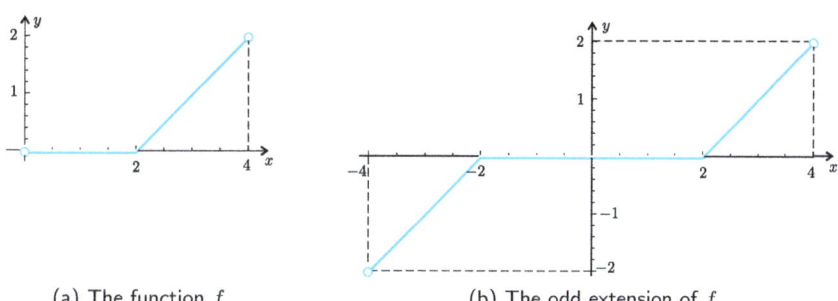

(a) The function f (b) The odd extension of f

Figure 3.56: Graphs of exercise 48

b) Since $\tilde{f}$ is an odd function, the Fourier series expansion is, in fact, a Fourier sine series. We can consider $\tilde{f}$ periodic with period $2l = 8$. The coefficients of the Fourier sine series are

$$b_n = \frac{2}{4}\int_0^4 f(x)\sin\left(\frac{n\pi x}{4}\right)dx = \frac{1}{2}\int_2^4 (x-2)\sin\left(\frac{n\pi x}{4}\right)dx$$

$$= \frac{1}{2}\left(\left[-\frac{4(x-2)}{n\pi}\cos\left(\frac{n\pi x}{4}\right)\right]_2^4 + \int_2^4 \frac{4}{n\pi}\cos\left(\frac{n\pi x}{4}\right)dx\right)$$

$$= \frac{1}{2}\left(-\frac{8}{n\pi}\cos(n\pi) + \left(\frac{4}{n\pi}\right)^2\left[\sin\left(\frac{n\pi x}{4}\right)\right]_2^4\right)$$

$$= \frac{1}{2}\left(-\frac{8(-1)^n}{n\pi} - \left(\frac{4}{n\pi}\right)^2\sin\left(\frac{n\pi}{2}\right)\right)$$

$$= -\frac{4(-1)^n}{n\pi} - \frac{8}{(n\pi)^2}\sin\left(\frac{n\pi}{2}\right),$$

and the series is

$$\frac{4}{\pi}\sum_{n=1}^{\infty}\left(-\frac{(-1)^n}{n} - \frac{2}{n^2\pi}\sin\left(\frac{n\pi}{2}\right)\right)\sin\left(\frac{n\pi x}{4}\right).$$

By Theorem 3.5.2, the series converges pointwise to the periodic function with period 8, illustrated in Fig. 3.57.

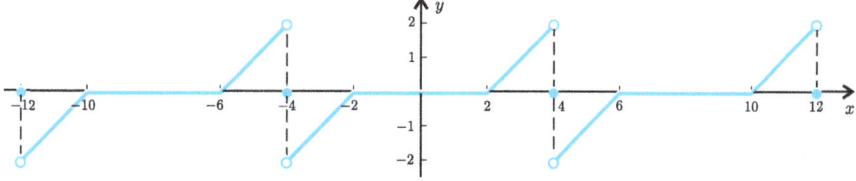

Figure 3.57: The sum of the series of exercise 48

49. If we want the Fourier cosine series of the function

$$f(x) = \begin{cases} 1, & \text{if } 0 \leq x < \pi \\ 0, & \text{if } \pi < x \leq 2\pi, \end{cases}$$

we must consider its extension as an even function, as illustrated in Fig. 3.58. By Proposition 3.5.7, the Fourier series of f is

$$\frac{a_0}{2} + \sum_{n=1}^{\infty} a_n \cos\left(\frac{nx}{2}\right),$$

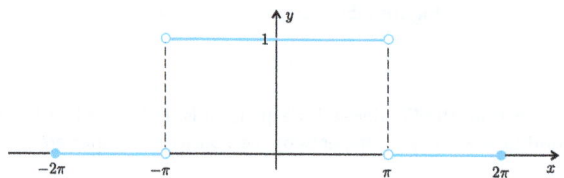

Figure 3.58: The even extension of the function f of exercise 49

where $a_n = \frac{2}{2\pi} \int_0^\pi f(x) \cos\left(\frac{nx}{2}\right) dx$, $n \in \mathbb{N}_0$, that is,

$$a_0 = \frac{2}{2\pi} \int_0^{2\pi} f(x)\, dx = \frac{1}{\pi} \int_0^{2\pi} f(x)\, dx = \frac{1}{\pi} \int_0^\pi 1\, dx = \frac{1}{\pi} \big[x\big]_0^\pi = 1;$$

$$a_n = \frac{1}{\pi} \int_0^{2\pi} f(x) \cos\left(\frac{nx}{2}\right) dx = \frac{1}{\pi} \int_0^\pi \cos\left(\frac{nx}{2}\right) dx = \frac{1}{\pi}\left[\frac{2}{n}\sin\left(\frac{nx}{2}\right)\right]_0^\pi = \frac{2}{n\pi}\sin\left(\frac{n\pi}{2}\right).$$

The Fourier series of f is

$$\frac{1}{2} + \frac{2}{\pi} \sum_{n=1}^{\infty} \frac{\sin\left(\frac{n\pi}{2}\right)}{n} \cos\left(\frac{nx}{2}\right) = \frac{1}{2} + \frac{2}{\pi} \sum_{n=1}^{\infty} \frac{(-1)^{n+1}}{2n-1} \cos\left(\frac{(2n-1)x}{2}\right).$$

50. Let us consider the function $f(x) = \pi - x$.

a) If we want the Fourier sine series of the function, we must consider its extension as an odd function, as illustrated in Fig. 3.59a.

By Proposition 3.5.8, the Fourier series of f is $\sum_{n=1}^{\infty} b_n \sin(nx)$, where $b_n = \frac{2}{\pi} \int_0^\pi f(x) \sin(nx)\, dx$, $n \in \mathbb{N}$, that is,

$$b_n = \frac{2}{\pi} \int_0^\pi f(x) \sin(nx)\, dx = \frac{2}{\pi} \int_0^\pi (\pi - x) \sin(nx)\, dx$$

$$= \frac{2}{\pi} \left(\left[-(\pi - x)\frac{\cos(nx)}{n}\right]_0^\pi - \int_0^\pi \frac{\cos(nx)}{n}\, dx \right)$$

$$= \frac{2}{\pi}\left(\frac{\pi}{n} - \left[\frac{\sin(nx)}{n^2}\right]_0^\pi\right) = \frac{2}{n}.$$

3.6. Solved Exercises

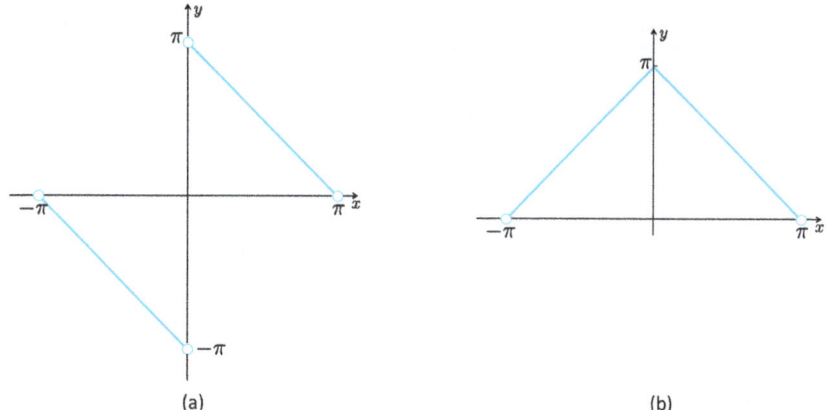

Figure 3.59: (a) The odd extension of $f(x) = \pi - x$. (b) The even extension of $f(x) = \pi - x$

The Fourier series of f is
$$\sum_{n=1}^{\infty} \frac{2}{n} \sin(nx).$$

By Theorem 3.5.1, the series converges pointwise, on the interval $]0, 2\pi[$, to the function $\tilde{f}(x) = \pi - x$. From this fact, it follows that
$$\sum_{n=1}^{\infty} \frac{\sin(nx)}{n} = \frac{\pi - x}{2}, \quad 0 < x < 2\pi.$$

b) The periodic function of period 2π defined on $]0, 2\pi[$ by $g(x) = \dfrac{\pi - x}{2}$ is piecewise continuous. By Theorem 3.5.7 we have
$$\int_0^x \frac{\pi - t}{2}\, dt = \int_0^x \sum_{n=1}^{\infty} \frac{\sin(nt)}{n}\, dt = \sum_{n=1}^{\infty} \int_0^x \frac{\sin(nt)}{n}\, dt, \quad 0 < x < 2\pi,$$

or,
$$\frac{\pi x}{2} - \frac{x^2}{4} = \sum_{n=1}^{\infty} \left(-\frac{\cos(nx)}{n^2} + \frac{1}{n^2} \right) = -\sum_{n=1}^{\infty} \frac{\cos(nx)}{n^2} + \sum_{n=1}^{\infty} \frac{1}{n^2}.$$

Taking $x = 0$ in exercise 43 b), we obtain
$$2\pi^2 = \frac{4\pi^2}{3} + 4 \sum_{n=1}^{\infty} \frac{1}{n^2};$$

therefore,
$$\sum_{n=1}^{\infty} \frac{1}{n^2} = \frac{\pi^2}{6}.$$

Then
$$\sum_{n=1}^{\infty} \frac{\cos(nx)}{n^2} = \frac{\pi^2}{6} - \frac{\pi x}{2} + \frac{x^2}{4} = \frac{3x^2 - 6\pi x + 2\pi^2}{12}.$$

c) If we want the Fourier cosine series of the function, we must consider its extension as an even function, as illustrated in Fig. 3.59b. By Proposition 3.5.7,

$$a_0 = \frac{1}{\pi}\int_{-\pi}^{\pi} f(x)\,dx = \frac{2}{\pi}\int_0^{\pi} f(x)\,dx = \frac{2}{\pi}\int_0^{\pi}(\pi - x)\,dx = \frac{2}{\pi}\left[\pi x - \frac{x^2}{2}\right]_0^{\pi} = \pi;$$

$$a_n = \frac{2}{\pi}\int_0^{\pi} f(x)\cos(nx)\,dx = \frac{2}{\pi}\int_0^{\pi}(\pi - x)\cos(nx)\,dx$$

$$= \frac{2}{\pi}\left(\left[\frac{(\pi - x)}{n}\sin(nx)\right]_0^{\pi} + \int_0^{\pi}\frac{1}{n}\sin(nx)\,dx\right)$$

$$= \frac{2}{n\pi}\left[-\frac{\cos(nx)}{n}\right]_0^{\pi} = \frac{2}{n^2\pi}(1 - (-1)^n).$$

The Fourier series of f is

$$\frac{\pi}{2} + \frac{2}{\pi}\sum_{n=1}^{\infty}\frac{1-(-1)^n}{n^2}\cos(nx) = \frac{\pi}{2} + \frac{4}{\pi}\sum_{n=1}^{\infty}\frac{1}{(2n-1)^2}\cos\left((2n-1)x\right).$$

d) As the even extension of the function f, $\tilde{f}(x) = \pi - |x|$, is a continuous function and piecewise C^1 on $]-\pi, \pi[$, we have

$$\pi - |x| = \frac{\pi}{2} + \frac{2}{\pi}\sum_{n=1}^{\infty}\frac{1-(-1)^n}{n^2}\cos(nx), \quad -\pi < x < \pi.$$

In particular, if $x = 0$,

$$\pi = \frac{\pi}{2} + \frac{2}{\pi}\sum_{n=1}^{\infty}\frac{1-(-1)^n}{n^2}.$$

Since the series $\sum_{n=1}^{\infty}\frac{1}{n^2}$ and $\sum_{n=1}^{\infty}\frac{(-1)^n}{n^2}$ are convergent, Theorem 2.2.2 allows us to write

$$\pi = \frac{\pi}{2} + \frac{2}{\pi}\sum_{n=1}^{\infty}\frac{1}{n^2} - \frac{2}{\pi}\sum_{n=1}^{\infty}\frac{(-1)^n}{n^2}.$$

By item b),

$$\pi = \frac{\pi}{2} + \frac{2}{\pi}\cdot\frac{\pi^2}{6} - \frac{2}{\pi}\sum_{n=1}^{\infty}\frac{(-1)^n}{n^2}.$$

Finally,

$$\sum_{n=1}^{\infty}\frac{(-1)^n}{n^2} = -\frac{\pi^2}{12}.$$

51. Let us consider the function $f(x) = x$.

a) If we want the Fourier sine series of the function, we must consider its extension as an odd function, as illustrated in Fig. 3.60a. By Proposition 3.5.8, the Fourier series of f is $\sum_{n=1}^{\infty} b_n \sin(nx)$, where

3.6. Solved Exercises

$b_n = \dfrac{2}{\pi} \int_0^\pi f(x) \, \sin(nx) \, dx$, $n \in \mathbb{N}$, that is,

$$\begin{aligned}
b_n &= \dfrac{2}{\pi} \int_0^\pi f(x) \, \sin(nx) \, dx = \dfrac{2}{\pi} \int_0^\pi x \, \sin(nx) \, dx \\
&= \dfrac{2}{\pi} \left(\left[-x \dfrac{\cos(nx)}{n} \right]_0^\pi + \int_0^\pi \dfrac{\cos(nx)}{n} \, dx \right) \\
&= \dfrac{2}{\pi} \left(-\dfrac{\pi(-1)^n}{n} + \left[\dfrac{\sin(nx)}{n^2} \right]_0^\pi \right) = \dfrac{2(-1)^{n+1}}{n}.
\end{aligned}$$

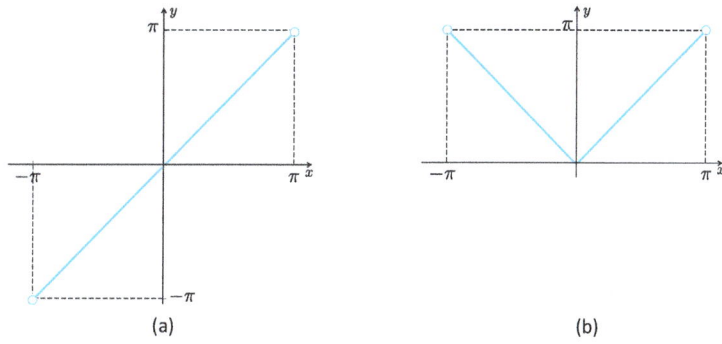

(a) (b)

Figure 3.60: (a) The odd extension of $f(x) = x$. (b) The even extension of $f(x) = x$

The Fourier series of f is

$$\sum_{n=1}^\infty \dfrac{2(-1)^{n+1}}{n} \sin(nx).$$

By Theorem 3.5.1, the series converges pointwise, on the interval $]-\pi, \pi[$, to the function $\tilde{f}(x) = x$. From this fact, it follows that

$$\sum_{n=1}^\infty (-1)^{n+1} \dfrac{\sin(nx)}{n} = \dfrac{x}{2}, \quad -\pi < x < \pi.$$

b) If we want the Fourier cosine series of the function, we must consider its extension as an even function, as illustrated in Fig. 3.60b. By Proposition 3.5.7,

$$\begin{aligned}
a_0 &= \dfrac{1}{\pi} \int_{-\pi}^\pi f(x) \, dx = \dfrac{2}{\pi} \int_0^\pi f(x) \, dx = \dfrac{2}{\pi} \int_0^\pi x \, dx = \dfrac{2}{\pi} \left[\dfrac{x^2}{2} \right]_0^\pi = \pi; \\
a_n &= \dfrac{2}{\pi} \int_0^\pi f(x) \cos(nx) \, dx = \dfrac{2}{\pi} \int_0^\pi x \cos(nx) \, dx \\
&= \dfrac{2}{\pi} \left(\left[\dfrac{x}{n} \sin(nx) \right]_0^\pi - \int_0^\pi \dfrac{1}{n} \sin(nx) \, dx \right) \\
&= \dfrac{2}{n\pi} \left[\dfrac{\cos(nx)}{n} \right]_0^\pi = \dfrac{2}{n^2 \pi} ((-1)^n - 1).
\end{aligned}$$

The Fourier series of f is

$$\frac{\pi}{2} + \frac{2}{\pi}\sum_{n=1}^{\infty}\frac{(-1)^n - 1}{n^2}\cos(nx) = \frac{\pi}{2} - \frac{4}{\pi}\sum_{n=1}^{\infty}\frac{1}{(2n-1)^2}\cos((2n-1)x).$$

52. The function $f(x) = |\sin(x)|$ is even (see Fig. 3.61); therefore, $b_n = 0$, $\forall n \in \mathbb{N}$. In addition, f is periodic with period π, so it also has period 2π. Let us calculate the Fourier coefficients.

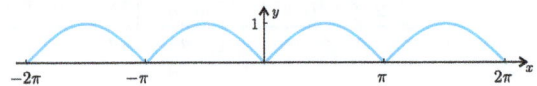

Figure 3.61: The function $f(x) = |\sin(x)|$

$$a_0 = \frac{2}{\pi}\int_0^{\pi} f(x)\,dx = \frac{2}{\pi}\int_0^{\pi}\sin(x)\,dx = \frac{2}{\pi}\big[-\cos(x)\big]_0^{\pi} = \frac{4}{\pi};$$

$$a_n = \frac{2}{\pi}\int_0^{\pi}\sin(x)\cos(nx)\,dx = \frac{2}{\pi}\int_0^{\pi}\frac{1}{2}\Big(\sin((n+1)x) + \sin((1-n)x)\Big)\,dx$$

$$= \begin{cases} \dfrac{1}{\pi}\left[-\dfrac{\cos((n+1)x)}{n+1} - \dfrac{\cos((1-n)x)}{1-n}\right]_0^{\pi}, & \text{if } n \ne 1 \\ \dfrac{1}{\pi}\left[-\dfrac{\cos(2x)}{2}\right]_0^{\pi}, & \text{if } n = 1 \end{cases}$$

$$= \begin{cases} \dfrac{1}{\pi}\left(\dfrac{(-1)^n + 1}{n+1} - \dfrac{(-1)^n + 1}{n-1}\right) = -\dfrac{2}{\pi}\cdot\dfrac{(-1)^n + 1}{n^2 - 1}, & \text{if } n \ne 1 \\ 0, & \text{if } n = 1. \end{cases}$$

The Fourier series of f is

$$\frac{2}{\pi} - \frac{2}{\pi}\sum_{n=2}^{\infty}\frac{(-1)^n + 1}{n^2 - 1}\cos(nx) = \frac{2}{\pi} - \frac{4}{\pi}\sum_{n=1}^{\infty}\frac{1}{4n^2 - 1}\cos(2nx).$$

Since f is a continuous function and piecewise C^1, Theorem 3.5.1 allows us to conclude that the series converges pointwise to f on $\mathbb{R}$.

Let $x = 0$. Then

$$f(0) = \frac{2}{\pi} - \frac{4}{\pi}\sum_{n=1}^{\infty}\frac{1}{4n^2 - 1}.$$

Since $f(0) = 0$, we have

$$\sum_{n=1}^{\infty}\frac{1}{4n^2 - 1} = \frac{1}{2}.$$

Let $x = \dfrac{\pi}{2}$. Then

$$f\left(\frac{\pi}{2}\right) = \frac{2}{\pi} - \frac{4}{\pi}\sum_{n=1}^{\infty}\frac{1}{4n^2 - 1}\cos(n\pi) = \frac{2}{\pi} - \frac{4}{\pi}\sum_{n=1}^{\infty}\frac{(-1)^n}{4n^2 - 1}.$$

3.6. Solved Exercises

Since $f\left(\dfrac{\pi}{2}\right) = 1$, we have

$$\frac{2}{\pi} - 1 = \frac{4}{\pi} \sum_{n=1}^{\infty} \frac{(-1)^n}{4n^2 - 1},$$

that is,

$$\sum_{n=1}^{\infty} \frac{(-1)^n}{4n^2 - 1} = \frac{1}{2} - \frac{\pi}{4}.$$

53. Since the function $f(x) = x^2(\pi - x)^2$ is defined on $]0, \pi[$ and we want to obtain a Fourier cosine series, we consider an even extension of f to the interval $]-\pi, \pi[$ (see Fig. 3.62). Since the function is even, the coefficients b_n are zero. Let us calculate a_0 and a_n.

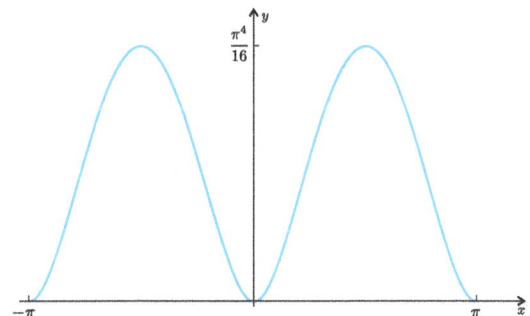

Figure 3.62: The even extension of the function of exercise 53

$$a_0 = \frac{2}{\pi} \int_0^\pi f(x)\,dx = \frac{2}{\pi} \int_0^\pi x^2(\pi - x)^2\,dx = \frac{2}{\pi} \int_0^\pi (x^4 - 2\pi x^3 + \pi^2 x^2)\,dx$$

$$= \frac{2}{\pi} \left[\frac{x^5}{5} - \frac{\pi x^4}{2} + \frac{\pi^2 x^3}{3} \right]_0^\pi = \frac{\pi^4}{15};$$

$$a_n = \frac{2}{\pi} \int_0^\pi x^2(\pi - x)^2 \cos(nx)\,dx = \frac{2}{\pi} \int_0^\pi (x^4 - 2\pi x^3 + \pi^2 x^2) \cos(nx)\,dx$$

$$= \frac{2}{\pi} \left(\left[(x^4 - 2\pi x^3 + \pi^2 x^2) \frac{\sin(nx)}{n} \right]_0^\pi - \int_0^\pi (4x^3 - 6\pi x^2 + 2\pi^2 x) \frac{\sin(nx)}{n}\,dx \right)$$

$$= \frac{2}{\pi n} \left(\left[(4x^3 - 6\pi x^2 + 2\pi^2 x) \frac{\cos(nx)}{n} \right]_0^\pi - \int_0^\pi (12x^2 - 12\pi x + 2\pi^2) \frac{\cos(nx)}{n}\,dx \right)$$

$$= \frac{2}{\pi n^2} \left(- \left[(12x^2 - 12\pi x + 2\pi^2) \frac{\sin(nx)}{n} \right]_0^\pi + \int_0^\pi (24x - 12\pi) \frac{\sin(nx)}{n}\,dx \right)$$

$$= \frac{2}{\pi n^3} \left(- \left[(24x - 12\pi) \frac{\cos(nx)}{n} \right]_0^\pi + \int_0^\pi 24 \frac{\cos(nx)}{n}\,dx \right)$$

$$= \frac{2}{\pi n^4} \left(-12\pi \cos(n\pi) - 12\pi + 24 \left[\frac{\sin(nx)}{n} \right]_0^\pi \right) = -\frac{24}{n^4}((-1)^n + 1)$$

The Fourier series of f is

$$\frac{\pi^4}{30} - 24 \sum_{n=2}^{\infty} \frac{(-1)^n + 1}{n^4} \cos(nx) = \frac{\pi^4}{30} - 3 \sum_{n=1}^{\infty} \frac{1}{n^4} \cos(2nx).$$

The extension, $\tilde{f}$, of f is a continuous function on $[-\pi, \pi]$, and Theorem 3.5.1 allows us to conclude that the series converges pointwise to $\tilde{f}$ on $[-\pi, \pi]$, that is,

$$\tilde{f}(x) = \frac{\pi^4}{30} - 3 \sum_{n=1}^{\infty} \frac{1}{n^4} \cos(2nx), \quad \forall x \in [-\pi, \pi].$$

The analytical expression of $\tilde{f}$ is

$$\tilde{f}(x) = \begin{cases} x^2(\pi + x)^2, & \text{if } -\pi \leq x \leq 0 \\ x^2(\pi - x)^2, & \text{if } 0 \leq x \leq \pi. \end{cases}$$

Let $x = 0$. Then

$$\tilde{f}(0) = \frac{\pi^4}{30} - 3 \sum_{n=1}^{\infty} \frac{1}{n^4}.$$

Since $\tilde{f}(0) = 0$, we have

$$\sum_{n=1}^{\infty} \frac{1}{n^4} = \frac{\pi^4}{90}.$$

Let $x = \dfrac{\pi}{2}$. Then

$$\tilde{f}\left(\frac{\pi}{2}\right) = \frac{\pi^4}{30} - 3 \sum_{n=1}^{\infty} \frac{1}{n^4} \cos(n\pi) = \frac{\pi^4}{30} - 3 \sum_{n=1}^{\infty} \frac{(-1)^n}{n^4}.$$

Since $\tilde{f}\left(\dfrac{\pi}{2}\right) = \dfrac{\pi^4}{16}$, we have

$$\tilde{f}\left(\frac{\pi}{2}\right) - \tilde{f}(0) = \frac{\pi^4}{16} = \frac{\pi^4}{30} - 3 \sum_{n=1}^{\infty} \frac{(-1)^n}{n^4} - \frac{\pi^4}{30} + 3 \sum_{n=1}^{\infty} \frac{1}{n^4} = 3 \sum_{n=1}^{\infty} \frac{1 - (-1)^n}{n^4},$$

that is, $\dfrac{\pi^4}{16} = 6 \sum_{n=1}^{\infty} \dfrac{1}{(2n-1)^4}$, which is equivalent to $\dfrac{\pi^4}{96} = \sum_{n=0}^{\infty} \dfrac{1}{(2n+1)^4}$.

54. a) The series $\sum_{n=1}^{\infty} n\, a_n \sin(nx)$ results from term by term differentiation of the original series. In fact, according to Theorem 3.5.8, by term by term differentiation of the series of $\tilde{f}$, we obtain the Fourier series of $\tilde{f}'$, that is,

$$\left(\frac{\pi}{2} + \frac{2}{\pi} \sum_{n=1}^{\infty} \frac{(-1)^n - 1}{n^2} \cos(nx)\right)' = \frac{2}{\pi} \sum_{n=1}^{\infty} \left(\frac{(-1)^n - 1}{n^2} \cos(nx)\right)' = -\frac{2}{\pi} \sum_{n=1}^{\infty} n \frac{(-1)^n - 1}{n^2} \sin(nx).$$

3.6. Solved Exercises

By the same theorem, this series converges to $\tilde{f}'$ at the points of continuity of $\tilde{f}'$. At the points of discontinuity the sum of the series is equal to the arithmetic mean of the left- and right-hand derivatives of the function $\tilde{f}'$ at these points. Thus,

$$-\frac{2}{\pi}\sum_{n=1}^{\infty} n \frac{(-1)^n - 1}{n^2} \sin(nx) = \begin{cases} \tilde{f}'(x), & \text{if } x \neq k\pi,\ k \in \mathbb{Z} \\ 0, & \text{if } x = k\pi,\ k \in \mathbb{Z}. \end{cases}$$

Therefore, the series $\sum_{n=1}^{\infty} n a_n \sin(nx)$ converges pointwise on $\mathbb{R}$.

b) Let $g(x) = -\sum_{n=1}^{\infty} n a_n \sin(nx)$. By the previous item,

$$g(x) = \begin{cases} \tilde{f}'(x), & \text{if } x \neq k\pi,\ k \in \mathbb{Z} \\ 0, & \text{if } x = k\pi,\ k \in \mathbb{Z} \end{cases}$$

$$= \begin{cases} 1, & \text{if } x \in\ ]2k\pi, (2k+1)\pi[,\ k \in \mathbb{Z} \\ -1, & \text{if } x \in\ ](2k+1)\pi, (2k+2)\pi[,\ k \in \mathbb{Z} \\ 0, & \text{if } x = k\pi,\ k \in \mathbb{Z}. \end{cases}$$

In Fig. 3.63 is illustrated the graph of g on the interval $[-2\pi, 2\pi]$.

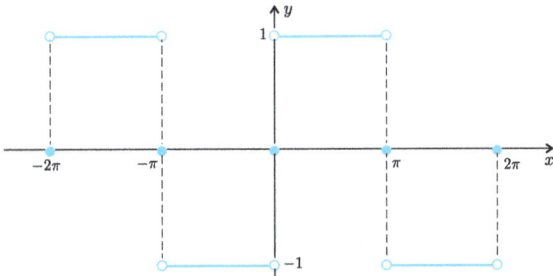

Figure 3.63: The graph of the function g of exercise 54 b)

c) The function $\tilde{f}$ is continuous and piecewise C^1 on $\mathbb{R}$; therefore,

$$\tilde{f}(x) = \frac{\pi}{2} + \frac{2}{\pi}\sum_{n=1}^{\infty} \frac{(-1)^n - 1}{n^2} \cos(nx) = \frac{\pi}{2} - \frac{4}{\pi}\sum_{n=1}^{\infty} \frac{1}{(2n-1)^2} \cos\bigl((2n-1)x\bigr),\ \forall x \in \mathbb{R}.$$

Let $x = 0$. Then, as $f(0) = 0$,

$$0 = \frac{\pi}{2} - \frac{4}{\pi}\sum_{n=1}^{\infty} \frac{1}{(2n-1)^2},$$

which implies that

$$\sum_{n=1}^{\infty} \frac{1}{(2n-1)^2} = \frac{\pi^2}{8}.$$

By Theorem 3.5.7, we have

$$\int_0^x f(t)\, dt = \int_0^x \frac{\pi}{2}\, dt - \frac{4}{\pi} \sum_{n=1}^{\infty} \frac{1}{(2n-1)^2} \int_0^x \cos((2n-1)t)\, dt,$$

that is, if $0 < x < \pi$,

$$\frac{x^2}{2} = \frac{\pi}{2} x - \frac{4}{\pi} \sum_{n=1}^{\infty} \frac{1}{(2n-1)^3} \sin((2n-1)x).$$

For $x = \frac{\pi}{2}$, we obtain

$$\frac{\pi^2}{8} = \frac{\pi^2}{4} + \frac{4}{\pi} \sum_{n=1}^{\infty} \frac{(-1)^n}{(2n-1)^3},$$

that is,

$$\sum_{n=1}^{\infty} \frac{(-1)^n}{(2n-1)^3} = -\frac{\pi^3}{32}.$$

55. Suppose that the function f has a Fourier sine series development, that is, $f(x) = \sum_{n=1}^{\infty} B_n \sin(n\pi x)$ (note that, in this case, the two conditions $f(0) = f(1) = 0$ are satisfied).

Assuming that f and its derivatives meet the conditions of Theorem 3.5.8, we have

$$f''(x) + 3f(x) = \sum_{n=1}^{\infty} \left(B_n \sin(n\pi x)\right)'' + 3 \sum_{n=1}^{\infty} B_n \sin(n\pi x) = -\sum_{n=1}^{\infty} n^2 \pi^2 B_n \sin(n\pi x) + 3 \sum_{n=1}^{\infty} B_n \sin(n\pi x).$$

Let $g(x) = 3x$, $0 < x < 1$. By extending g to an odd function, we obtain the Fourier sine series of g on the interval $]-1, 1[$

$$g(x) \sim \sum_{n=1}^{\infty} b_n \sin(n\pi x),$$

where

$$b_n = 2 \int_0^1 3x \sin(n\pi x)\, dx = 2 \left(\left[-\frac{3x}{n\pi} \cos(n\pi x) \right]_0^1 + \int_0^1 \frac{3}{n\pi} \cos(n\pi x)\, dx \right) = -\frac{6(-1)^n}{n\pi}.$$

Since $g(x) = 3x$ is a C^1 function on $]-1, 1[$, we have the equality

$$3x = \sum_{n=1}^{\infty} \frac{6(-1)^{n+1}}{n\pi} \sin(n\pi x), \quad \forall x \in\,]-1, 1[.$$

Thus,

$$f''(x) + 3f(x) = 3x \Leftrightarrow -\sum_{n=1}^{\infty} n^2 \pi^2 B_n \sin(n\pi x) + 3 \sum_{n=1}^{\infty} B_n \sin(n\pi x) = \sum_{n=1}^{\infty} \frac{6(-1)^{n+1}}{n\pi} \sin(n\pi x)$$

$$\Leftrightarrow \sum_{n=1}^{\infty} \left(-n^2 \pi^2 B_n + 3 B_n\right) \sin(n\pi x) = \sum_{n=1}^{\infty} \frac{6(-1)^{n+1}}{n\pi} \sin(n\pi x).$$

3.6. Solved Exercises

From the uniqueness of the Fourier series, it follows that

$$-n^2\pi^2 B_n + 3B_n = \frac{6(-1)^{n+1}}{n\pi}, \quad \forall n \in \mathbb{N},$$

that is,

$$B_n = \frac{6(-1)^{n+1}}{n\pi(3-n^2\pi^2)}, \quad \forall n \in \mathbb{N}.$$

The desired function is $f(x) = \sum_{n=1}^{\infty} \frac{6(-1)^{n+1}}{n\pi(3-n^2\pi^2)} \sin(n\pi x)$.

56. a) The Fourier coefficients of f are

$$a_0 = \int_0^2 f(x)\,dx = \int_0^2 x^2\,dx = \left[\frac{x^3}{3}\right]_0^2 = \frac{8}{3};$$

$$a_n = \int_0^2 x^2 \cos(n\pi x)\,dx = \left[\frac{x^2}{n\pi}\sin(n\pi x)\right]_0^2 - \int_0^2 \frac{2x}{n\pi}\sin(n\pi x)\,dx$$

$$= -\frac{2}{n\pi}\int_0^2 x\sin(n\pi x)\,dx = -\frac{2}{n\pi}\left(\left[-\frac{x}{n\pi}\cos(n\pi x)\right]_0^2 + \int_0^2 \frac{\cos(n\pi x)}{n\pi}\,dx\right)$$

$$= -\frac{2}{n\pi}\left(-\frac{2}{n\pi} + \frac{1}{n\pi}\int_0^2 \cos(n\pi x)\,dx\right) = \frac{4}{(n\pi)^2} - \frac{2}{(n\pi)^2}\left[\frac{\sin(n\pi x)}{n\pi}\right]_0^2 = \frac{4}{(n\pi)^2};$$

$$b_n = \int_0^2 x^2 \sin(n\pi x)\,dx = \left[-\frac{x^2}{n\pi}\cos(n\pi x)\right]_0^2 - \int_0^2 -\frac{2x}{n\pi}\cos(n\pi x)\,dx$$

$$= -\frac{4}{n\pi} + \frac{2}{n\pi}\int_0^2 x\cos(n\pi x)\,dx = -\frac{4}{n\pi} + \frac{2}{n\pi}\left(\left[\frac{x}{n\pi}\sin(n\pi x)\right]_0^2 - \int_0^2 \frac{\sin(n\pi x)}{n\pi}\,dx\right)$$

$$= -\frac{4}{n\pi} + \frac{2}{(n\pi)^2}\left[\frac{\cos(n\pi x)}{n\pi}\right]_0^2 = -\frac{4}{n\pi}.$$

The Fourier series of f is

$$\frac{4}{3} + \sum_{n=1}^{\infty}\left(\frac{4}{(n\pi)^2}\cos(n\pi x) - \frac{4}{n\pi}\sin(n\pi x)\right)$$

and converges pointwise to the function

$$S(x) = \begin{cases} (x-2k)^2, & \text{if } 2k < x < 2k+2, \ k \in \mathbb{Z} \\ 2, & \text{if } x = 2k, \ k \in \mathbb{Z}. \end{cases}$$

b) Term by term differentiation of the series obtained in item a) results in the series

$$\sum_{n=1}^{\infty}\left(-\frac{4}{(n\pi)^2}n\pi\sin(n\pi x) - \frac{4}{n\pi}n\pi\cos(n\pi x)\right) = \sum_{n=1}^{\infty}\left(-\frac{4}{n\pi}\sin(n\pi x) - 4\cos(n\pi x)\right),$$

which is divergent because if $x \in\,]0,2[$, the general term of the series does not have a limit. Therefore, the sum of the series does not exist and cannot define the function f'.

Note that the function f does not verify the conditions of Theorem 3.5.8. In fact, the periodic extension of f to the interval $]-1,1[$

$$g(x) = \begin{cases} (x+2)^2, & \text{if } -1 < x \leq 0, \\ x^2, & \text{if } 0 < x < 1 \end{cases}$$

is not continuous at 0.

57. a) Let us calculate the Fourier coefficients of f.

$$a_0 = \frac{1}{3}\int_{-3}^{3} f(x)\,dx = \frac{1}{3}\int_{-3}^{3} (x-x^2)\,dx = \frac{1}{3}\left[\frac{x^2}{2} - \frac{x^3}{3}\right]_{-3}^{3} = -6;$$

$$a_n = \frac{1}{3}\int_{-3}^{3} (x-x^2)\cos\left(\frac{n\pi x}{3}\right)dx = \frac{1}{3}\left(\left[\frac{3(x-x^2)}{n\pi}\sin\left(\frac{n\pi x}{3}\right)\right]_{-3}^{3} - \int_{-3}^{3} \frac{3(1-2x)}{n\pi}\sin\left(\frac{n\pi x}{3}\right)dx\right)$$

$$= \frac{1}{n\pi}\int_{-3}^{3} (2x-1)\sin\left(\frac{n\pi x}{3}\right)dx = \frac{1}{n\pi}\left(\left[-\frac{3(2x-1)}{n\pi}\cos\left(\frac{n\pi x}{3}\right)\right]_{-3}^{3} + \int_{-3}^{3}\frac{6}{n\pi}\cos\left(\frac{n\pi x}{3}\right)dx\right)$$

$$= \frac{1}{n\pi}\left(-\frac{36(-1)^n}{n\pi} + \frac{18}{(n\pi)^2}\left[\sin\left(\frac{n\pi x}{3}\right)\right]_{-3}^{3}\right) = -\frac{36(-1)^n}{(n\pi)^2};$$

$$b_n = \frac{1}{3}\int_{-3}^{3} (x-x^2)\sin\left(\frac{n\pi x}{3}\right)dx = \frac{1}{3}\left(\left[\frac{3(x^2-x)}{n\pi}\cos\left(\frac{n\pi x}{3}\right)\right]_{-3}^{3} + \int_{-3}^{3}\frac{3(1-2x)}{n\pi}\cos\left(\frac{n\pi x}{3}\right)dx\right)$$

$$= -\frac{1}{n\pi}\left(6(-1)^n - \left[\frac{3(1-2x)}{n\pi}\sin\left(\frac{n\pi x}{3}\right)\right]_{-3}^{3} - \int_{-3}^{3}\frac{6}{n\pi}\sin\left(\frac{n\pi x}{3}\right)dx\right)$$

$$= -\frac{1}{n\pi}\left(6(-1)^n + \frac{18}{(n\pi)^2}\left[\cos\left(\frac{n\pi x}{3}\right)\right]_{-3}^{3}\right) = -\frac{6(-1)^n}{n\pi}.$$

The Fourier series of f is

$$-3 + \sum_{n=1}^{\infty}\left(-\frac{36(-1)^n}{(n\pi)^2}\cos\left(\frac{n\pi x}{3}\right) - \frac{6(-1)^n}{n\pi}\sin\left(\frac{n\pi x}{3}\right)\right)$$

and, on the interval $]-3,3[$, its sum is equal to f because f is continuous and of class C^1.

b) Since

$$\int_0^x \sin\left(\frac{n\pi t}{3}\right)dt = \frac{3}{n\pi}\left[-\cos\left(\frac{n\pi t}{3}\right)\right]_0^x = \frac{3}{n\pi}\left(1 - \cos\left(\frac{n\pi x}{3}\right)\right)$$

and

$$\int_0^x \cos\left(\frac{n\pi t}{3}\right)dt = \frac{3}{n\pi}\left[\sin\left(\frac{n\pi t}{3}\right)\right]_0^x = \frac{3}{n\pi}\sin\left(\frac{n\pi x}{3}\right),$$

the series

$$\frac{a_0 x}{2} + \sum_{n=1}^{\infty}\frac{3}{n\pi}b_n + \sum_{n=1}^{\infty}\frac{3}{n\pi}\left(a_n\sin\left(\frac{n\pi x}{3}\right) - b_n\cos\left(\frac{n\pi x}{3}\right)\right),$$

3.6. Solved Exercises

where a_0, a_n, and b_n are the Fourier coefficients of f, results from the term by term integration between 0 and x of the series of f. By Theorem 3.5.7, with $a = 0$ and $b = x$, the sum of the series is equal to $\int_0^x f(t)\,dt$, that is,

$$S(x) + \sum_{n=1}^{\infty} \frac{3}{n\pi} b_n = \int_0^x f(t)\,dt = \int_0^x (t - t^2)\,dt = \left[\frac{t^2}{2} - \frac{t^3}{3}\right]_0^x = \frac{x^2}{2} - \frac{x^3}{3}, \quad x \in\,]-3, 3[.$$

Therefore, using the result of exercise 50d),

$$S(x) = \frac{x^2}{2} - \frac{x^3}{3} + \sum_{n=1}^{\infty} \frac{18(-1)^n}{n^2 \pi^2} = \frac{x^2}{2} - \frac{x^3}{3} - \frac{3}{2}, \quad x \in\,]-3, 3[.$$

3.7 Proposed Exercises

1. For each value of x, test the convergence of the series $\sum_{n=1}^{\infty} e^{\frac{x}{2}+nx^3}$ and, in case of convergence, find its sum.

2. For each value of x, test the convergence of the series $\sum_{n=1}^{\infty} (x^n + n^{-3x})$.

3. Considering that $e^x = \sum_{n=0}^{\infty} \frac{x^n}{n!}$, show that
$$\lim_{x \to 0} \frac{e^x - 1}{x} = 1.$$

4. For each value of $x \neq 0$, examine the convergence of the series $\sum_{n=1}^{\infty} \frac{n^\alpha}{x^n}$, $\alpha \in \mathbb{R}$, and show that it is uniformly convergent on the interval $[2,3]$.

5. Show that the series of functions $\sum_{n=2}^{\infty} \left(x^{\frac{1}{n}} - x^{\frac{1}{n-1}} \right)$ converges pointwise but not uniformly on $[0,1]$.

6. For each $n \in \mathbb{N}$, let $f_n(x) = \frac{n^{n+1}}{n!} x^n e^{-nx}$.

   a) Study the pointwise convergence of the series $\sum_{n=1}^{\infty} f_n$ on the interval $]1, 2]$.

   b) Show that the series $\sum_{n=1}^{\infty} f_n$ is uniformly convergent on $[2, +\infty[$. What can be said about the continuity of the sum of the series on this interval?

7. Consider the function $f(x) = \sum_{n=1}^{\infty} \frac{1}{n^x}$.

   a) Determine the domain of f.

   b) Prove that f is continuous on its domain.

8. Consider the function $f(x) = 1 + \sum_{n=1}^{\infty} \frac{e^{nx}}{n^3 + 1}$. Show that $\int_{-1}^{0} f(x)\, dx = 1 + \sum_{n=1}^{\infty} \frac{1 - e^{-n}}{n(n^3 + 1)}$.

9. Consider the series of functions $\sum_{n=1}^{\infty} \frac{\sin(nx)}{n^2}$.

   a) Show that the series converges uniformly on $\mathbb{R}$, but that there are points in $\mathbb{R}$ where the series of derivatives diverges.

   b) Prove that $f(x) = \sum_{n=1}^{\infty} \frac{\sin(nx)}{n^2}$ is integrable on $[0,1]$ and express $\int_{0}^{1} f(x)dx$ as the sum of a series.

10. Consider the series $\sum_{n=1}^{\infty} \frac{n}{\sqrt{n^5 x + 1}}$.

    a) Study the uniform convergence of the series on the interval $[2, 4]$.

    b) If f is the sum of the series, compute $\int_{2}^{4} f(x)\, dx$.

11. Consider f, a real-valued function of a real variable, defined by $f(x) = \sum_{n=1}^{\infty} \frac{x}{n(x+n)}$, $x \in \mathbb{R}_0^+$.

    a) Prove that f is continuous on $[0,1]$.

    b) Calculate $\int_{0}^{1} f(x)dx$.

12. Show that the series of functions $\sum_{n=0}^{\infty} \frac{\cos(nx)}{2^n}$ converges uniformly on $\mathbb{R}$ and that the sum is a differentiable function.

13. Consider f, a real-valued function of a real variable, defined by $f(x) = \sum_{n=1}^{\infty} \frac{\arctan(nx) \sin\left(\frac{1}{n^2}\right)}{n}$.

    a) Find the domain of f. Give a clear explanation of the answer.

    b) Prove that f is differentiable on $\mathbb{R}$ and write its derivative as the sum of a series.

14. Indicate the real values of x for which the following series are absolutely convergent, conditionally convergent, and divergent. Explain the findings for each question.

3.7. Proposed Exercises

a) $\sum_{n=1}^{\infty} n^n x^n$

b) $\sum_{n=1}^{\infty} \left(1+\frac{1}{n}\right)^{2n} \cdot \left(\frac{x}{x+1}\right)^n$

c) $\sum_{n=1}^{\infty} \frac{(-1)^n}{2n-1} \left(\frac{1-x}{1+x}\right)^n$

d) $\sum_{n=1}^{\infty} \frac{(-1)^n n}{2^n (n^2+7)} x^n$

e) $\sum_{n=0}^{\infty} \frac{(x+3)^n}{(n+1)2^n}$

f) $\sum_{n=1}^{\infty} \frac{(-1)^n \sqrt{n+3}}{5^n n^4} (x-3)^n$

g) $\sum_{n=0}^{\infty} \frac{(1-x^2)^n}{\sqrt{n+2}}$

h) $\sum_{n=1}^{\infty} \frac{2^{1/n} \pi^n}{n(n+1)(n+2)} (x-2)^n$

i) $\sum_{n=0}^{\infty} \left(1+\frac{(-1)^n}{2}\right)^n (2x+1)^n$

j) $\sum_{n=1}^{\infty} \frac{(n+1)^n}{(2n^2)^{n/2}} (x+1)^n$

k) $\sum_{n=1}^{\infty} \frac{n}{2^n(3n-1)} (x-1)^n$

l) $\sum_{n=1}^{\infty} \frac{(-1)^n}{3^n \sqrt[3]{n}} (x-4)^n$

m) $\sum_{n=1}^{\infty} \frac{2^n}{n+2} \left(\frac{1}{x-1}\right)^n$

n) $\sum_{n=0}^{\infty} \frac{n^3+3^n}{2^n} (x-1)^n$

o) $\sum_{n=0}^{\infty} \frac{n+2}{n^2+1} (x+1)^n$

p) $\sum_{n=0}^{\infty} \frac{2^n \sqrt{n}}{n^2+1} \left(\frac{x+2}{4}\right)^n$

q) $\sum_{n=0}^{\infty} \left(\frac{1}{e^n} - \frac{1}{e^{n+1}}\right) (x-e)^n$

r) $\sum_{n=0}^{\infty} \frac{(-1)^n}{4^n \sqrt[3]{n^2+1}} (x-2)^n$

s) $\sum_{n=0}^{\infty} \frac{3^n}{1+9^n} (x-2)^n$

t) $\sum_{n=0}^{\infty} \frac{(-1)^n}{2^n(n^2+3)} (x-2)^n$

15. Let $a, b \in \mathbb{R}^+$. Determine the values of x for which the following series are absolutely convergent:

    a) $\sum_{n=0}^{\infty} a^{n^2} x^n$, $a < 1$

    b) $\sum_{n=0}^{\infty} \frac{x^n}{a^n + b^n}$

16. Expand the following functions into a Maclaurin series and indicate the interval on which the development is valid. Give a justification for the answers provided.

    a) a^x, $a > 0$
    b) $x e^{x^2}$
    c) $\frac{1}{a^2+x^2}$, $a \neq 0$
    d) $\arctan(x)$
    e) $\sinh(x)$
    f) $\cosh(x)$
    g) $2^x + \frac{1}{2+x}$
    h) $\int_0^x e^{-t^2} dt$
    i) $\frac{1}{x^2+x-2}$
    j) $\int_0^x \frac{e^{3t}-1}{t} dt$
    k) $\frac{2}{x^2-4x+3}$
    l) $\frac{x+1}{(1-x)^2}$
    m) $\int_0^x \frac{1-\cos(t)}{t^2} dt$
    n) $\frac{1}{x^2-3x+2}$
    o) $\frac{3}{x-5} + \frac{x}{(x-5)^2}$

17. Expand the function
$$f(x) = x \log(1+x^3)$$
into a Maclaurin series and use the development to prove that the function has a minimum at $x=0$.

18. Consider f, a real-valued function of a real variable, defined by $f(x) = x e^x$.

    a) Develop f into a Maclaurin series and indicate for which values of x the expansion is valid. Explain the response given.

b) Using the series of derivatives from the result of the previous item, calculate the sum of the series
$$\sum_{n=0}^{\infty} \frac{n+1}{n!}.$$

19. Consider the function $f : \mathbb{R} \to \mathbb{R}$ defined by
$$f(x) = x \sin(2x).$$

a) Write Maclaurin series of f and indicate for which values of x the development is valid.

b) Using the result of the previous item and by term by term differentiation, compute the sum of the series
$$\sum_{n=0}^{\infty} (-1)^n \frac{(2n+2)\pi^{2n+1}}{(2n+1)!}.$$
Justify the answer.

20. Consider the function $f : \mathbb{R} \to \mathbb{R}$ defined by
$$f(x) = \sin^2(x) + \log(1 - x^2).$$

a) Using term by term integration, develop f into a Maclaurin series and indicate the real values of x for which the development is valid.
Note: $2\sin(x)\cos(x) = \sin(2x)$.

b) Using the result from the previous item, indicate the value of $f^{(15)}(0)$ and $f^{(16)}(0)$.

21. Develop the function
$$f(x) = \frac{2}{4x+5}$$
into a power series of $x+3$ and determine the radius of convergence of the series.

22. Develop the function
$$f(x) = \frac{1}{x^2 - 6x + 5}$$
into a power series of $x-3$ and determine the interval of convergence of the series.

23. Obtain the Maclaurin series of the function
$$f(x) = (1+x)^{-2}$$
by two different processes. What is the radius of convergence of the series?

24. Determine two power series that represent the function
$$f(x) = \frac{1}{2-x}$$
on the interval $\left[\frac{1}{2}, \frac{3}{2}\right]$. Justify the answer.

25. Develop the function
$$f(x) = \log(x)$$
into a power series of $x-2$ and indicate an open interval on which the function coincides with the sum of the series.

26. Expand the function
$$f(x) = e^{-x+4} + 3^x$$
into a power series of $x-1$ and indicate the largest open interval on which the development represents the function.

27. Let f be the function defined by $f(x) = x^2 \log(x^2)$ on $\mathbb{R} \setminus \{0\}$. Develop f into a power series of $x - 1$ and indicate the largest open interval where this expansion represents the function.

28. Determine whether the following functions are even or odd and sketch their graphs:

a) $x^2 - x$ g) $x|x| + 3$

b) $x^2 \sin(3x)$ h) $\sin^2(x)$

c) $\tan(6x)$ i) $e^{-|x|}$

d) $x^4 + 20\cos(5x)$ j) $\dfrac{x^3 e^{-x^2}}{\cos(x)}$

e) $x^5 - 2x$ k) $x + 5\cos(x)$

f) $\sin(x) + \cos(2x)$ l) $\log(x^4)$

3.7. Proposed Exercises

29. Sketch the graph of the following functions, check if they are periodic, and, if so, indicate a period:

    a) $\sin(7x)$
    b) $\sinh(2x)$
    c) $\tan(\pi x)$
    d) $\sin\left(\dfrac{\pi x}{4}\right)$
    e) $\sin\left(\dfrac{x}{3}\right) + \cos(2x)$
    f) $\sin^2(x)$
    g) $\sin\left(\dfrac{x}{\pi}\right)\cos(x)$
    h) $x + \cos(8x)$
    i) $\begin{cases} 0, & \text{if } 2n-1 \le x < 2n,\ n \in \mathbb{Z} \\ 1, & \text{if } 2n \le x < 2n+1,\ n \in \mathbb{Z} \end{cases}$

30. Determine the Fourier series related to the functions periodic with period 2π, defined on the interval $]-\pi, \pi[$ as follows:

    a) $\begin{cases} -1, & \text{if } -\pi < x \le 0 \\ \dfrac{1}{2}, & \text{if } 0 < x < \pi \end{cases}$
    b) $\begin{cases} 1, & \text{if } -\pi < x \le 0 \\ 3, & \text{if } 0 < x < \pi \end{cases}$
    c) $\begin{cases} x^2, & \text{if } -\pi < x \le 0 \\ -x^2, & \text{if } 0 < x < \pi \end{cases}$
    d) $x + |x|$
    e) $e^{-|x|}$
    f) $\cos^2(2x)$

31. Show that, on the interval $]0, \pi[$, we have

    a) $(\pi - x)\sin(x) =$
    $= 1 + \dfrac{1}{2}\cos(x) - 2 \sum_{n=2}^{\infty} \dfrac{1}{n^2 - 1}\cos(nx)$

    b) $\pi x\,(\pi - x) = 8 \sum_{n=1}^{\infty} \dfrac{1}{(2n-1)^3}\sin\big((2n-1)x\big)$

32. Considering that the following functions, defined on the indicated intervals, are periodic with period 2π, determine their respective Fourier series and the sum of the obtained series.

    a) $|\cos(x)|,\ 0 < x < 2\pi$
    b) $3x(\pi^2 - x^2),\ -2\pi < x < 0$
    c) $\begin{cases} -2x, & \text{if } 0 < x \le \pi \\ 2x, & \text{if } \pi < x < 2\pi \end{cases}$
    d) $\begin{cases} 0, & \text{if } -\dfrac{\pi}{2} < x \le 0 \\ \pi, & \text{if } 0 < x < \dfrac{3\pi}{2} \end{cases}$
    e) $\begin{cases} 0, & \text{if } -\dfrac{3\pi}{2} < x \le 0 \\ \sin(x), & \text{if } 0 < x < \dfrac{\pi}{2} \end{cases}$

33. The following functions are periodic and defined on the interval corresponding to one period. Determine the Fourier series and the respective sum for each function.

    a) $x,\ 0 < x < 3$
    b) $2 - x,\ 0 < x < 2$
    c) $\begin{cases} 1, & \text{if } -2 \le x \le -1 \\ 0, & \text{if } -1 < x < 1 \\ 1, & \text{if } 1 \le x \le 2 \end{cases}$
    d) $\begin{cases} x+1, & \text{if } -1 \le x \le 0 \\ 1-x, & \text{if } 0 < x \le 1 \end{cases}$
    e) $\begin{cases} 0, & \text{if } -3 \le x \le 0 \\ x^2(3-x), & \text{if } 0 < x \le 3 \end{cases}$

34. Let $L > 0$. Develop the following functions into Fourier series on the indicated intervals:

    a) $|x|,\ ]-L, L[$
    b) $e^x,\ ]-L, L[$
    c) $\cosh(|x| - \pi L),\ ]-2\pi L, 2\pi L[$

35. Determine the Fourier sine series and the respective sum for each function.

    a) $1,\ 0 < x < \pi$
    b) $\begin{cases} 0, & \text{if } 0 < x < \pi \\ 1, & \text{if } \pi < x < 2\pi \\ 2, & \text{if } 2\pi < x < 3\pi \end{cases}$
    c) $2 - x^2,\ 0 < x < 2$
    d) $\begin{cases} x, & \text{if } 0 < x \le 1 \\ 2-x, & \text{if } 1 < x < 2 \end{cases}$
    e) $x^3,\ 0 < x < 1$

36. Determine the Fourier cosine series and its respective sum for each function.

a) $x\cos(x)$, $0 < x < \pi$

b) $x^2 - 2x$, $0 < x < 4$

c) $\begin{cases} x, & \text{if } 0 < x \leq 1 \\ 2-x, & \text{if } 1 < x < 2 \end{cases}$

37. Let $h \in \,]0, \pi[$. Determine the Fourier cosine series of the function defined by

$$f(x) = \begin{cases} 1, & \text{if } 0 < x < h \\ 0, & \text{if } h < x < \pi. \end{cases}$$

38. Let $h \in \,]0, \dfrac{\pi}{2}[$. Determine the Fourier cosine series of the function defined by

$$f(x) = \begin{cases} 1, & \text{if } x = 0 \\ \dfrac{2h-x}{2h}, & \text{if } 0 < x < 2h \\ 0, & \text{if } 2h < x < \pi, \end{cases}$$

where f is a periodic function with period 2π.

39. Let $f(x) = \dfrac{\pi}{4} - \dfrac{x}{2}$, $x \in \,]0, \pi[$.

    a) Develop f into a Fourier sine series.

    b) Develop f into a Fourier cosine series.

40. Let $f : \mathbb{R} \to \mathbb{R}$ be a function such that

$$f(x) = \begin{cases} 1, & \text{if } 0 \leq x \leq \dfrac{1}{2} \\ -1, & \text{if } \dfrac{1}{2} < x \leq 1. \end{cases}$$

    a) Determine the Fourier cosine series of the function f on $[0, 1]$ and indicate the values for which the series converges. Explain the reasoning behind the answer.

    b) Sketch the graph of the sum of the Fourier cosine series of f.

41.
    a) Determine the Fourier cosine series of the function $f(x) = \sin(x)$, $0 \leq x < \pi$.

    b) Determine the Fourier sine series of the function $g(x) = \cos(x)$, $0 < x < \pi$.

    c) Can the function f be developed into a Fourier cosine series on $-\pi \leq x \leq \pi$?

42. Let $f : \mathbb{R} \to \mathbb{R}$ be a periodic function with period 2π such that

$$f(x) = \begin{cases} \pi, & \text{if } -\pi < x < 0 \\ -x + \pi, & \text{if } 0 \leq x < \pi. \end{cases}$$

    a) Represent f as the sum of a Fourier series on the interval $\,]-\pi, \pi[$ and sketch, on $\mathbb{R}$, the graph of its sum.

    b) Show that $\displaystyle\sum_{n=1}^{\infty} \dfrac{1}{(2n-1)^2} = \dfrac{\pi^2}{8}$.

43. Let $f : \mathbb{R} \to \mathbb{R}$ be a periodic function with period 4 such that

$$f(x) = \begin{cases} -x - 2, & \text{if } -2 < x \leq -1 \\ \dfrac{|x|}{x}, & \text{if } |x| < 1 \text{ and } x \neq 0 \\ -x + 2, & \text{if } 1 \leq x < 2 \end{cases}$$

and $S(x)$ be the Fourier series associated with f.

    a) Without calculating $S(x)$, indicate the value of $S(6^{12345})$ and $S(65)$. Provide a clear explanation of the answer.

    b) Determine $S(x)$.

44. Let f be the function defined by

$$f(x) = \begin{cases} \dfrac{\pi}{4}, & \text{if } -\pi < x < 0 \\ \dfrac{\pi}{4} - x, & \text{if } 0 \leq x < \pi. \end{cases}$$

    a) Determine the Fourier series of f.

    b) Let

    $$S(x) = \dfrac{a_0 x}{2} + \sum_{n=1}^{\infty} \dfrac{1}{n}\left(a_n \sin(nx) - b_n \cos(nx)\right),$$

    where a_0, a_n, and b_n are the Fourier coefficients of f. Determine the expression of $S(x)$, $x \in \,]-\pi, \pi[$.

45. Use the Fourier series of

$$f(x) = \begin{cases} -1, & \text{if } -3 \leq x < 0 \\ 1, & \text{if } 0 \leq x < 3 \end{cases}$$

to determine the Fourier series of

$$F(x) = \begin{cases} -x - 3, & \text{if } -3 \leq x < 0 \\ x - 3, & \text{if } 0 \leq x < 3. \end{cases}$$

3.7. Proposed Exercises

46. Using the equality

$$x = 2\sum_{n=1}^{\infty} \frac{(-1)^{n+1}}{n} \sin(nx), \quad -\pi < x < \pi,$$

obtain the Fourier series development of $f(x) = x^4$.

47. Determine, using Fourier series, the function f such that

$$\begin{cases} f''(x) + 10f(x) = 5x, \quad -2 < x < 2 \\ f(-2) = f(2) = 0. \end{cases}$$

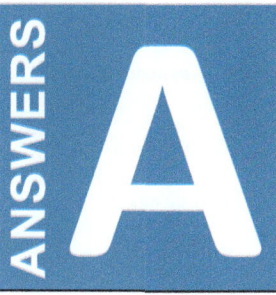

Answers to Proposed Exercises

Chapter 1

3. a) $-\frac{1}{5}$ c) 1   e) $-\frac{1}{4}$
    b) $+\infty$ d) 0

4. a) 0   c) 0   e) 1
    b) -1 d) $\frac{1}{2}$

5. c) n^n, $n!$, e^n, 2^n, n^3, $2n$, $\sqrt{10n}$, $\log(n)$

6. $\frac{1}{n^n}$, $\frac{1}{n!}$, $\frac{1}{e^n}$, $\frac{1}{2^n}$, $\frac{1}{n^3}$, $\frac{1}{2n}$, $\frac{1}{\sqrt{10n}}$, $\frac{1}{\log(n)}$

7. a) e^4 c) $+\infty$
    b) 0   d) 1

8. a) $\frac{1}{2e}$ b) $\frac{4}{e}$

9. $p = \frac{1}{3e}$

10. a) 1 if $x = 2k\pi$, $k \in \mathbb{Z}$
       limit does not exist if $x = (2k+1)\pi$, $k \in \mathbb{Z}$
       0 if $x \in \mathbb{R} \setminus \{k\pi, k \in \mathbb{Z}\}$
    b) 0   e) 2   h) $+\infty$
    c) $+\infty$ f) $\frac{1}{e}$ i) 0
    d) $\frac{2}{e}$ g) $+\infty$ j) 1

13. -1

15. c) $\sqrt{2}$

16. a) True   b) False   c) False

17. a) $\overline{\lim}\, u_n = +\infty$; $\underline{\lim}\, u_n = 0$
    b) $\overline{\lim}\, u_n = 1$; $\underline{\lim}\, u_n = -1$
    c) $\overline{\lim}\, u_n = +\infty$; $\underline{\lim}\, u_n = -1$
    d) $\overline{\lim}\, u_n = 1$; $\underline{\lim}\, u_n = -1$
    e) $\overline{\lim}\, u_n = +\infty$; $\underline{\lim}\, u_n = 0$
    f) $\overline{\lim}\, u_n = 1$; $\underline{\lim}\, u_n = 0$
    g) $\overline{\lim}\, u_n = +\infty$; $\underline{\lim}\, u_n = -\infty$
    h) If $a \in \left[\frac{\pi}{4} + 2k\pi, \frac{3\pi}{4} + 2k\pi\right]$, $k \in \mathbb{Z}$,
       $\overline{\lim}\, u_n = \sin(a)$; $\underline{\lim}\, u_n = -\sin(a)$
       If $a \in \left[\frac{5\pi}{4} + 2k\pi, \frac{7\pi}{4} + 2k\pi\right]$, $k \in \mathbb{Z}$,
       $\overline{\lim}\, u_n = -\sin(a)$; $\underline{\lim}\, u_n = \sin(a)$
       If $a \in \left[-\frac{\pi}{4} + 2k\pi, \frac{\pi}{4} + 2k\pi\right]$, $k \in \mathbb{Z}$,
       $\overline{\lim}\, u_n = \cos(a)$; $\underline{\lim}\, u_n = -\cos(a)$
       If $a \in \left[\frac{3\pi}{4} + 2k\pi, \frac{5\pi}{4} + 2k\pi\right]$, $k \in \mathbb{Z}$,
       $\overline{\lim}\, u_n = -\cos(a)$; $\underline{\lim}\, u_n = \cos(a)$
    i) $\overline{\lim}\, u_n = +\infty$; $\underline{\lim}\, u_n = 0$
    j) $\overline{\lim}\, u_n = +\infty$; $\underline{\lim}\, u_n = -\infty$

22. a) $\left(\frac{8}{9}\right)^n$

Chapter 2

1. a) $\dfrac{1}{(n+1)^2 - 1} = \dfrac{1}{n(n+2)}; \dfrac{3}{4}$

   b) $\dfrac{1}{n(n+1)(n+2)}; \dfrac{1}{4}$

   c) $\dfrac{1}{n(n+1)(n+2)(n+3)}; \dfrac{1}{18}$

2. a) -1

   b) $-\cot(a) + \dfrac{1}{a}$

   c) 1

   d) $-\dfrac{17}{4}$

   e) $-e^{-1}$

   f) $\dfrac{6}{7}$

   g) $\dfrac{1}{\arctan(2)}$

4. a) $x \in \,]-\infty, -3\,[\,\cup\,]1, +\infty\,[\,;\; \dfrac{(x+1)^3}{(x+1)^3 - 8}$

   b) $x \in \,]-2, 0[\,\cup\,]0, 2\,[\,;\; \dfrac{|x| - 1}{2 - |x|}$

   c) $x \in \,]-1, 1\,[\,;\; \dfrac{x}{1 + x^2}$

6. a) Conditionally convergent
   b) Absolutely convergent
   c) Divergent

7. b) $S_{98} = \displaystyle\sum_{n=1}^{98} \dfrac{(-1)^n}{(n+1)^2}$

8. a) Convergent
   b) Convergent or divergent, depending on the series
   c) Convergent
   d) Divergent
   e) Convergent or divergent, depending on the series
   f) Divergent
   g) Convergent or divergent, depending on the series
   h) Convergent or divergent, depending on the series

9. a) Convergent e) Divergent
   b) Divergent   f) Convergent
   c) Convergent g) Divergent
   d) Convergent h) Divergent

10. a) Convergent d) Convergent
    b) Convergent e) Divergent
    c) Convergent f) Convergent

11. a) Convergent e) Divergent
    b) Divergent   f) Convergent
    c) Convergent g) Convergent
    d) Convergent h) Convergent

12. a) Convergent e) Convergent
    b) Divergent
    c) Convergent f) Convergent
    d) Divergent   g) Convergent

13. a) Convergent c) Divergent
    b) Convergent d) Divergent

14. a) Divergent
    b) Absolutely convergent
    c) Divergent
    d) Absolutely convergent
    e) Absolutely convergent
    f) Absolutely convergent
    g) Absolutely convergent
    h) Conditionally convergent
    i) Divergent
    j) Divergent
    k) Absolutely convergent

Answers to Proposed Exercises 329

l) Absolutely convergent

m) Divergent

n) Divergent

o) Absolutely convergent

p) Absolutely convergent

q) Divergent

r) Absolutely convergent

s) Conditionally convergent

15. a) Absolutely convergent

b) Absolutely convergent

c) Divergent

d) Conditionally convergent

e) Conditionally convergent

f) Absolutely convergent

g) Absolutely convergent

h) Absolutely convergent

i) Absolutely convergent

j) Absolutely convergent

k) Conditionally convergent

l) Absolutely convergent if $a > -\frac{3}{2}$; divergent if $a \leq -\frac{3}{2}$

m) Absolutely convergent

n) Divergent

o) Divergent

p) Absolutely convergent

q) Absolutely convergent

r) Divergent

s) Absolutely convergent

t) Absolutely convergent

u) Absolutely convergent

16. a) Absolutely convergent if $(2k+1)\pi < \alpha < (2k+2)\pi$, $k \in \mathbb{Z}$
diverges if $2k\pi \leq \alpha \leq (2k+1)\pi$, $k \in \mathbb{Z}$

b) Absolutely convergent if $\alpha > 1$
conditionally convergent if $0 < \alpha \leq 1$
divergent if $\alpha \leq 0$

17. a) Absolutely convergent

b) 0

19. Convergent

20. a) Divergent

b) Convergent

23. a) Absolutely convergent

b) i. Absolutely convergent if $\beta - \alpha > 1$
divergent if $\beta - \alpha \leq 1$

ii. It is a sum with a finite number of terms.

25. a) $\sum_{n=0}^{2} \left(\sum_{k=0}^{n} \frac{(-1)^k}{k!} \cdot \frac{1}{(n-k+1)\sqrt{n-k+1}} \right) = \frac{1}{3\sqrt{3}} + \frac{1}{2}$

b) Convergent

26. $\frac{8}{7}$

27. 32

Chapter 3

1. Convergent on $]-\infty, 0[$, $S = e^{\frac{x}{2}+x^3}\left(1-e^{x^3}\right)^{-1}$, divergent on $[0, +\infty[$

2. Convergent on $\left]\frac{1}{3}, 1\right[$,
   divergent on $\left]-\infty, \frac{1}{3}\right] \cup [1, +\infty[$

4. Absolutely convergent on $]-\infty, -1[\cup]1, +\infty[$ and divergent on $]-1, 1[$, $\forall \alpha \in \mathbb{R}$
   if $x = 1$: absolutely convergent if $\alpha < -1$ and divergent if $\alpha \geq -1$
   if $x = -1$: absolutely convergent if $\alpha < -1$, conditionally convergent if $-1 \leq \alpha < 0$ and divergent if $\alpha \geq 0$

6. a) Convergent b) Is continuous

7. a) $]1, +\infty[$

9. b) $\sum_{n=1}^{\infty} \frac{1-\cos(n)}{n^3}$

10. a) Converges uniformly
    b) $\sum_{n=1}^{\infty} \frac{2(\sqrt{4n^5+1} - \sqrt{2n^5+1})}{n^4}$

11. b) $\sum_{n=1}^{\infty} \left(\frac{1}{n} - \log\left(\frac{n+1}{n}\right)\right)$

13. a) $\mathbb{R}$ b) $\sum_{n=1}^{\infty} \frac{\sin\left(\frac{1}{n^2}\right)}{1+(nx)^2}$

14. a) Divergent on $\mathbb{R} \setminus \{0\}$

    b) Absolutely convergent on $\left]-\frac{1}{2}, +\infty\right[$
       Divergent on $\left]-\infty, -\frac{1}{2}\right] \setminus \{-1\}$

    c) Absolutely convergent on $]0, +\infty[$
       Conditionally convergent at $x = 0$
       Divergent on $]-\infty, 0[\setminus \{-1\}$

    d) Absolutely convergent on $]-2, 2[$
       Conditionally convergent at $x = 2$
       Divergent on $]-\infty, -2] \cup]2, +\infty[$

    e) Absolutely convergent on $]-5, -1[$
       Conditionally convergent at $x = -5$
       Divergent on $]-\infty, -5[\cup [-1, +\infty[$

    f) Absolutely convergent on $[-2, 8]$
       Divergent on $]-\infty, -2[\cup]8, +\infty[$

    g) Absolutely convergent on $]-\sqrt{2}, \sqrt{2}[\setminus \{0\}$
       Conditionally convergent at $x = -\sqrt{2}$ and at $x = \sqrt{2}$
       Divergent on $]-\infty, -\sqrt{2}[\cup]\sqrt{2}, +\infty[\cup \{0\}$

    h) Absolutely convergent on $\left[2-\frac{1}{\pi}, 2+\frac{1}{\pi}\right]$
       Divergent on $\left]-\infty, 2-\frac{1}{\pi}\right[\cup \left]2+\frac{1}{\pi}, +\infty\right[$

    i) Absolutely convergent on $\left]-\frac{5}{6}, -\frac{1}{6}\right[$
       Divergent on $\left]-\infty, -\frac{5}{6}\right] \cup \left[-\frac{1}{6}, +\infty\right[$

    j) Absolutely convergent on $]-\sqrt{2}-1, \sqrt{2}-1[$
       Divergent on $]-\infty, -\sqrt{2}-1] \cup [\sqrt{2}-1, +\infty[$

    k) Absolutely convergent on $]-1, 3[$
       Divergent on $]-\infty, -1] \cup [3, +\infty[$

    l) Absolutely convergent on $]1, 7[$
       Conditionally convergent at $x = 7$
       Divergent on $]-\infty, 1] \cup]7, +\infty[$

    m) Absolutely convergent on $]-\infty, -1[\cup]3, +\infty[$
       Conditionally convergent at $x = -1$
       Divergent on $]-1, 3] \setminus \{1\}$

    n) Absolutely convergent on $\left]\frac{1}{3}, \frac{5}{3}\right[$
       Divergent on $\left]-\infty, \frac{1}{3}\right] \cup \left[\frac{5}{3}, +\infty\right[$

    o) Absolutely convergent on $]-2, 0[$
       Conditionally convergent at $x = -2$
       Divergent on $]-\infty, -2[\cup [0, +\infty[$

    p) Absolutely convergent on $[-4, 0]$
       Divergent on $]-\infty, -4[\cup]0, +\infty[$

    q) Absolutely convergent on $]0, 2e[$
       Divergent on $]-\infty, 0] \cup [2e, +\infty[$

    r) Absolutely convergent on $]-2, 6[$
       Conditionally convergent at $x = 6$
       Divergent on $]-\infty, -2] \cup]6, +\infty[$

Answers to Proposed Exercises

s) Absolutely convergent on $]-1,5[$
Divergent on $]-\infty,-1] \cup [5,+\infty[$

t) Absolutely convergent on $[0,4]$
Divergent on $]-\infty,0[\cup]4,+\infty[$

15. a) Absolutely convergent on $\mathbb{R}$

b) Absolutely convergent on
$]-a,a[$, if $a \geq b$
$]-b,b[$, if $b > a$

16. a) $\sum_{n=0}^{\infty} \frac{(\log(a))^n}{n!} x^n$, for all $x \in \mathbb{R}$

b) $\sum_{n=0}^{\infty} \frac{x^{2n+1}}{n!}$, for all $x \in \mathbb{R}$

c) $\sum_{n=0}^{\infty} \frac{(-1)^n}{a^{2n+2}} x^{2n}$, $|x| < |a|$

d) $\sum_{n=0}^{\infty} \frac{(-1)^n}{2n+1} x^{2n+1}$, $|x| < 1$

e) $\sum_{n=0}^{\infty} \frac{1}{(2n+1)!} x^{2n+1}$, for all $x \in \mathbb{R}$

f) $\sum_{n=0}^{\infty} \frac{1}{(2n)!} x^{2n}$, for all $x \in \mathbb{R}$

g) $\sum_{n=0}^{\infty} \left(\frac{(\log(2))^n}{n!} + \frac{(-1)^n}{2^{n+1}} \right) x^n$, $|x| < 2$

h) $\sum_{n=0}^{\infty} \frac{(-1)^n}{n!(2n+1)} x^{2n+1}$, for all $x \in \mathbb{R}$

i) $\sum_{n=0}^{\infty} \frac{1}{3} \left(\frac{(-1)^{n+1}}{2^{n+1}} - 1 \right) x^n$, $|x| < 1$

j) $\sum_{n=1}^{\infty} \frac{3^n}{n! \, n} x^n$, for all $x \in \mathbb{R}$

k) $\sum_{n=0}^{\infty} \left(1 - \frac{1}{3^{n+1}}\right) x^n$, $|x| < 1$

l) $\sum_{n=0}^{\infty} (2n+1) x^n$, $|x| < 1$

m) $\sum_{n=1}^{\infty} \frac{(-1)^{n+1}}{(2n)!(2n-1)} x^{2n-1}$, $\forall x \in \mathbb{R}$

n) $\sum_{n=0}^{\infty} \left(1 - \frac{1}{2^{n+1}}\right) x^n$, $|x| < 1$

o) $\sum_{n=0}^{\infty} \frac{n-3}{5^{n+1}} x^n$, $|x| < 5$

17. $\sum_{n=0}^{\infty} \frac{(-1)^n}{n+1} x^{3n+4}$, $|x| < 1$

18. a) $\sum_{n=0}^{\infty} \frac{x^{n+1}}{n!}$, $\forall x \in \mathbb{R}$

b) $2e$

19. a) $\sum_{n=0}^{\infty} \frac{(-1)^n \, 2^{2n+1}}{(2n+1)!} x^{2n+2}$, $\forall x \in \mathbb{R}$

b) $-\pi$

20. a) $\sum_{n=0}^{\infty} \left(\frac{(-1)^n \, 2^{2n+1}}{(2n+2)!} - \frac{1}{n+1} \right) x^{2n+2}$, $|x| < 1$

b) $f^{(15)}(0) = 0$; $f^{(16)}(0) = -2^{15} - 2 \times 15!$

21. $\sum_{n=0}^{\infty} \frac{-2^{2n+1}}{7^{n+1}} (x+3)^n$, $-\frac{19}{4} < x < -\frac{5}{4}$, $r = \frac{7}{4}$

22. $\sum_{n=0}^{\infty} \frac{-1}{2^{2n+2}} (x-3)^{2n}$, $1 < x < 5$

23. $\sum_{n=0}^{\infty} (-1)^n (n+1) x^n$, $|x| < 1$, $r = 1$

24. $\sum_{n=0}^{\infty} (x-1)^n$, $0 < x < 2$

$\sum_{n=0}^{\infty} \frac{1}{2^{n+1}} x^n$, $|x| < 2$

25. $\log(2) + \sum_{n=0}^{\infty} \frac{(-1)^n}{(n+1)2^{n+1}} (x-2)^{n+1}$, $0 < x < 4$

26. $\sum_{n=0}^{\infty} \frac{e^3 (-1)^n + 3 \log^n(3)}{n!} (x-1)^n$, $\forall x \in \mathbb{R}$

27. $2(x-1) + 3(x-1)^2 + 4 \sum_{n=0}^{\infty} \frac{(-1)^n (x-1)^{n+3}}{(n+1)(n+2)(n+3)}$, $0 < x < 2$

28. a) Neither even nor odd

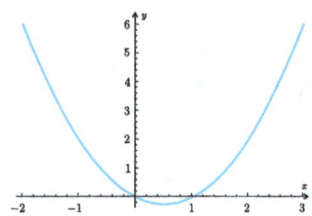

b) Odd

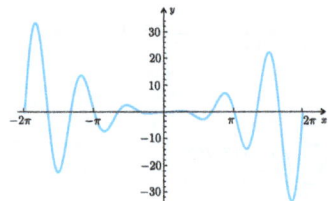

c) Odd

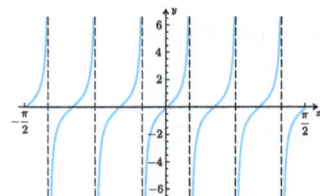

d) Even

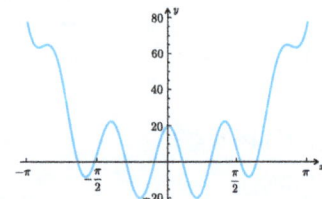

e) Odd

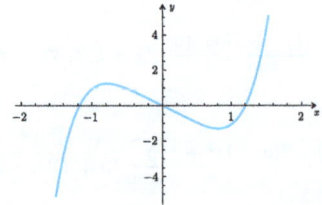

f) Neither even nor odd

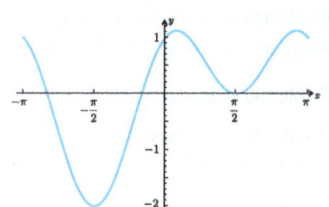

g) Neither even nor odd

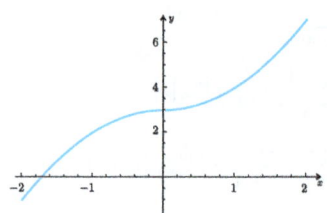

h) Even

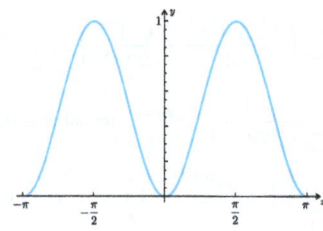

i) Even

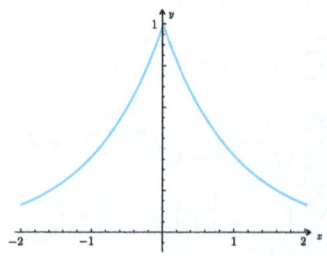

Answers to Proposed Exercises

j) Odd

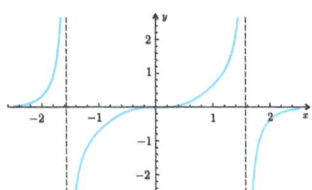

k) Neither even nor odd

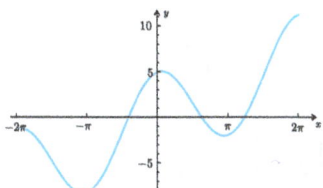

l) Even

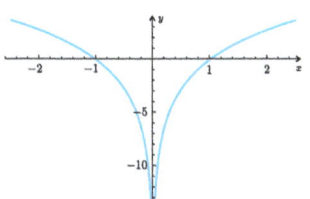

29. a) Periodic with period $\dfrac{2\pi}{7}$

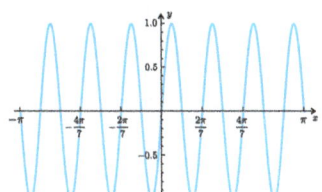

b) Not periodic

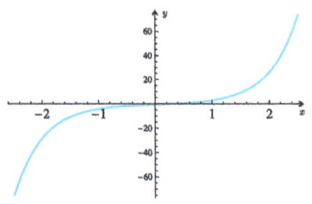

c) Periodic with period 1

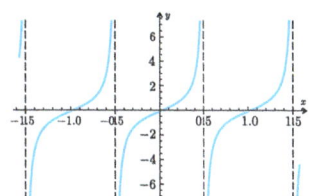

d) Periodic with period 8

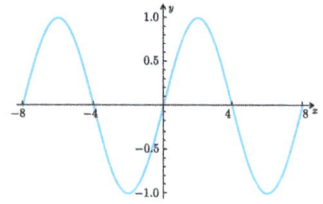

e) Periodic with period 6π

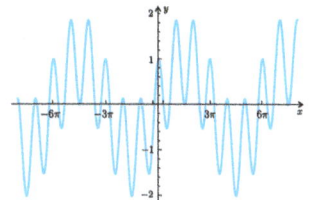

f) Periodic with period π

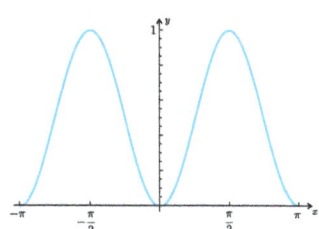

g) Not periodic

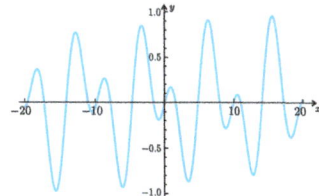

h) Not periodic

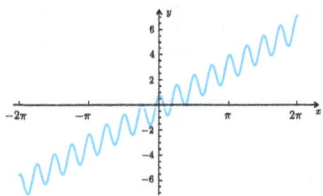

i) Periodic with period 2

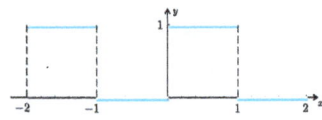

30. a) $-\dfrac{1}{4} + \dfrac{3}{\pi}\displaystyle\sum_{n=1}^{\infty} \dfrac{1}{2n-1}\sin((2n-1)x)$

b) $2 + \dfrac{4}{\pi}\displaystyle\sum_{n=1}^{\infty} \dfrac{1}{2n-1}\sin((2n-1)x)$

c) $\dfrac{2}{\pi}\displaystyle\sum_{n=1}^{\infty} \dfrac{2+(-1)^n(\pi^2 n^2 - 2)}{n^3}\sin(nx)$

d) $\dfrac{\pi}{2} + 2\displaystyle\sum_{n=1}^{\infty}\left(\dfrac{(-1)^n - 1}{\pi n^2}\cos(nx) - \dfrac{(-1)^n}{n}\sin(nx)\right)$

e) $\dfrac{1-e^{-\pi}}{\pi} + \dfrac{2}{\pi}\displaystyle\sum_{n=1}^{\infty} \dfrac{1-(-1)^n e^{-\pi}}{n^2+1}\cos(nx)$

f) $\dfrac{1}{2} + \dfrac{1}{2}\cos(4x)$

32. a) $\dfrac{2}{\pi} - \dfrac{4}{\pi}\displaystyle\sum_{n=1}^{\infty} \dfrac{(-1)^n}{4n^2 - 1}\cos(2nx)$

$|\cos(x)|,\ \forall x \in \mathbb{R}$

b) $3\pi^3 + 18\displaystyle\sum_{n=1}^{\infty}\left(\dfrac{2\pi}{n^2}\cos(nx) + \dfrac{n^2\pi^2 - 2}{n^3}\sin(nx)\right)$

$\begin{cases} 3(x-2k\pi)(\pi^2 - (x-2k\pi)^2), \\ \qquad\qquad\text{if } x \in](2k-2)\pi, 2k\pi[\\ 9\pi^3, \qquad \text{if } x = 2k\pi \end{cases}$

$k \in \mathbb{Z}$

c) $\pi + 8\displaystyle\sum_{n=1}^{\infty}\left(\dfrac{\cos((2n-1)x)}{(2n-1)^2\pi} - \dfrac{\sin((2n-1)x)}{2n-1}\right)$

$\begin{cases} -2(x - 2k\pi), & \text{if } x \in]2k\pi, (2k+1)\pi[\\ 0, & \text{if } x = (2k+1)\pi \\ 2(x - 2k\pi), & \text{if } x \in](2k+1)\pi, (2k+2)\pi[\\ 2\pi, & \text{if } x = 2k\pi \end{cases}$

$k \in \mathbb{Z}$

d) $\dfrac{3\pi}{4} + \displaystyle\sum_{n=1}^{\infty}\left(\dfrac{\sin\left(\frac{3n\pi}{2}\right)}{n}\cos(nx) + \dfrac{2\sin^2\left(\frac{3n\pi}{4}\right)}{n}\sin(nx)\right)$

$\begin{cases} 0, & \text{if } x \in](4k-1)\frac{\pi}{2}, 2k\pi[\\ \frac{\pi}{2}, & \text{if } x = (4k-1)\frac{\pi}{2} \vee x = 2k\pi \\ \pi, & \text{if } x \in]2k\pi, (4k+3)\frac{\pi}{2}[\end{cases}$

$k \in \mathbb{Z}$

e) $\dfrac{1}{2\pi} + \dfrac{1}{2\pi}\cos(x) + \dfrac{1}{4}\sin(x)$

$+\dfrac{1}{\pi}\displaystyle\sum_{n=2}^{\infty}\left(\dfrac{n\sin\left(\frac{n\pi}{2}\right) - 1}{n^2 - 1}\cos(nx) - \dfrac{n\cos\left(\frac{n\pi}{2}\right)}{n^2 - 1}\sin(nx)\right)$

$\begin{cases} \sin(x), & \text{if } x \in [2k\pi, (4k+1)\frac{\pi}{2}[\\ \frac{1}{2}, & \text{if } x = (4k+1)\frac{\pi}{2} \\ 0, & \text{if } x \in](4k+1)\frac{\pi}{2}, (2k+2)\pi[\end{cases}$

$k \in \mathbb{Z}$

33. a) $\dfrac{3}{2} - \dfrac{3}{\pi}\displaystyle\sum_{n=1}^{\infty} \dfrac{1}{n}\sin\left(\dfrac{2n\pi x}{3}\right)$

$\begin{cases} x - 3k, & \text{if } x \in]3k, 3k+3[\\ \frac{3}{2}, & \text{if } x = 3k \end{cases}$

$k \in \mathbb{Z}$

b) $1 + \dfrac{2}{\pi}\displaystyle\sum_{n=1}^{\infty} \dfrac{1}{n}\sin(n\pi x)$

$\begin{cases} 2 - (x - 2k), & \text{if } x \in]2k, 2k+2[\\ 1, & \text{if } x = 2k \end{cases}$

$k \in \mathbb{Z}$

Answers to Proposed Exercises

c) $\dfrac{1}{2} + \dfrac{2}{\pi} \sum_{n=1}^{\infty} \dfrac{(-1)^n}{2n-1} \cos\left(\dfrac{(2n-1)\pi x}{2}\right)$

$\begin{cases} 1, & \text{if } x \in\,]1+4k, 3+4k[\\ 0, & \text{if } x \in\,]3+4k, 5+4k[\\ \frac{1}{2}, & \text{if } x = 2k+1 \end{cases}$

$k \in \mathbb{Z}$

d) $\dfrac{1}{2} + \dfrac{4}{\pi^2} \sum_{n=1}^{\infty} \dfrac{1}{(2n-1)^2} \cos((2n-1)\pi x)$

$\begin{cases} x - 2k + 1, & \text{if } x \in [2k-1, 2k] \\ -x + 2k + 1, & \text{if } x \in [2k, 2k+1] \end{cases}$

$k \in \mathbb{Z}$

e) $\dfrac{9}{8} - \dfrac{27}{\pi^4} \sum_{n=1}^{\infty} \left(\dfrac{6 + (-1)^n(n^2\pi^2 - 6)}{n^4} \cos\left(\dfrac{n\pi x}{3}\right) \right.$
$\left. + \dfrac{2\pi(1 + 2(-1)^n)}{n^3} \sin\left(\dfrac{n\pi x}{3}\right) \right)$

$\begin{cases} (x-6k)^2(3+6k-x), & \text{if } x \in [6k, 3+6k] \\ 0, & \text{if } x \in [3+6k, 6+6k] \end{cases}$

$k \in \mathbb{Z}$

34. a) $\dfrac{L}{2} - \dfrac{4L}{\pi^2} \sum_{n=1}^{\infty} \dfrac{1}{(2n-1)^2} \cos\left(\dfrac{(2n-1)\pi x}{L}\right)$

b) $\sinh(L) \left(\dfrac{1}{L} + 2 \sum_{n=1}^{\infty} \dfrac{L(-1)^n}{L^2 + n^2\pi^2} \cos\left(\dfrac{n\pi x}{L}\right) \right.$
$\left. - \dfrac{n\pi(-1)^n}{L^2 + n^2\pi^2} \sin\left(\dfrac{n\pi x}{L}\right) \right)$

c) $\sinh(\pi L) \left(\dfrac{1}{\pi L} + \dfrac{2L}{\pi} \sum_{n=1}^{\infty} \dfrac{1}{L^2 + n^2} \cos\left(\dfrac{nx}{L}\right) \right)$

35. a) $\dfrac{4}{\pi} \sum_{n=1}^{\infty} \dfrac{1}{2n-1} \sin((2n-1)x)$

$\begin{cases} 1, & \text{if } x \in\,]2k\pi, (2k+1)\pi[\\ -1, & \text{if } x \in\,](2k+1)\pi, (2k+2)\pi[\\ 0, & \text{if } x = k\pi \end{cases}$

$k \in \mathbb{Z}$

b) $\dfrac{4}{\pi} \sum_{n=1}^{\infty} \dfrac{\cos\left(\frac{n\pi}{2}\right)\cos\left(\frac{n\pi}{6}\right) - (-1)^n}{n} \sin\left(\dfrac{nx}{3}\right)$

$\begin{cases} -2, & \text{if } x \in\,]-3\pi + 6k\pi, -2\pi + 6k\pi[\\ -1, & \text{if } x \in\,]-2\pi + 6k\pi, -\pi + 6k\pi[\\ 0, & \text{if } x \in\,]-\pi + 6k\pi, \pi + 6k\pi[\\ 1, & \text{if } x \in\,]\pi + 6k\pi, 2\pi + 6k\pi[\\ 2, & \text{if } x \in\,]2\pi + 6k\pi, 3\pi + 6k\pi[\\ \frac{1}{2}, & \text{if } x = \pi + 6k\pi \\ -\frac{1}{2}, & \text{if } x = -\pi + 6k\pi \\ \frac{3}{2}, & \text{if } x = 2\pi + 6k\pi \\ -\frac{3}{2}, & \text{if } x = -2\pi + 6k\pi \\ 0, & \text{if } x = 3\pi + 6k\pi \end{cases}$

$k \in \mathbb{Z}$

c) $\dfrac{4}{\pi} \sum_{n=1}^{\infty} \left(\dfrac{(-1)^n + 1}{n} - \dfrac{4((-1)^n - 1)}{n^3\pi^2} \right) \sin\left(\dfrac{n\pi x}{2}\right)$

$\begin{cases} 2 - (x-4k)^2, & \text{if } x \in\,]4k, 4k+2[\\ (x-4k)^2 - 2, & \text{if } x \in\,]4k-2, 4k[\\ 0, & \text{if } x = 2k \end{cases}$

$k \in \mathbb{Z}$

d) $\dfrac{8}{\pi^2} \sum_{n=1}^{\infty} \dfrac{(-1)^{n+1}}{(2n-1)^2} \sin\left(\dfrac{(2n-1)\pi x}{2}\right)$

$\begin{cases} x - 4k, & \text{if } x \in [-1+4k, 1+4k] \\ 2 - (x-4k), & \text{if } x \in [1+4k, 3+4k] \end{cases}$

$k \in \mathbb{Z}$

e) $\dfrac{2}{\pi^3} \sum_{n=1}^{\infty} (-1)^n \dfrac{6 - n^2\pi^2}{n^3} \sin(n\pi x)$

$\begin{cases} (x-2k)^3, & \text{if } x \in\,]2k-1, 2k+1[\\ 0, & \text{if } x = 2k+1 \end{cases}$

$k \in \mathbb{Z}$

36. a) $-\dfrac{2}{\pi} + \dfrac{\pi}{2}\cos(x) - \dfrac{4}{\pi} \sum_{n=1}^{\infty} \dfrac{4n^2 + 1}{(4n^2 - 1)^2} \cos(2nx)$

$\begin{cases} -(x - 2k\pi)\cos(x), & \text{if } x \in [(2k-1)\pi, 2k\pi] \\ (x - 2k\pi)\cos(x), & \text{if } x \in [2k\pi, (2k+1)\pi] \end{cases}$

$k \in \mathbb{Z}$

b) $\dfrac{4}{3} + \dfrac{16}{\pi^2} \sum_{n=1}^{\infty} \dfrac{1 + 3(-1)^n}{n^2} \cos\left(\dfrac{n\pi x}{4}\right)$

$(x - 8k)^2 - 2|x - 8k|$, if $x \in [8k-4, 8k+4]$

$k \in \mathbb{Z}$

c) $\dfrac{1}{2} - \dfrac{4}{\pi^2} \sum_{n=1}^{\infty} \dfrac{1+(-1)^n - 2\cos\left(\frac{n\pi}{2}\right)}{n^2} \cos\left(\dfrac{n\pi x}{2}\right)$

$= \dfrac{1}{2} - \dfrac{4}{\pi^2} \sum_{n=1}^{\infty} \dfrac{1}{(2n-1)^2} \cos((2n-1)\pi x)$

$\begin{cases} x-2k, & \text{if } x \in [2k, 2k+1] \\ 2-x+2k, & \text{if } x \in [2k+1, 2k+2] \end{cases}$

$k \in \mathbb{Z}$

37. $\dfrac{h}{\pi} + \dfrac{2}{\pi} \sum_{n=1}^{\infty} \dfrac{\sin(nh)}{n} \cos(nx)$

38. $\dfrac{h}{\pi} + \dfrac{2}{\pi h} \sum_{n=1}^{\infty} \dfrac{\sin^2(nh)}{n^2} \cos(nx)$

39. a) $\sum_{n=1}^{\infty} \dfrac{\sin(2nx)}{2n}$

   b) $\dfrac{2}{\pi} \sum_{n=1}^{\infty} \dfrac{\cos((2n-1)x)}{(2n-1)^2}$

40. a) $\dfrac{4}{\pi} \sum_{n=1}^{\infty} \dfrac{(-1)^{n+1}}{2n-1} \cos((2n-1)\pi x)$

$\begin{cases} 1, & \text{if } x \in \,]-\frac{1}{2}+2k, \frac{1}{2}+2k[\\ -1, & \text{if } x \in \,]\frac{1}{2}+2k, \frac{3}{2}+2k[\\ 0, & \text{if } x = \frac{2k+1}{2} \end{cases}$

$k \in \mathbb{Z}$

b)
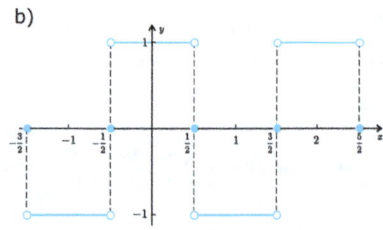

41. a) $\dfrac{2}{\pi} + \dfrac{4}{\pi} \sum_{n=1}^{\infty} \dfrac{1}{1-4n^2} \cos(2nx)$

   b) $\dfrac{8}{\pi} \sum_{n=1}^{\infty} \dfrac{n}{4n^2-1} \sin(2nx)$

   c) No

42. a) $\dfrac{3\pi}{4} + \sum_{n=1}^{\infty} \left(\dfrac{1-(-1)^n}{\pi n^2} \cos(nx) + \dfrac{(-1)^n}{n} \sin(nx) \right)$

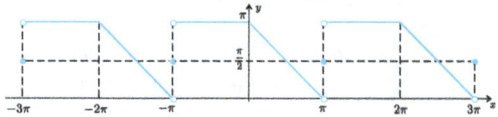

43. a) $S(6^{12345}) = 0,\ S(65) = 1$

   b) $\dfrac{2}{\pi} \sum_{n=1}^{\infty} \left(\dfrac{1}{n} + \dfrac{2}{n^2 \pi} \sin\left(\dfrac{n\pi}{2}\right) \right) \sin\left(\dfrac{n\pi x}{2}\right)$

44. a) $\sum_{n=1}^{\infty} \left(\dfrac{1-(-1)^n}{\pi n^2} \cos(nx) + \dfrac{(-1)^n}{n} \sin(nx) \right)$

   b) $S(x) = \begin{cases} \dfrac{\pi}{4}x + \dfrac{\pi^2}{12}, & \text{if } -\pi < x < 0 \\ \dfrac{\pi}{4}x - \dfrac{x^2}{2} + \dfrac{\pi^2}{12}, & \text{if } 0 \leq x < \pi \end{cases}$

45. $f \sim \sum_{n=1}^{\infty} \dfrac{4}{(2n-1)\pi} \sin\left(\dfrac{(2n-1)\pi x}{3}\right)$

$F(x) = -\dfrac{3}{2} - \sum_{n=1}^{\infty} \dfrac{12}{(2n-1)^2 \pi^2} \cos\left(\dfrac{(2n-1)\pi x}{3}\right)$

46. $x^4 = \dfrac{\pi^4}{5} + 8 \sum_{n=1}^{\infty} \dfrac{(-1)^n}{n^4} (\pi^2 n^2 - 6) \cos(nx)$

47. $\dfrac{80}{\pi} \sum_{n=1}^{\infty} \dfrac{(-1)^n}{n((n\pi)^2 - 40)} \sin\left(\dfrac{n\pi x}{2}\right)$

Bibliography

1. Anton, H., Bivens, I., Davis, S.: Calculus, 12th edn. Wiley, Hoboken, New Jersey (2022)

2. Apostol, T.: Calculus, 2nd edn. Wiley, New Delhi (2003)

3. Bourchtein, L., Bourchtein, A.: Theory of Infinite Sequences and Series. Birkhäuser, Cham, Switzerland (2022)

4. Ellis, R., Gulick, D.: Calculus with Analytic Geometry, 5th edn. Saunders College Publishing, Harcourt Brace College Publishers, Fort Worth (1994)

5. Hunt, R.A.: Calculus, 2nd edn. HarperCollins College Publishers, New York (1994)

6. Kreyszig, E.: Advanced Engineering Mathematics, 10th edn. Wiley, New York (2011)

7. Laczkovich, M., Sós, V.T.: Real Analysis, Series, Functions of Several Variables, and Applications. Springer, New York (2017)

8. Larson, R., Hostetler, R., Edwards, B.H.: Calculus with Analytic Geometry, 7th edn. Houghton Mifflin, Boston (2001)

9. Little, C.H.C., Teo, K.L., Van Brunt, B.: Real Analysis via Sequences and Series. Springer, New York (2015)

10. Pinkus, A., Zafrany, S.: Fourier Series and Integral Transforms. Cambridge University Press, Cambridge (1997)

11. Mcquarrie, D.A.: Mathematical Methods for Scientists and Engineers, University Science Books, Sausalito, California (2003)

12. Salas, S., Hille, E., Etgen, G.: Calculus, One and Several Variables, 10th edn. Wiley, Hoboken, New Jersey (2007)

13. Spivak, M.: Calculus, 4th edn. Publish or Perish, Inc, Houston, Texas (2008)

14. Stewart, J.: Calculus, 9th edn. Cengage Learning, Boston, MA (2021)

15. Stewart, J.: Calculus, Concepts and Contexts, 5th edn. Cengage Learning, Boston, MA (2023)

16. Taylor, A., Mann, R.: Advanced Calculus, 3rd edn. Wiley, New York (1983)

17. Tolstov, G.P.: Fourier Series. Dover Publications, New York (1962)

Index

Symbols
Int(x), integer part of x, viii
$\overline{\mathbb{R}}$, extended real line, 15

A
Alternating series, 82
arithmetic progression, *see* arithmetic sequence
arithmetic sequence, 6
    common difference, 6
    definition, 6
    general term, 7
    sum of the first n terms, 7

B
bounded, viii
bounded from above, viii
bounded from below, viii

C
Cauchy
    Cauchy product, 113
    Cauchy's Criterion, 78
    Cauchy's Root Test, 106
    Condensation Test, 90
    sequence, 35
Cauchy, Augustin Louis, 36

D
D'Alembert's Test, 102
D'Alembert, Jean Le Rond, 102
difference of sequences, 4
Dirichlet's Test, 83
Dirichlet, Peter Gustave Lejeune, 83

E
Euler's constant, 30
Euler, Leonhard, 29
extended real line, 15

F
Fibonacci
    sequence, 2
Fibonacci, Leonardo, 2
Fourier series
    coefficients, 205, 223
    convergence, 225
    cosine series, 216
    definition, 205
    differentiation term by term, 226
    even function, ix
    integration term by term, 226
    odd function, ix
    period, 201
    periodic extension, 206
    periodic function, 201
    piecewise continuous, 211
    piecewise C^1, 212
    pointwise convergence, 212
    sine series, 217
    trigonometric polynomial, 202
    trigonometric series, 202
        coefficients, 202
    uniform convergence, 225
Fourier, Joseph, 205
function
    even, ix, 214
    odd, ix, 214
    of class C^1, ix

of class C^p, ix
piecewise continuous, 211
piecewise C^1, 212

G
Gauss, Carl Friedrich, 7
General Comparison Test, 95
general term of a sequence, 1
geometric progression, *see* geometric sequence
geometric sequence, 8
    common ratio, 8
    definition, 8
    general term, 9
    sum of the first n terms, 9

I
indeterminate forms, ix
infimum, viii
integer part, viii
Integral Test, 93
interval, vii

K
Kummer's Test, 109
Kummer, Ernst Eduard, 109

L
L'Hôpital's Rule, ix
Leibniz's Test, 82
Leibniz, Gottfried Wilhelm, 82
lower bound, viii

M
Maclaurin
    binomial series, 194
        Maclaurin series of $\sin(x)$, 196
        Maclaurin series of e^x, 197
Maclaurin, Collin, 194
maximum, viii

Method of Mathematical Induction, viii
minimum, viii

N
neighborhood
    of a, vii, 15
    of $+\infty$, 15
    of $-\infty$, 15
Neper's number, 30
Neper, Jhone, 30
Newton Binomial, viii
Numerical series
    absolutely convergent, 87
    alternating harmonic, 86
    alternating series, 82
    associative property, 80
    Cauchy Condensation Test, 90
    Cauchy product, 113
    Cauchy's Criterion, 78
    Cauchy's Root Test, 106
    conditionally convergent, 87
    convergent, 69
    definition, 69
    divergent, 69
    General Comparison Test, 95
    general term, 69
    geometric, 70
    harmonic, 71
    Integral Test, 93
    Kummer's Test, 109
    p-series, 91
    partial sums, 69
    product of series, 113
    Ratio Test, 101
    rearrangement, 87

Index

remainder of order p, 79
Root Test, 106
sum, 69
telescopic, 74
terms of the series, 69
Theorem of Mertens, 115

P
period, 201
periodic extension, 206
periodic function, 201
piecewise continuous, 211
piecewise C^1, 212
Power series
    absolute convergence, 186
    convergence at endpoints, 189
    definition, 186
    divergence, 186
    interval of convergence, 187
    Maclaurin series, 193
    radius of convergence, 187
    Taylor series, 193
    term by term differentiation, 191
    term by term integration, 190
    uniform convergence, 190
product of sequences, 4

Q
quotient of sequences, 4

R
Raabe's Test, 111
Raabe, Joseph L., 111
Ratio Test, 101
Root Test, 106

S
Sequence of functions, 167
    convergence at a point, 167
    limit function, 167
    pointwise convergence, 167
    uniform convergence, 169
Sequence of real numbers, 1
    arithmetic progression, *see* arithmetic sequence
    arithmetic sequence, 6
        common difference, 6
        definition, 6
        general term, 7
        sum of the first n terms, 7
    bounded, 2
    bounded above, 2
    bounded below, 2
    Cauchy sequence, 35
    convergent, 13
    decreasing, 4
    definition, 1
    divergent, 13
    Fibonacci, 2
    general infinite limit, 11
    general term, 1
    geometric progression, *see* geometric sequence
    geometric sequence, 8
        common ratio, 8
        definition, 8
        general term, 9
        sum of the first n terms, 9
    graph, 1
    increasing, 4
    indeterminate form
        division by the highest power, 41
    inferior limit, 35
    infinite limit, 11
    limit, 13
    lower limit, 35
    monotonic, 4
    null sequence, 17
    operations with limits, 16

range, 1
recursive relation, 2
Squeeze Theorem, 19
strictly decreasing, 4
strictly increasing, 4
strictly monotonic, 4
sublimit, 33
subsequence, 31
superior limit, 35
tends to, 11, 13
terms, 1
uniqueness of the limit, 16
upper limit, 35

Sequences of real numbers
difference, 4
product, 4
quotient, 4
sum, 4

Series of functions
convergence at a point, 172
differentiable term by term, 183
Fourier series, 201
  coefficients, 205
  definition, 205
  periodic extension, 206
general term, 172
integrable term by term, 180, 183
Maclaurin series, 193
partial sum, 172
pointwise convergence, 172
power series, 186
  absolute convergence, 186
  convergence at endpoints, 189
  definition, 186
  divergence, 186
  interval of convergence, 187
  radius of convergence, 187
  term by term differentiation, 191
  term by term integration, 190
  uniform convergence, 190
sum function of the series, 172
Taylor series, 193
uniform convergence, 174
Weierstrass' Test, 175

sum of sequences, 4
supremum, viii

T

Taylor series, 193
  Maclaurin series, 193
  uniqueness, 197
Taylor's formula, 193
  Lagrange remainder, 193
Taylor, Brook, 193
triangle inequality, vii
trigonometric polynomial, 202
trigonometric series, 202
  coefficients, 202

U

upper bound, viii

W

Weierstrass' Test, 175
Weierstrass, Karl Theodor Wilhelm, 175

Z

Zeno, 68
Zeno's Paradoxes, 68

SPRINGER NATURE

GPSR Compliance

The European Union's (EU) General Product Safety Regulation (GPSR) is a set of rules that requires consumer products to be safe and our obligations to ensure this.

If you have any concerns about our products, you can contact us on ProductSafety@springernature.com

In case Publisher is established outside the EU, the EU authorized representative is:

Springer Nature Customer Service Center GmbH
Europaplatz 3
69115 Heidelberg, Germany

The manufacturer's authorised representative in the EU is Springer Nature Customer Service Centre GmbH, Europaplatz 3, 69115 Heidelberg, Germany. If you have any concerns regarding our products, please contact ProductSafety@springernature.com

Printed and bound by CPI Group (UK) Ltd, Croydon, CR0 4YY

26/03/2026

02078955-0001